GROUND WATER
POLLUTION CONTROL

By Larry W. Canter and Robert C. Knox

CRC Press
Taylor & Francis Group
Boca Raton London New York

CRC Press is an imprint of the
Taylor & Francis Group, an **informa** business

First published 1985 by Lewis Publishers, Inc

Published 2019 by CRC Press
Taylor & Francis Group
6000 Broken Sound Parkway NW, Suite 300
Boca Raton, FL 33487-2742

First issued in paperback 2019

No claim to original U.S. Government works

ISBN 13: 978-0-367-45171-4 (pbk)
ISBN 13: 978-0-87371-014-5 (hbk)

Visit the Taylor & Francis Web site at
http://www.taylorandfrancis.com

and the CRC Press Web site at
http://www.crcpress.com

Library of Congress Cataloging in Publication Data

Canter, Larry W.
    Ground water pollution control

    Bibliography: p.
    Includes index.
    1. Water, Underground – Pollution. I. Knox, R.C.
II. Title.
TD426.C36   1985   363.7'394   84-28927

PREFACE

Ground water has become a natural resource of national and international prominence within the last several years, and indications are that this resource will receive still more attention due to increasing usage, frequent revelations of pollution, and new legislative initiatives and management strategies. Ground water is perceived by many as the most critical natural resource problem in the decade of the 1980s, and there is technical information which suggests that ground water pollution problems and associated concerns will heighten and possibly reach crisis proportions within the next 10 to 20 years.

Ground water pollution problems can be considered in terms of resource extraction and natural and man-made pollution. Resource extraction may lead to ground water mining or depletion, salt or saline water intrusion, and land subsidence. Ground water quality must be considered relative to potential ground water usage and the associated quality requirements for usage. Natural geological conditions can cause poor ground water quality via the introduction of inorganics and metals; examples include chlorides, sulfates, hardness, and iron and manganese. There are numerous categories of man-made sources of ground water pollution. Leachates containing potentially toxic constituents from municipal and chemical landfills and abandoned dump sites are receiving attention as a result of legal requirements and special programs such as "Superfund." Placement of septic tank systems in hydrologically and geologically unsuitable locations can also lead to nitrate and bacterial pollution of ground water. Many other categories of man-made sources of ground water pollution could be cited.

As a result of the growing recognition of ground water pollution problems, there is a growing interest in, and implementation of, ground water pollution control programs. These programs are an outgrowth of: (1) the federal "Superfund" program administered by the U.S. Environmental Protection Agency (EPA); (2) the Installation Restoration Program administered by the Department of Defense; (3) the increasing number of state-level "Superfund" programs; (4) public pressures to clean up accidental spills and leakage from underground storage tanks; and (5) the expanding knowledge base on technical alternatives which are potentially

iii

available for usage in these programs. In addition, with the issuance of the EPA Ground Water Protection Strategy, state and local governmental programs directed toward ground water protection and cleanup are expected to increase in both number and institutional coverage. As experience is gained in the planning and implementation of aquifer restoration/remedial action programs, there is a growing realization that these measures can: (1) be very expensive to plan, construct, and operate; (2) require considerable time to develop information on the problem to be addressed and the technically feasible measures which can be used to address the problem; and (3) be less than complete in terms of maintaining their structural stability and operational capability over time.

This book has been written to summarize the state-of-knowledge on technical alternatives available for usage in ground water pollution control programs, and to delineate decision-making techniques which can be used in the selection process for these alternatives. The book is organized into three major headings: Technologies; Decision-Making; and Case Studies and Applications. Following an introductory chapter, three categories of technologies are addressed: physical control measures in Chapter 2, treatment of ground water in Chapter 3, and in situ chemical and biological measures in Chapter 4. Chapter 2 includes information on source control strategies, stabilization/solidification methods, well systems, interceptor systems, capping and liners, sheet piling, grouting and slurry walls. Treatment of ground water via air stripping, carbon adsorption, biological treatment, chemical precipitation, and other treatment techniques is summarized in Chapter 3. Chapter 4 summarizes the advantages and disadvantages of in situ chemical treatment, and presents extensive information on in situ biological stabilization via enhancement of the indigenous microbial population or addition of acclimated microorganisms.

Decision-making is addressed in Chapters 5 through 8. Chapter 5 presents a protocol for aquifer restoration decision-making and includes a summary of preliminary study activities and the development and evaluation of alternatives. An extensive summary of techniques for decision-making is in Chapter 6, including information on importance weighting techniques and scaling/rating/ranking of alternatives. Chapter 7 addresses risk assessment as related to aquifer restoration planning and includes summaries of two methods applicable to hazardous waste disposal sites. Chapter 8 focuses on public participation and highlights approaches for identifying publics and selecting public participation techniques.

Case studies and applications of ground water pollution control are addressed in Appendices A and B. Appendix A summarizes information on ground water cleanup programs for Tar Creek in Oklahoma, the

Gilson Road Site in New Hampshire, and Rocky Mountain Arsenal in Colorado. Highlighted case studies of aquifer restoration via biological processes are also in Appendix A. Appendix B contains an annotated bibliography of 225 selected references related to ground water pollution control.

The authors gratefully acknowledge the contributions made to this book by faculty colleagues at Oklahoma State University and Rice University. Drs. D. F. Kincannon and E. L. Stover of the School of Civil Engineering at Oklahoma State University prepared Chapter 3 and information on the Gilson Road case study in Appendix A. Dr. C. H. Ward and M. D. Lee of the Department of Environmental Science and Engineering at Rice University prepared Chapter 4 and the highlighted case studies of aquifer restoration via biological processes in Appendix A.

The authors wish to express their appreciation to several additional persons instrumental in the development of this book. First, James McNabb and Dick Scalf of the U.S. Environmental Protection Agency served as project officers on a research study of the state-of-the-art of aquifer restoration. In addition, a panel of experts provided periodic advice, consultation, and review comments during the conduction of the research study. The panel included: William Althoff, New Jersey Department of Environmental Protection; Rich Barteldt, Region V Superfund Coordinator, EPA; Richard Conway, Union Carbide Corporation; George Dixon, Office of Solid Waste, EPA; Lawrence Greenfield, Office of Solid Waste, EPA; Jerry Kotas, Ground Water Protection Branch, EPA; Roy Murphy, Office of Waste Programs Enforcement, EPA; and Dr. Rick Standford, Office of Emergency Response, EPA. Second, Debby Fairchild of the Environmental and Ground Water Institute at the University of Oklahoma conducted the computer-based literature searches basic to this effort. The able typing assistance of Mrs. Wilma Clark and Mrs. Mittie Durham is gratefully acknowledged. Finally, and most important, the authors are indebted to Mrs. Leslie Rard of the Environmental and Ground Water Institute for her typing skills and dedication to the preparation of this manuscript.

The authors also wish to express their appreciation to the University of Oklahoma College of Engineering for its basic support of faculty writing endeavors, and to their families for their understanding and patience.

Larry W. Canter
Sun Company Professor
  of Ground Water Hydrology
University of Oklahoma
Norman, Oklahoma

Robert C. Knox
Assistant Professor of
  Civil Engineering
McNeese State University
Lake Charles, Louisiana

February, 1985

To Donna, Doug, Steve, and Greg

To Ruth

Larry W. Canter

LARRY W. CANTER, P.E., is the Sun Company Professor of Ground Water Hydrology, and Director, Environmental and Ground Water Institute, at the University of Oklahoma, Norman, Oklahoma, in the USA. Dr. Canter received his Ph.D. in Environmental Health Engineering from the University of Texas in 1967, MS in Sanitary Engineering from the University of Illinois in 1962, and BE in Civil Engineering from Vanderbilt University in 1961. Before joining the faculty of the University of Oklahoma in 1969, he was on the faculty at Tulane University and was a sanitary engineer in the U.S. Public Health Service. He served as Director of the School of Civil Engineering and Environmental Science at the University of Oklahoma from 1971 to 1979.

Dr. Canter has published several books and has written chapters in other books; he is also the author or co-author of numerous papers and research reports. His research interests include environmental impact assessment and ground water pollution control. In 1982 he received the Outstanding Faculty Achievement in Research Award from the College of Engineering, and in 1983 the Regent's Award for Superior Accomplishment in Research.

Dr. Canter currently serves on the U.S. Army Corps of Engineers Environmental Advisory Board. He has conducted research, presented short courses, or served as advisor to institutions in Mexico, Panama, Colombia, Venezuela, Peru, Scotland, The Netherlands, France, Germany, Italy, Greece, Turkey, Kuwait, Thailand, and the People's Republic of China.

Robert C. Knox

ROBERT C. KNOX is an Assistant Professor of Civil Engineering at McNeese State University in Lake Charles, Louisiana. Dr. Knox received his BS, MS and Ph.D. degrees in Civil Engineering from the University of Oklahoma. Prior to joining the faculty at McNeese, he was a research engineer at the Environmental and Ground Water Institute at the University of Oklahoma.

Dr. Knox's research interests include ground water contamination and pollution control, environmental impact assessment, and wastewater treatment. Dr. Knox's dissertation research involved one of the first assessments of a ground water pollution control technology focusing on subsurface impermeable barriers. Dr. Knox has published several technical reports and articles concerning ground water pollution control.

Contents

DECISION-MAKING IN
AQUIFER RESTORATION PROJECTS

List of Figures

List of Tables

List of Tables

CHAPTER 1

Introduction

Pollution of ground water can result from many activities, including leaching from municipal and chemical landfills and abandoned dump sites, accidental spills of chemicals or waste materials, improper underground injection of liquid wastes, and placement of septic tank systems in hydrologically and geologically unsuitable locations. In recent years incidents of aquifer pollution from man's waste disposal activities have been discovered with increasing regularity. Concurrently, demands for ground water usage have been increasing due to population growth and diminishing opportunities to economically develop surface water supplies. Until recently the general viewpoint held by many ground water professionals and policy-makers was that once an aquifer had become polluted its water usage must be curtailed or possibly eliminated. However, this viewpoint is changing as a result of increasing needs for ground water utilization and the development of appropriate methodologies for ground water pollution control (aquifer cleanup). The focus on methodologies has been heightened by current hazardous waste site cleanup efforts funded by "Superfund."

CLASSIFICATION OF METHODOLOGIES

Table 1.1 lists methodologies for ground water quality protection and treatment organized according to whether the pollution problem is acute or chronic. Acute pollution may occur from inadvertent spills of chemicals or releases of undesirable materials and chemicals during a transportation accident. Acute pollution events are unplanned and are characterized by their emergency nature. Chronic aquifer pollution may occur from numerous point and area sources and may involve traditional pollutants such as nitrates and bacteria, or unique pollutants such as gasoline, metals, and synthetic organic chemicals.

1

Table 1.1   Methodologies for Ground Water Quality Protection and Treatment

Pollution Problem	Goal	Methodologies
Acute	Abatement	1. In situ chemical fixation. 2. Excavation of contaminated soil with subsequent backfilling with "clean" soil.
	Restoration	1. Removal wells, treatment of contaminated ground water, and recharge. 2. Removal wells, treatment of contaminated ground water, and discharge to surface drainage. 3. Removal wells and discharge to surface drainage.
Chronic	Abatement	1. In situ chemical fixation. 2. Excavation of contaminated soil with subsequent backfilling with "clean" soil. 3. Interceptor trenches to collect polluted water as it moves laterally away from site. 4. Surface capping with impermeable material to inhibit infiltration of leachate-producing precipitation. 5. Subsurface barriers of impermeable materials to restrict hydraulic flow from sources. 6. Modify pumping patterns at existing wells. 7. Inject fresh water in a series of wells placed around source or contaminant plume to develop pressure ridge to restrict movement of pollutants.
Chronic	Restoration	1. Removal wells, treatment of contaminated ground water, and recharge. 2. Removal wells, treatment of contaminated ground water, and discharge to surface drainage. 3. Removal wells and discharge to surface drainage. 4. In situ chemical treatment. 5. In situ biological treatment.

Methodologies for ground water pollution control can also be considered in terms of the goals of abatement and restoration. Abatement means "to put an end to"; therefore, abatement as a goal refers to the application of methodologies which will aid in preventing pollutant movement into ground water, or preventing contaminated plume movement into usable aquifer zones. The latter example of abatement is also called plume management. Aquifer restoration as a goal refers to the restoration of water quality to its normal quality, usually by removing both the source of pollution and renovating the polluted ground water. If the pollution source has already been dissipated by time, restoration may only involve renovation of the polluted ground water.

It should be noted that a given aquifer cleanup project may involve usage of several methodologies in combination. For example, for an acute problem, excavation and backfilling may be used in conjunction with removal wells, treatment of contaminated ground water, and discharge to surface drainage. A chronic pollution cleanup project may include surface capping, subsurface barriers, and in situ chemical treatment.

OVERVIEW STUDIES

In 1977, Lindorff and Cartwright surveyed the nation for case histories of aquifer clean-up. Information on 116 cases of aquifer pollution was summarized, with most of the pollution either caused by industrial wastes or by leaching from municipal landfills. Table 1.2 summarizes the cleanup methodologies used in 32 of the cases. (Lindorff and Cartwright, 1977). Removal wells are best applied when the pollution has not traveled far from the source, and they have most often been used to remove oil and oil products. This technique has also been used to control water-soluble chemicals. For oil pollution two or more pumping locations in each well may be used, a deep pump to maintain a drawdown cone and a shallow skimming pump to remove the oil (Todd, 1973). For pollutants that have traveled very far from the source or have dispersed, the use of interceptor trenches is often more effective than the drawdown cone in a removal well. The interceptor trench (or pit or ditch) is placed across the pollution plume to capture the ground water flow. The water is then pumped out and treated. This method is obviously most useful for shallow aquifers (Committee on Environmental Affairs, 1972).

From May to October, 1980, a survey of on-going and completed remedial action projects was conducted by Neely et al. A total of 169 hazardous waste sites were identified; however, because controls for ground water cleanup are more expensive and time consuming than those for surface water, complete remedial measures were implemented at only a few sites. Details of remedial measures were reported for nine selected sites, with the measures appearing to be effective at only two sites (Neely et al., 1981).

An expert seminar was held in Paris, France in November, 1980, under sponsorship of the Organization for Economic Co-operation and Development (OECD). Table 1.3 summarizes remedial actions used at thirteen sites including five from the USA, four from Germany, and one each from France, the Netherlands, Sweden, and the United Kingdom (Organization for Economic Co-operation and Development, 1983). One con-

Table 1.2   Aquifer Cleanup Survey (Lindorff and Cartwright, 1977)

Remedial Action	State	Ground Water Pollutant or Source
Removal wells	California	gasoline
	Delaware	landfilling
	Georgia	gasoline
	Illinois	acrylonitrile
		chemical wastes
		cyanide
	Indiana	acetone cyanohydrin
		oil
	Kentucky	hydrochloric acid
	Michigan	industrial waste pond
	Minnesota	solvents and acids
	New Jersey	chromium
		heavy metals
	New Mexico	chlorides
	New York	gasoline
	Ohio	brines
		chlorides
	Oregon	aluminum and mine tailings
	Pennsylvania	fuel oil
		gasoline
		insecticides
		terpenes and acids
	Wisconsin	organic wastes
		sulfite liquor
Removal wells and treatment	California	phenols [a]
	Connecticut	styrene [b]
	New Jersey	petrochemicals [b]
Interception trench	Georgia	gasoline
	Idaho	cadmium, lead and zinc
	Pennsylvania	fuel oil
		landfill
	South Dakota	landfill

[a]Treatment with chlorine dioxide.
[b]Treatment with activated carbon filters.

clusion from the seminar was that a careful evaluation of the cost-effectiveness of possible remedial measures is essential before decisions are made with regard to implementation of a particular methodology. All components of the cost of a remedial program must be taken into account. These include, in addition to the actual engineering cost, the physicochemical site evaluation required prior to implementation of

Table 1.3 Examples of Aquifer Clean-up Projects (Organization for Economic Cooperation and Development, 1983)

Brief Site Description	Nature of "Problem"	Remedial Action	Comments
1. *Hamburg, West Germany* Contamination, mainly heavy metals detected in 1978 at a central decontamination plant in an industrial area. The 100-m² area contained 100 m³ of contaminated material.	Soil and groundwater contamination shown by drilling and sampling.	Removal of contaminated soil to hazardous waste disposal site. Pumping of groundwater.	
2. *La Bounty Dump, Charles City, Iowa, USA* Chemical waste dump operated in river flood plain over major aquifer, 1952–1977. Contained over 3 million tons of feed additives and veterinary pharmaceuticals including 28 priority pollutants.	Pollution of river.	Monitoring and closure of site with a "capping". Some in situ stabilization may follow.	If remedial action is not effective enough, an up gradient cut-off wall may be installed to reduce water flow.
3. *Windham, Connecticut, USA* Landfill to Willimatic Township	Groundwater pollution and threatened public water supply reservoir.	PVC Cap installed to prevent rain water inflow to refuse. Site closed.	90% of water to be intercepted and routed away.
4. *Olin Corporation Site, Saltville, PA., USA* Mercury wastes from nearby alkali industry.	Mercury globules visible at surface. River polluted. Vapor hazard.	Temporary cosmetic treatment to reduce immediate hazard by covering exposed waste near river with plastic and sand to reduce pollution to river.	Only satisfactory solution is a $23 million removal operation. Recovery of mercury currently not economical because of price slump.

Table 1.3, continued

Brief Site Description	Nature of "Problem"	Remedial Action	Comments
5. *Firestone Tire and Rubber Site, Pottstown, PA., USA* Tires, scrubber waste, organic waste, pigments, and PVC sludge.	Unknown.	Recovery wells to intercept polluted water and recycle it through plant.	100% efficient solution of the problem anticipated. Length of pumping and costs not quoted.
6. *Various locations in USA* Sites containing solvents, oils, paint wastes, heavy metals, acid sludges and PCBs.	Toxic liquid lagoons. Leachate surface migration. Leachate spread.	Techniques include: removal of waste at the site; containment of waste at the site; emptying of pits/lagoons; oil absorption in fuller's earth; mobile active carbon unit.	Many cases had received temporary remedial treatment and were awaiting funds before complete remedial work could be carried out.
7. *City of Hamburg, Germany* Toluene contamination from 700 m³ of waste reached 100 m² of subway in the marketing area of the city.	Damage to concrete of subway. Soil and groundwater contamination. Toluene vapor.	Reconstruction of part of tunnel shown by drilling and sampling to be in contaminated area. Removal of soil to proper landfill site. Three years ground water pumping.	
8. *Grenzach Wyllen, Germany* Disposal of chlorinated hydrocarbons, chemical warfare agents in gravel pit.	Chlorinated hydrocarbons in drinking water. Highly toxic material.		Development of special investigation and analysis methods.
9. *Marals de Ponteau, France* 30,000 m³ of oil waste on salt.	Surface water contamination aesthetic insult.	Removal, incineration and solidification.	Cost—6 million francs. Successful solidification.

10. *Lekkerkerk, Netherlands* Houses built on chemical waste with high solvent content.	Contamination of drinking water and air.	Total removal, incineration and disposal.	High cost, evacuation of residents necessary.
11. *R. T. Kemi, Sweden* Large-scale disposal of pesticides in agricultural region near surface water.	Surface water contamination, odors, and possible public health impacts.	Water management, excavation and incineration.	Experience in groundwater purging. Heavy participation by public media.
12. *Malkins Bank, United Kingdom* General industrial waste tip in rural area.	Noxious odors, surface water contamination.	Encapsulation and restoration to public golf course.	Private land acquired by government.
13. *Gendorf, Germany* Organics and other pesticides on production site.	Soil, groundwater and surface water contamination.	Encapsulation, surface sealing and solidification.	Excavation too dangerous. Permanent restrictions on land use.

remedial measures and the subsequent monitoring. It is possible that the cost of prior evaluation and subsequent monitoring may equal that of the remedial methodology used.

As a result of these overview studies, three generalizations about ground water pollution control strategies can be made. First, they are costly. Restoration costs may run into tens of millions of dollars. Second, they are time consuming. Because most problems have been recognized only after the pollutant has been moving and spreading for a number of years, the areal extent of pollution is often quite large. Consequently, it takes a long time to clean up these large problems. Third, aquifer restoration strategies are not always effective. Although the success ratio of recent years has increased, there are a number of instances where the cleanup strategy has been ineffective and second and third attempts to restore the aquifer have been made.

ORGANIZATION OF BOOK

This book addresses ground water quality protection and treatment in accordance with three major headings: Technologies; Decision-Making; and Case Studies and Applications. Following this introductory chapter, three chapters address Technologies: physical control measures; treatment of ground water; and in situ measures. Decision-making is addressed in chapters on a protocol for aquifer restoration decision-making, techniques for decision-making, risk assessment as related to aquifer restoration planning, and public participation. Appendix A summarizes a number of case studies of ground water pollution control, while Appendix B contains a topically organized annotated bibliography of 225 selected references.

SELECTED REFERENCES

Committee on Environmental Affairs, "The Migration of Petroleum Products in Soil and Ground Water — Principles and Countermeasures", Publ No. 4149, 1972, American Petroleum Institute, Washington, D.C.

Lindorff, D.E., and Cartwright, K., "Ground Water Contamination: Problems and Remedial Actions", Report No. 81, May 1977, Illinois State Geological Survey, Urbana, Illinois.

Neely, D., et al., "Remedial Actions at Hazardous Waste Sites, Survey and Case Studies", EPA 430/9-81-05, Jan. 1981, U.S. Environmental Protection Agency, Cincinnati, Ohio.

Organization for Economic Co-operation and Development, "Hazardous Waste Problem Sites", 1983, Paris, France.

Todd, D.K., "Ground Water Pollution in Europe—A Conference Summary", Contract No. EPA-68-01-0759, 1973, Center for Advanced Studies, General Electric Co., Santa Barbara, California.

TECHNOLOGIES FOR GROUND WATER POLLUTION CONTROL

CHAPTER 2

Physical Control Measures

This chapter is primarily oriented to those measures for preventing/ minimizing the occurrence of ground water pollution. Included herein is information on source control strategies, stabilization/solidification strategies, well systems, interceptor systems, surface capping and liners, sheet piling, grouting and slurry walls.

SOURCE CONTROL STRATEGIES

Source control strategies represent attempts to minimize or prevent ground water pollution before a potential polluting activity is initiated. The objectives of source control strategies are to reduce the volume of waste to be handled, or reduce the threat a certain waste poses by altering its physical or chemical makeup.

Source control strategies are most applicable in the design of new facilities as a component of one or more ground water pollution prevention steps. However, some of the strategies can be used as abatement measures for existing facilities.

The development and design of each of the strategies listed below are site and situation specific, depending very heavily on the type of waste to be treated. Probably the only criterion universally applicable to these strategies is that they will require extensive monitoring and recording. Most certainly, indiscriminate dumping must become a thing of the past. Some general advantages of source control strategies include:

(1) reduces the threat to the ground water environment;
(2) accelerates the time for "stabilization" of waste disposal facilities; and
(3) offers opportunities for economic recovery.

Two general disadvantages of source control strategies are:

(1) increased capital and maintenance costs; and
(2) monitoring and skilled operator requirements.

Table 2.1 lists some of the source control strategies that can be applied for ground water pollution prevention, usually at waste disposal facilities. The list is not totally comprehensive. It should be emphasized, however, that the possibility for implementation of source control strategies should be examined for all future waste disposal facilities. The following section contains additional information on stabilization and solidification strategies.

Table 2.1   Potential Source Control Strategies

I. **Volume Reduction Strategies**
   A. Recycling
   B. Resource Recovery
      1. materials recovery
      2. waste-to-energy conversion
   C. Centrifugation
   D. Filtration
   E. Sand Drying Beds
II. **Physical/Chemical Alteration Strategies**
   A. Chemical Fixation
      1. neutralization
      2. precipitation
      3. chelation
      4. cementation
      5. oxidation-reduction
      6. biodegradation
   B. Detoxification
      1. thermal
      2. chemical—ion-exchange, pyrolysis, etc.
      3. biological—activated sludge, aerated lagoons, etc.
   C. Degradation
      1. hydrolysis
      2. dechlorination
      3. photolysis
      4. oxidation
   D. Encapsulation
   E. Waste Segregation
   F. Co-Disposal
   G. Leachate Recirculation

STABILIZATION/SOLIDIFICATION STRATEGIES

One means of reducing the threat to ground water from land disposal of waste materials is to structurally isolate the waste material in a solid matrix prior to landfilling. This process, known as stabilization/solidification, is becoming increasingly popular for hazardous and radioactive waste disposal (U.S. Environmental Protection Agency, 1979). The objective of solidification/stabilization processes is to chemically fix the waste in a solid matrix. This reduces the exposed surface area and minimizes leaching of toxic constituents. Effective immobilization includes reacting toxic components chemically to form compounds immobile in the environment and/or entrapping the toxic material in an inert stable solid. Thus stabilization and solidification have different meanings, although the terms are often used interchangeably. From a definitional perspective, stabilization refers to immobilization by chemical reaction or entrapping (watertight inert polymer or crystal lattice), while solidification means the production of a solid, monolithic mass with sufficient integrity to be easily transported. It will become apparent that these processes may overlap or take place within one operation. An example is cementation where the process both stabilizes by producing insoluble heavy metal compounds, and solidifies into a formed mass while entrapping the pollutants.

Chemical stabilization is designed to provide a substance which is more resistant to leaching, and also more amenable to the solidification process. By chemically fixing the hazardous waste constituents, their release will be minimized in the event of a breakdown of the solid matrix. Probably the simplest stabilization process is pH adjustment. In most industrial sludges, toxic metals are precipitated as amorphous hydroxides that are insoluble at an elevated pH. By carefully selecting a stabilization system of suitable pH, the solubility of any metal hydroxide can be minimized. Certain metals can also be stabilized by forming insoluble carbonates or sulfides. Care should be taken to ensure that these metals are not remobilized because of changes in pH or redox conditions after they have been introduced into the environment. Where possible, it is desirable to co-dispose of wastes which stabilize without the addition of extraneous chemicals.

Stabilized wastes can be solidified into a solid mass by microencapsulation or macroencapsulation. Microencapsulation refers to the dispersion and chemical reaction of the toxic materials within a solid matrix. Therefore, any breakdown of the solid material only exposes the material located at the surface to potential release to the environment. Macroencapsulation is the sealing of the waste in a thick, relatively impermeable coating layer. Plastic and asphalt coatings, or secured land-filling, are considered to be macroencapsulation methods. Breakdown of the protective layer with macroencapsulation could result in a significant release of toxic material to the environment.

Stabilization techniques have concentrated on the containment of toxic inorganic compounds. This is because many of the techniques originated as methods for treating radioactive wastes which consist primarily of inorganic isotopes. Also, organic compounds may interfere with the stabilization/solidification process, although small amounts may be incorporated under test conditions. Chemical oxidation or incineration have been found to be the most successful treatment methods for the majority of dangerous organic chemicals.

Stabilization/solidification processes should produce a material whose physical placement will not render the land on which it is disposed unusable for other purposes. The resultant stabilized/solidified material should be impervious, with good dimensional stability and load-bearing characteristics. It should also have satisfactory wet-dry and freeze-thaw weathering resistance.

Present solidification/stabilization systems can be grouped into seven classes or processes:

(1) solidification through cement addition;
(2) solidification through the addition of lime and other pozzolanic materials;
(3) techniques involving embedding wastes in thermoplastic materials such as bitumen, paraffin, or polyethylene;
(4) solidification by addition of an organic polymer;
(5) encapsulation of wastes in an inert coating;
(6) treatment of the wastes to produce a cementitious product with major additons of other constituents; and
(7) formation of a glass by fusion of wastes with silica.

WELL SYSTEMS

Well systems for ground water pollution control are based on manipulating the subsurface hydraulic gradient through injection or withdrawal of water. Well systems are designed to control the movement of the water phase directly, and subsurface pollutants indirectly. This approach is often referred to as plume management. There are three general classes of well systems. Two of these systems, which involve withdrawal of water, are well point systems and deep well systems. The other technology involves injection of water and can be referred to as a pressure ridge system. All three of the systems may require the installation of several wells at selected sites. These wells are then pumped (or injected) at specified rates so as to dictate the movement of the water phase and associated pollutants.

Construction

The single most important construction aspect of well systems is the installation of the wells. The sequence for installation of wells involves the following steps:

(1) setting up the drilling equipment;
(2) drilling the well hole;
(3) installing casings and liners;
(4) grouting and sealing the annular spaces between casings and boreholes;
(5) installing well screens and fittings;
(6) gravel-packing and placing materials; and
(7) developing the well.

Not all of the above steps are required for each well. Conversely, in addition to the above steps, the process could also require construction of aboveground facilities (pumphouses, etc.) and testing (pump tests).

Several methods exist for the initial drilling of the borehole, and these methods can usually be listed under one of the headings shown in Table 2.2. After the borehole is drilled, the next series of steps involves setting of casings, grouting and sealing of casings, and installing well screens. Excellent discussions are available on these procedures (Campbell and Lehr, 1977; Johnson Division, UOP Inc., 1975).

The next step in well installation is well development or completion. Development is the process by which the finer particles of the water-bearing formation around the well are removed, thus resulting in enlarged passages for moisture travel. Three beneficial results of well development are:

(1) correction of damage to, or clogging of, water-bearing formation which occurs as a side effect from drilling;
(2) increases the porosity and permeability of the formation in vicinity of the well; and
(3) stabilizes the sand formation around a screened well so that the well will yield water free of sand (Johnson Division, UOP Inc., 1975).

The final activity would be the construction of the aboveground facilities associated with the well. This could include pumphouses, pumps, and piping networks.

Table 2.2   Water Well Drilling Methods

Equipment Type	Procedure Description
1. Cable Tool	The cable tool drills by lifting and dropping a string of tools suspended on a cable. A bit at the bottom of the tool string strikes the bottom of the hold, crushing and breaking the formation material. Cuttings are removed by bailing or recirculation of a slurry. Casing is usually driven concurrent with drilling operations (Campbell and Lehr, 1977).
2. Hydraulic Rotary Drilling	Hydraulic rotary drilling consists of cutting a borehole by means of a rotating bit and removing the cuttings by continuous circulation of a drilling fluid as the bit penetrates the formation materials (Johnson Division, UOP Inc., 1975). Drilling fluid is usually a water-based slurry containing some fine material (clay) in suspension.
3. Reverse Circulation Drilling	Same as (2) except the fluid is forced downhole outside of the casing and returns up the inside of the casing. For (2) the route is down inside the casing and returning up outside the casing.
4. Air Rotary Drilling	Same as (2) only forced air becomes the drilling fluid.
5. Air-Percussion Rotary Drilling	Rotary technique in which the main source of energy for fracturing rock is obtained from a percussion machine connected directly to the bit.
6. Hollow-Rod Drilling	Similar to (1) except with more rapid and shorter strokes. Ball and check valve on bit allows water and cuttings to enter and travel up the casing (hollow-rod).
7. Jet Drilling	Similar to (6) except water is pumped down inside casing (hollow-rod) and returns up outside the casing.
8. Driven Wells	Pipe with a driving plug on the end is driven into the ground by repeated blows from a driving weight.
9. Hollow Stem Auger Drilling	Wells can be drilled in unconsolidated material and sampled by coring, with no drilling fluid or air required. Limited to shallow (less than 100 ft) unconsolidated material.

Design

The design of any of the three well systems should be preceded by a preliminary investigation. This preliminary investigation should involve a hydrogeologic study of the area. The hydrogeologic study should generate two categories of information. First, the contaminant plume dimensions (width, length, depth and general shape) and the hydraulic gradient across the plume should be determined (Lundy and Mahan,

1982). Second, the hydrogeologic characteristics of the aquifer should be identified. This information can be used along with the concepts of well hydraulics to establish the number and spacing of needed wells.

Some equations available for analyzing the flow to wells under different aquifer conditions are given in Tables 2.3 and 2.4 (Cohen and Miller, 1983). Table 2.5 summarizes the methods that can be used to calculate the radius of influence under various aquifer conditions (Kufs et al., 1983). Simplistically, the design of a well system involves an iterative procedure of utilizing a set pumping rate, calculating the radius of influence associated with that rate, and checking the radius of influence against the plume dimensions.

Lundy and Mahan (1982) identify three constraints which well recovery systems must satisfy:

(1) the recovery system must be located near enough to the downgradient plume boundary to reverse the hydraulic gradient at that boundary;

(2) total discharge is large enough to create limiting flowlines that bound the plume up to its widest part upgradient of the well system; and

(3) drawdowns of water levels resulting from withdrawals from the system do not exceed limits that are a function of aquifer saturated thickness.

Well point systems involve a number of closely spaced, shallow wells each connected to a main pipe (header) which is connected to a centrally located suction lift pump. Well point systems are used only for shallow water table aquifers because the maximum drawdown obtainable by suction life is the difference between the suction limit of the pump and the distance from the static water level to the center of the pump. The upper limit for lifting water by suction lift is around 25 ft. Pumping by suction lift is depicted in Figure 2.1 (Johnson Division, UOP Inc., 1975). Well point systems should be designed so that the radius of influence (drawdown) of the system completely intercepts the plume of contaminants, as shown in Figure 2.2. Deep well systems are similar to well point systems, except they are used for greater depths and are usually pumped individually. The deep well system should be designed so that its radius of influence completely intercepts the contaminant plume, as shown in Figure 2.2.

Pressure ridge systems can be conceptualized as the inverse of the previous two systems. Pressure ridges use the injection of water to form an upconing of the water table which acts as a barrier to ground water flow. Figure 2.3 depicts the phenomena of a pressure ridge system. These systems have found their widest application in coastal areas to prevent saltwater intrusion.

The basis of design for all three of the well systems is that movement of the ground water flow and pollutant be completely controlled.

Table 2.3   Analytical Solutions of Steady Flow Applicable to Corrective Actions (Cohen and Miller, 1983)

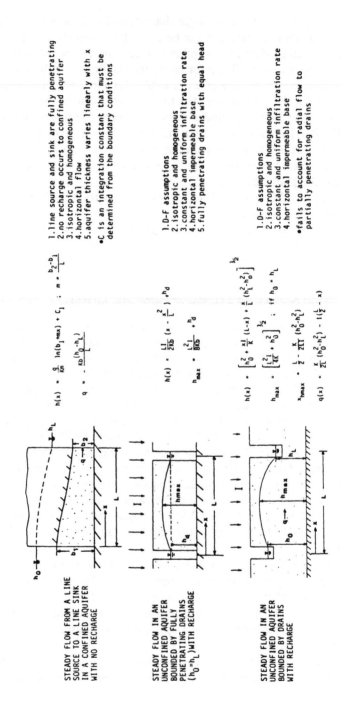

STEADY FLOW FROM A LINE SOURCE TO A LINE SINK IN A CONFINED AQUIFER WITH NO RECHARGE

$$h(x) = \frac{q}{Kn} \ln(b_1 \cdot m_x) + C_1 \; ; \quad m = \frac{b_2 - b_1}{L}$$

$$q = -\frac{Kb'(h_0 - h_L)}{L}$$

1. line source and sink are fully penetrating
2. no recharge occurs to confined aquifer
3. isotropic and homogeneous
4. horizontal flow
5. aquifer thickness varies linearly with x

● C is an integration constant that must be determined from the boundary conditions

STEADY FLOW IN AN UNCONFINED AQUIFER BOUNDED BY FULLY PENETRATING DRAINS $(h_0 = h_L)$ WITH RECHARGE

$$h(x) = \frac{LI}{2Kb}\left(x - \frac{x^2}{L}\right) + h_d$$

$$h_{max} = \sqrt{\frac{L^2 I}{8Kb}} + h_d$$

1. D-F assumptions
2. isotropic and homogeneous
3. constant and uniform infiltration rate
4. horizontal impermeable base
5. fully penetrating drains with equal head

STEADY FLOW IN AN UNCONFINED AQUIFER BOUNDED BY DRAINS WITH RECHARGE

$$h(x) = \left[h_0^2 + \frac{xI}{K}(L-x) + \frac{x}{L}(h_L^2 - h_0^2)\right]^{\frac{1}{2}}$$

$$x_{hmax} = \left[\frac{L^2 I}{4K} + h_0^2\right]^{\frac{3}{2}} \; ; \; \text{if } h_0 = h_L$$

$$x_{hmax} = \frac{L}{2} - \frac{K}{2LI}(h_0^2 - h_L^2)$$

$$q(x) = \frac{K}{2L}(h_0^2 - h_L^2) - I\left(\frac{L}{2} - x\right)$$

1. D-F assumptions
2. isotropic and homogeneous
3. constant and uniform infiltration rate
4. horizontal impermeable base

● fails to account for radial flow to partially penetrating drains

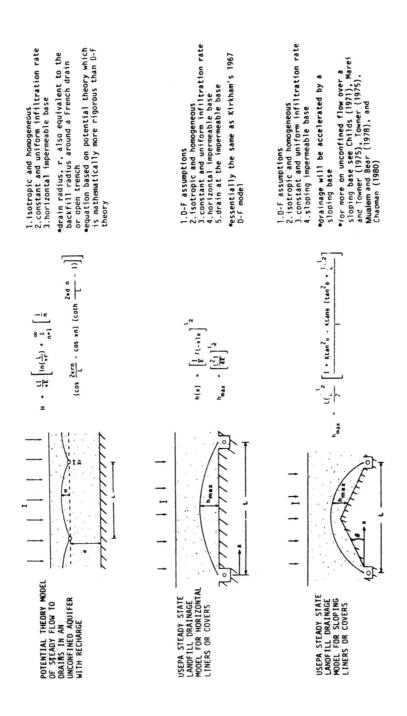

POTENTIAL THEORY MODEL OF STEADY FLOW TO DRAINS IN AN UNCONFINED AQUIFER WITH RECHARGE

1. isotropic and homogeneous
2. constant and uniform infiltration rate
3. horizontal impermeable base

- drain radius, r, also equivalent to the backfill radius around a French drain or open trench
- equation based on potential theory which is mathematically more rigorous than D-F theory

$$H = \frac{LI}{\pi K}\left[\ln\left(\frac{L}{\pi r}\right) + \sum_{n=1}^{\infty}\frac{1}{n}\left(\cos\frac{2\pi rn}{L} - \cos \pi n\right)\left(\coth\frac{2\pi d\,n}{L} - 1\right)\right]$$

USEPA STEADY STATE LANDFILL DRAINAGE MODEL FOR HORIZONTAL LINERS OR COVERS

1. D-F assumptions
2. isotropic and homogeneous
3. constant and uniform infiltration rate
4. horizontal impermeable base
5. drain at the impermeable base

- essentially the same as Kirkham's 1967 D-F model

$$h(x) = \left[\frac{I}{K}(L-x)x\right]^{\frac{1}{2}}$$

$$h_{max} = \left[\frac{L^2 I}{4K}\right]^{\frac{1}{2}}$$

USEPA STEADY STATE LANDFILL DRAINAGE MODEL FOR SLOPING LINERS OR COVERS

1. D-F assumptions
2. isotropic and homogeneous
3. constant and uniform infiltration rate
4. sloping impermeable base

- drainage will be accelerated by a sloping base
- for more on unconfined flow over a sloping base see Childs (1971), Marei and Towner (1975), Towner (1975), Mualem and Bear (1978), and Chapman (1980)

$$h_{max} = L\left(\frac{I}{K}\right)^{\frac{1}{2}}\left[\frac{1 + K\tan^2\theta_v - K\tan\theta\,(\tan^2\theta + \frac{I}{K})^{\frac{1}{2}}}{A}\right]$$

Table 2.3, continued

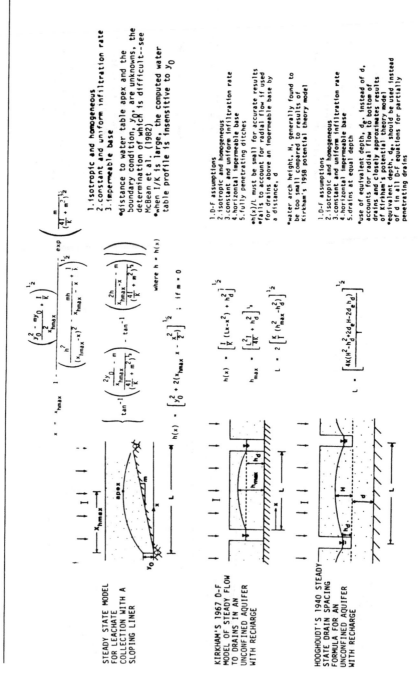

STEADY STATE MODEL FOR LEACHATE COLLECTION WITH A SLOPING LINER

1. isotropic and homogeneous
2. constant and uniform infiltration rate
3. impermeable base

• distance to water table apex and the boundary condition, y_0, are unknowns, the determination of which is difficult--see McBean et al. (1982)
• when I/K is large, the computed water table profile is insensitive to y_0

KIRKHAM'S 1967 D-F MODEL OF STEADY FLOW TO DRAINS IN AN UNCONFINED AQUIFER WITH RECHARGE

1. D-F assumptions
2. isotropic and homogeneous
3. constant and uniform infiltration rate
4. horizontal impermeable base
5. fully penetrating ditches

• h(x)/L must be small for accurate results
• fails to account for radial flow if used for drains above an impermeable base by a distance, d
• water arch height, H, generally found to be too small compared to results of Kirkham's 1958 potential theory model

HOOGHOUDT'S 1940 STEADY STATE DRAIN SPACING FORMULA FOR AN UNCONFINED AQUIFER WITH RECHARGE

1. D-F assumptions
2. isotropic and homogeneous
3. constant and uniform infiltration rate
4. horizontal impermeable base
5. drains at equal depth

• use of equivalent depth, d_e, instead of d, accounts for radial flow to bottom of drains and closely approximates results of Kirkham's potential theory model
• equivalent depth, d_e, should be used instead of d in all D-F equations for partially penetrating drains

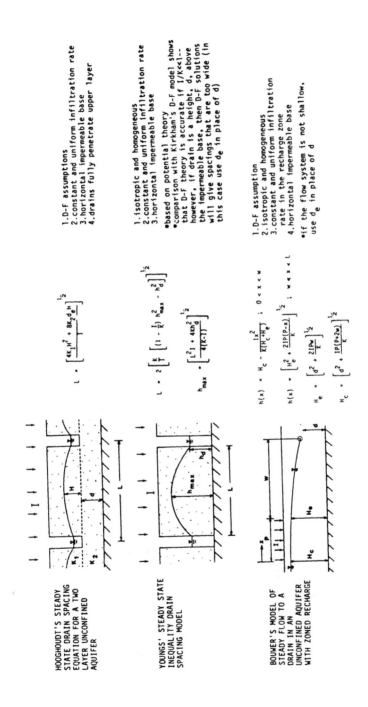

HOOGHOUDT'S STEADY STATE DRAIN SPACING EQUATION FOR A TWO LAYER UNCONFINED AQUIFER

$$L = \left[\frac{4K_1 H^2 + 8K_2 d_e H}{I} \right]^{\frac{1}{2}}$$

1. D-F assumptions
2. constant and uniform infiltration rate
3. horizontal impermeable base
4. drains fully penetrate upper layer

YOUNGS' STEADY STATE INEQUALITY DRAIN SPACING MODEL

$$L = 2 \left[\frac{K}{I} \left[1 - \frac{I}{K} \right] h_{max}^2 - h_d^2 \right]^{\frac{1}{2}}$$

$$h_{max} = \left[\frac{L^2 I + 4Kh_d^2}{4(K-I)} \right]^{\frac{1}{2}}$$

1. isotropic and homogeneous
2. constant and uniform infiltration rate
3. horizontal impermeable base

• based on potential theory
• comparison with Kirkham's D-F model shows that D-F theory is accurate if I/K<<1--however, if drain is a height, d, above the impermeable base, then D-F solutions will give spacings that are too wide (in this case use d_e in place of d)

BOUWER'S MODEL OF STEADY FLOW TO A DRAIN IN AN UNCONFINED AQUIFER WITH ZONED RECHARGE

$$h(x) = H_c - \frac{Ix^2}{K(H_c + H_e)} \quad ; \quad 0 < x < w$$

$$h(x) = \left[H_e^2 + \frac{2IP(P-x)}{K} \right]^{\frac{1}{2}} \quad ; \quad w \le x \le L$$

$$H_e = \left[d^2 + \frac{2IPw}{K} \right]^{\frac{1}{2}}$$

$$H_c = \left[d^2 + \frac{IP(P+2w)}{K} \right]^{\frac{1}{2}}$$

1. D-F assumption
2. isotropic and homogeneous
3. constant and uniform infiltration rate in the recharge zone
4. horizontal impermeable base

• if the flow system is not shallow, use d_e in place of d

Table 2.3, continued

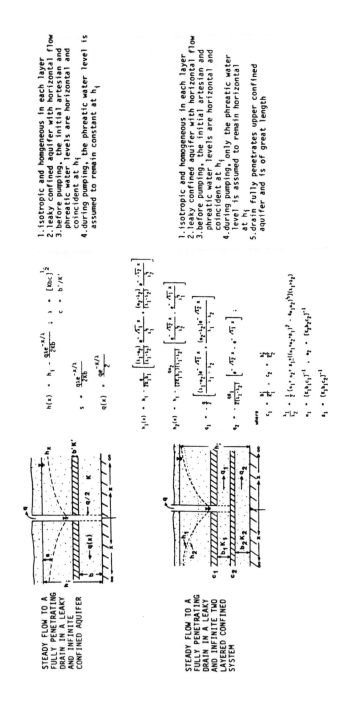

STEADY FLOW TO A FULLY PENETRATING DRAIN IN A LEAKY AND INFINITE CONFINED AQUIFER

$$h(x) = h_i - \frac{q\lambda e^{-x/\lambda}}{2Kb} \quad ; \quad \lambda = [Kbc]^{\frac{1}{2}} \quad ; \quad c = b'/K'$$

$$s = \frac{q\lambda e^{-x/\lambda}}{2Kb}$$

$$q(x) = \frac{qe^{-x/\lambda}}{2}$$

1. isotropic and homogeneous in each layer
2. leaky confined aquifer with horizontal flow
3. before pumping, the initial artesian and phreatic water levels are horizontal and coincident at h_i
4. during pumping, the phreatic water level is assumed to remain constant at h_i

STEADY FLOW TO A FULLY PENETRATING DRAIN IN A LEAKY AND INFINITE TWO LAYERED CONFINED SYSTEM

$$h_1(x) = h_i - \frac{q\sigma}{2K_1b_1}\left[\frac{(1_1+z_2)e^{-\sqrt{z_1}x}}{z_1-z_2}\cdot\frac{\sqrt{z_1}x}{z_1} - \frac{(z_2+z_2)}{z_1-z_2}e^{-\sqrt{z_2}x}\cdot\frac{\sqrt{z_2}x}{z_2}\right]$$

$$h_2(x) = h_i - \frac{q\sigma}{2K_2b_2}\left[\frac{(1_1+z_2)e^{-\sqrt{z_1}x}}{z_1-z_2} - \frac{(z_2+z_2)e^{-\sqrt{z_2}x}}{z_1-z_2}\right]$$

$$q_1 = -\frac{q}{2}\frac{\sigma}{z_1[z_1-z_2]}\left[(1_1+z_2)e^{-\sqrt{z_1}x}-(z_2+z_2)e^{-\sqrt{z_2}x}\right]$$

$$q_2 = -\frac{q\sigma}{2[z_1-z_2]}\left[-e^{-\sqrt{z_1}x}-e^{-\sqrt{z_2}x}\right];$$

where
$$c_1 = \frac{b_1'}{K_1'} \quad ; \quad c_2 = \frac{b_2'}{K_2'}$$

$$\frac{z_1}{z_2} = \frac{1}{2}\{\alpha_1+\alpha_2+\beta_1\pm[(\alpha_1+\alpha_2+\beta_1)^2-4\alpha_1\beta_2]^{\frac{1}{2}}\}(\beta_1+\beta_2)$$

$$\alpha_1 = [c_1b_1c_1]^{-1} \quad , \quad \alpha_2 = [c_2b_2c_2]^{-1}$$

$$\beta_1 = [c_1b_2c_2]^{-1}$$

1. isotropic and homogeneous in each layer
2. leaky confined aquifer with horizontal flow
3. before pumping, the initial artesian and phreatic water levels are horizontal and coincident at h_i
4. during pumping, only the phreatic water level is assumed to remain horizontal at h_i
5. drain fully penetrates upper confined aquifer and is of great length

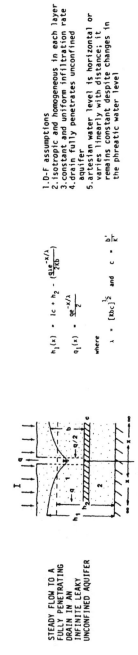

STEADY FLOW TO A FULLY PENETRATING DRAIN IN AN INFINITE LEAKY UNCONFINED AQUIFER

$$h_1(x) = lc + h_2 - \left(\frac{qle^{-x/\lambda}}{2Kb}\right)$$

$$q_1(x) = \frac{qe^{-x/\lambda}}{2}$$

where

$$\lambda = [Kbc]^{1/2} \quad \text{and} \quad c = \frac{b'}{K'}$$

1. D-F assumptions
2. isotropic and homogeneous in each layer
3. constant and uniform infiltration rate
4. drain fully penetrates unconfined aquifer
5. artesian water level is horizontal or varies linearly with distance; it remains constant despite changes in the phreatic water level

STEADY FLOW TO PARTIALLY PENETRATING DRAINS IN A LEAKY UNCONFINED AQUIFER

$$h_1(x) = h_2 + C_1 e^{-x/\lambda} + C_2 e^{x/\lambda} + \frac{l\lambda^2}{Kb}$$

$$\lambda = [Kbc]^{1/2}$$

$$c = \frac{b'}{K'}$$

1. D-F assumptions
2. isotropic and homogeneous in each layer
3. constant and uniform infiltration rate
4. artesian water level remains constant
5. horizontal impermeable base

* C_1 and C_2 are integration constants that must be determined from the boundary conditions

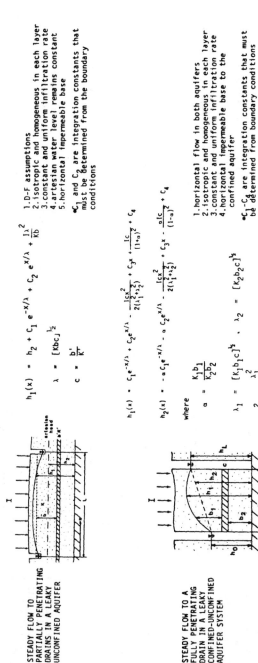

STEADY FLOW TO A FULLY PENETRATING DRAIN IN A LEAKY CONFINED-UNCONFINED AQUIFER SYSTEM

$$h_1(x) = C_1 e^{-x/\lambda} + C_2 e^{x/\lambda} - \frac{lcx^2}{2(\lambda_1^2+\lambda_2^2)} + C_3 x + \frac{lc}{(1+\alpha)}\frac{x^2}{2} + C_4$$

$$h_2(x) = -\alpha C_1 e^{-x/\lambda} - \alpha C_2 e^{x/\lambda} - \frac{lcx^2}{2(\lambda_1^2+\lambda_2^2)} + C_3 x - \frac{\alpha lc}{(1-\alpha)}\frac{x^2}{2} + C_4$$

where

$$\alpha = \frac{K_1 b_1}{K_2 b_2}$$

$$\lambda_1 = [K_1 b_1 c]^{1/2}, \quad \lambda_2 = [K_2 b_2 c]^{1/2}$$

$$\lambda^2 = \frac{\lambda_1^2}{1+\alpha}$$

$$c = \frac{b'}{K'}$$

1. horizontal flow in both aquifers
2. isotropic and homogeneous in each layer
3. constant and uniform infiltration rate
4. horizontal impermeable base to the confined aquifer

* C_1-C_4 are integration constants that must be determined from boundary conditions

Table 2.4   Analytical Solutions of Transient Flow Applicable to Corrective Actions (Cohen and Miller, 1983)

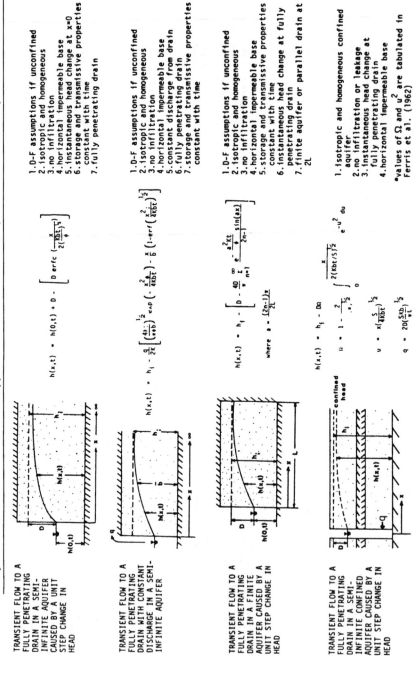

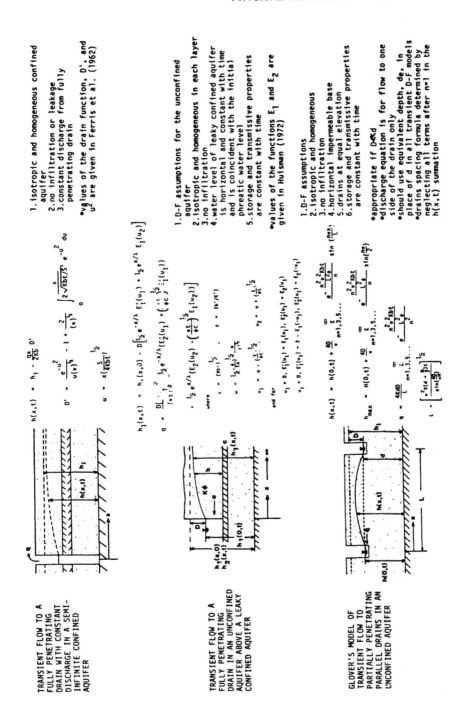

TRANSIENT FLOW TO A FULLY PENETRATING DRAIN WITH CONSTANT DISCHARGE IN A SEMI-INFINITE CONFINED AQUIFER

$$h(x,t) = h_i - \frac{Qx}{2KbS}\, D'$$

$$D' = \frac{e^{-u^2}}{u(\pi)^{\frac{1}{2}}} - 1 + \frac{2}{(\pi)^{\frac{1}{2}}} \int_0^{\frac{x}{2\sqrt{Kbt/S}}} e^{-u^2}\, du$$

$$u = x\left(\frac{S}{Kbt}\right)^{\frac{1}{2}}$$

1. isotropic and homogeneous confined aquifer
2. no infiltration or leakage
3. constant discharge from fully penetrating drain

• values of the drain function, D', and u^2 are given in Ferris et al. (1962)

TRANSIENT FLOW TO A FULLY PENETRATING DRAIN IN AN UNCONFINED AQUIFER ABOVE A LEAKY CONFINED AQUIFER

$$h_1(x,t) = h_i(x,0) - D\left[\tfrac{1}{2}e^{-x/\lambda} E_1'(u_1) + \tfrac{1}{2}e^{x/\lambda} E_1'(u_2)\right]$$

$$q = \frac{D(\cdot)}{(\pm)^{1/2}}\left[\tfrac{1}{2}e^{-x/\lambda}\big(E_2'(u_2) - (\tfrac{\pi t}{4c})^{\frac{1}{2}} E_1'(u_2)\big)\right]$$

$$+ \tfrac{1}{2}e^{x/\lambda}\big(E_2'(u_2) - (\tfrac{\pi t}{4c})^{\frac{1}{2}} E_1'(u_2)\big)\big]$$

where

$$\lambda = (cb)^{\frac{1}{2}} \qquad c = (b'/K')$$

$$u_1 = \tfrac{1}{2}(\tfrac{S}{Kbt})^{\frac{1}{2}} \cdot \tfrac{x}{t^{\frac{1}{2}}}$$

$$u_2 = u + (\tfrac{t}{4c})^{\frac{1}{2}}$$

and for

$$u_1 \ll 0, \quad E_1'(u_1), \; E_2'(u_1), \; E_2'(u_1)$$

$$u_1 \ll 0, \quad E_1'(u_1) = 2 - E_1'(-u_1), \; E_2'(u_1), \; E_2'(-u_1)$$

1. D-F assumptions for the unconfined aquifer
2. isotropic and homogeneous in each layer
3. no infiltration
4. water level of leaky confined aquifer is horizontal and constant with time and is coincident with the initial phreatic water level
5. storage and transmissive properties are constant with time

• values of the functions E_1 and E_2 are given in Huisman (1972)

GLOVER'S MODEL OF TRANSIENT FLOW TO PARTIALLY PENETRATING PARALLEL DRAINS IN AN UNCONFINED AQUIFER

$$h(x,t) = H(0,t) + \frac{40}{\pi}\sum_{n=1,3,5,\dots}^{\infty}\frac{e^{-\frac{n^2\pi^2 Kbt}{L^2}}}{n} \sin\left(\frac{n\pi x}{L}\right)$$

$$h_{max} = H(0,t) + \frac{40}{\pi}\sum_{n=1,3,5,\dots}^{\infty}\frac{e^{-\frac{n^2\pi^2 Kbt}{L^2}}}{n}\sin\frac{n\pi}{2}$$

$$q = \frac{4KdD}{L}\sum_{n=1,3,5,\dots}^{\infty} e^{-\frac{n^2\pi^2 Kbt}{L^2}}$$

$$L = \left[\frac{2c(c-\tfrac{B}{2})t}{S\ln(\frac{d}{h_i})}\right]^{\frac{1}{2}}$$

1. D-F assumptions
2. isotropic and homogeneous
3. no infiltration
4. horizontal impermeable base
5. drains at equal elevation
6. storage and transmissive properties are constant with time

• appropriate if $D < d$
• discharge equation is for flow to one side of the drain only
• should use equivalent depth, d_e, in place of d in all transient D-F models
• drains spacing formula determined by neglecting all terms after $n=1$ in the $h(x,t)$ summation

Table 2.4, continued

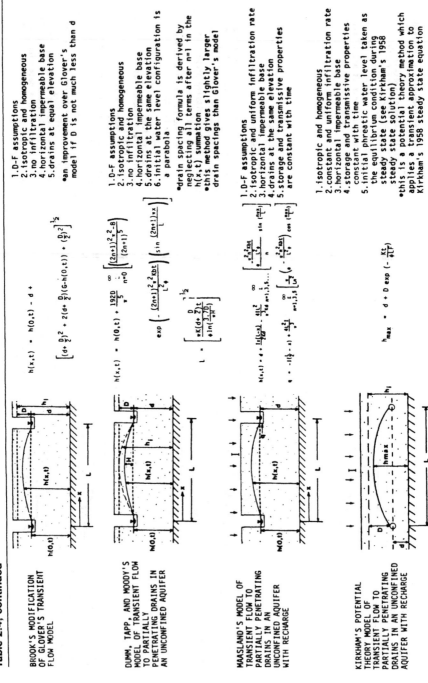

BROOK'S MODIFICATION OF GLOVER'S TRANSIENT FLOW MODEL

$$h(x,t) = h(0,t) - d + \left[(d+\tfrac{D}{2})^2 + 2(d+\tfrac{D}{2})(G-h(0,t)) + (\tfrac{D}{2})^2\right]^{\frac{1}{2}}$$

1. D-F assumptions
2. isotropic and homogeneous
3. no infiltration
4. horizontal impermeable base
5. drains at equal elevation

• an improvement over Glover's model if D is not much less than d

DUMM, TAPP, AND MOODY'S MODEL OF TRANSIENT FLOW TO PARTIALLY PENETRATING DRAINS IN AN UNCONFINED AQUIFER

$$h(x,t) = h(0,t) + \frac{192D}{\pi^5} \sum_{n=0}^{\infty} \left[\frac{(2n+1)^2 \pi^2 - 8}{(2n+1)^5}\right] \left(\sin\frac{(2n+1)\pi x}{L}\right) \exp\left(-\frac{(2n+1)^2 \pi^2 kbt}{L^2}\right)$$

$$L = \left[\frac{\pi K(d+\tfrac{D}{2})t}{\phi \ln\left(\frac{3.70}{\pi H}\right)}\right]^{\frac{1}{2}}$$

1. D-F assumptions
2. isotropic and homogeneous
3. no infiltration
4. horizontal impermeable base
5. drains at the same elevation
6. initial water level configuration is a parabola

• drain spacing formula is derived by neglecting all terms after n=1 in the h(x,t) summation
• this method gives slightly larger drain spacings than Glover's model

MAASLAND'S MODEL OF TRANSIENT FLOW TO PARTIALLY PENETRATING DRAINS IN AN UNCONFINED AQUIFER WITH RECHARGE

$$h(x,t) = d + \frac{i\frac{L}{4}(\frac{L}{4}-x)}{kd} - \frac{4iL^2}{\pi^3 kd} \sum_{n=1,3,5\ldots} \frac{1}{n^3}\left(e^{-\frac{\pi^2 kd t}{\phi L^2}}\right)\sin\frac{n\pi x}{L}$$

$$q = -i(\frac{L}{4}-x) - \frac{4iL}{\pi^2}\sum_{n=1,3,5}\frac{1}{n^2}\left(e^{-\frac{\pi^2 kd t}{\phi L^2}}\right)\cos\frac{n\pi x}{L}$$

1. D-F assumptions
2. isotropic and uniform infiltration rate
3. horizontal impermeable base
4. drains at the same elevation
5. storage and transmissive properties are constant with time

KIRKHAM'S POTENTIAL THEORY MODEL OF TRANSIENT FLOW TO PARTIALLY PENETRATING DRAINS IN AN UNCONFINED AQUIFER WITH RECHARGE

$$h_{max} = d + D \exp\left(-\frac{Kt}{\phi L^2}\right)$$

1. isotropic and homogeneous
2. constant and uniform infiltration rate
3. horizontal impermeable base
4. storage and transmissive properties constant with time
5. initial phreatic water level taken as the equilibrium condition during steady state (see Kirkham's 1958 steady state solution)

• this is a potential theory method which applies a transient approximation to Kirkham's 1958 steady state equation

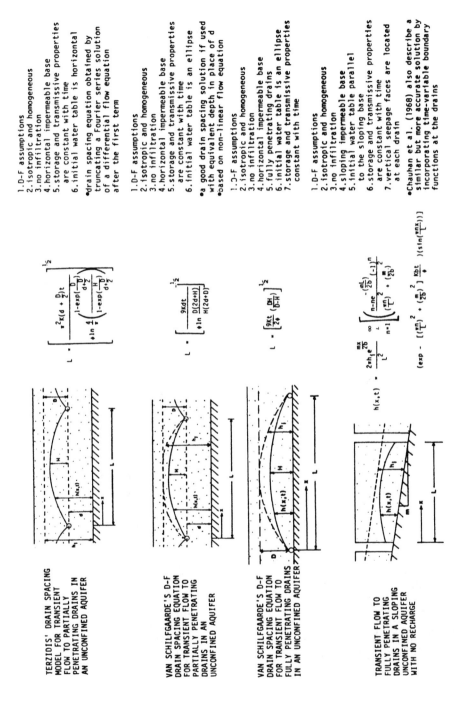

TERZIDIS' DRAIN SPACING MODEL FOR TRANSIENT FLOW TO PARTIALLY PENETRATING DRAINS IN AN UNCONFINED AQUIFER

1. D-F assumptions
2. isotropic and homogeneous
3. no infiltration
4. horizontal impermeable base
5. storage and transmissive properties are constant with time
6. initial water table is horizontal

• drain spacing equation obtained by truncating a Fourier series solution of a differential flow equation after the first term

$$L = \left[\frac{\pi^2 k (d + \frac{D}{2})t}{\phi \ln \frac{4}{\pi} \frac{1-\exp(-\frac{D}{d+\frac{D}{2}})}{1-\exp(-\frac{H}{d+\frac{D}{2}})}} \right]^{\frac{1}{2}}$$

VAN SCHILFGAARDE'S D-F DRAIN SPACING EQUATION FOR TRANSIENT FLOW TO PARTIALLY PENETRATING DRAINS IN AN UNCONFINED AQUIFER

1. D-F assumptions
2. isotropic and homogeneous
3. no infiltration
4. horizontal impermeable base
5. storage and transmissive properties are constant with time
6. initial water table is an ellipse

• a good drain spacing solution if used with equivalent depth in place of d
• based on non-linear flow equation

$$L = \left[\frac{9kdt}{\phi \ln \frac{D(2d+H)}{H(2d+D)}} \right]^{\frac{1}{2}}$$

VAN SCHILFGAARDE'S D-F DRAIN SPACING EQUATION FOR TRANSIENT FLOW TO FULLY PENETRATING DRAINS IN AN UNCONFINED AQUIFER

1. D-F assumptions
2. isotropic and homogeneous
3. no infiltration
4. horizontal impermeable base
5. fully penetrating drains
6. initial water table is an ellipse
7. storage and transmissive properties constant with time

$$L = \left[\frac{9kt}{2\phi} \left(\frac{DH}{D-H} \right) \right]^{\frac{1}{2}}$$

TRANSIENT FLOW TO FULLY PENETRATING DRAINS IN A SLOPING UNCONFINED AQUIFER WITH NO RECHARGE

1. D-F assumptions
2. isotropic and homogeneous
3. no infiltration
4. sloping impermeable base
5. initial water table parallel to the sloping base
6. storage and transmissive properties are constant with time
7. vertical seepage faces are located at each drain

• Chauhan et al. (1968) also describe a similar but more accurate solution by incorporating time-variable boundary functions at the drains

$$h(x,t) = \frac{2\pi h_i}{L^2}\frac{mx}{2b} + \frac{2\pi h_i}{L^2}\sum_{n=1}^{\infty}\left[\frac{n-ne^{-(\frac{mL}{2b})}(-1)^n}{(\frac{n\pi}{L})^2 + (\frac{m}{2b})^2} \right]\left(\exp - \left[(\frac{n\pi}{L})^2 + (\frac{m}{2b})^2\right]\frac{kbt}{\phi}\right)\left(\sin(\frac{n\pi x}{L})\right)$$

Table 2.4, continued

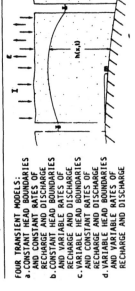

FOUR TRANSIENT MODELS:
a. CONSTANT HEAD BOUNDARIES AND CONSTANT RATES OF RECHARGE AND DISCHARGE
b. CONSTANT HEAD BOUNDARIES AND VARIABLE RATES OF RECHARGE AND DISCHARGE
c. VARIABLE HEAD BOUNDARIES AND CONSTANT RATES OF RECHARGE AND DISCHARGE
d. VARIABLE HEAD BOUNDARIES AND VARIABLE RATES OF RECHARGE AND DISCHARGE

refer to Singh and Jacob (1977)

refer to Singh and Jacob (1977)

HANTUSH'S 1967 D-F MODEL OF THE GROWTH AND DECAY OF WATER TABLE MOUNDS IN RESPONSE TO UNIFORM PERCOLATION

$$h(x,y,t) = \left[h_i^2 + \frac{I\upsilon t}{2K} \left\{ S^*\left(\frac{\ell+x}{(4\upsilon t)^{\frac{1}{2}}} , \frac{a+y}{(4\upsilon t)^{\frac{1}{2}}} \right) + S^*\left(\frac{\ell+x}{(4\upsilon t)^{\frac{1}{2}}} , \frac{a-y}{(4\upsilon t)^{\frac{1}{2}}} \right) + S^*\left(\frac{\ell-x}{(4\upsilon t)^{\frac{1}{2}}} , \frac{a+y}{(4\upsilon t)^{\frac{1}{2}}} \right) + S^*\left(\frac{\ell-x}{(4\upsilon t)^{\frac{1}{2}}} , \frac{a-y}{(4\upsilon t)^{\frac{1}{2}}} \right) \right\} \right]$$

where

$$\upsilon = \frac{K\bar{b}}{\phi}$$

$$\bar{b} = 0.5\ h_i(x,y,o) + h(x,y,t)$$

$$S^*(\alpha,\beta) = \int_0^1 erf\left(\frac{\alpha}{\tau^{\frac{1}{2}}}\right) erf\left(\frac{\beta}{\tau^{\frac{1}{2}}}\right) d\tau$$

$$h_{max} = h_i^2 + \frac{2I\upsilon t}{K} S^*\left(\frac{\ell}{(4\upsilon t)^{\frac{1}{2}}} , \frac{a}{(4\upsilon t)^{\frac{1}{2}}} \right)$$

1. D-F assumptions
2. isotropic and homogeneous
3. infinite in areal extent
4. horizontal impermeable base
5. rectangular recharge area receiving constant and uniform infiltration
6. storage and transmissive properties remain constant with time

• S^* has been tabulated for different values of α and β by Hantush (1967)
• the water table mound decay caused by reducing recharge with a low permeability cover can be estimated by superimposing a uniform discharge rate equal to the reduction in the infiltration rate and subtracting this solution from the build-up solution
• Hantush (1967) also presents a solution for a circular recharge basin

Table 2.5   Methods for Calculating Radius of Influence (Kufs et al., 1983)

Pumping Condition	Method[a]
Equilibrium	
—Exact	$\ln R_o = \dfrac{K(H^2 - hw^2)}{458 \, Q} + \ln R_o = \dfrac{T(H - hw)}{229 \, Q} + \ln r_w$
—Approximate	$R_o = 3(H - h_w)(0.47K)^{1/2}$
Non-Equilibrium	
—Exact	Drawdown vs. log distance plots
—Approximate	$R_o = r_w + \left(\dfrac{Tt}{4790 \, S}\right)^{1/2}$

[a] R_o = radius of influence (ft)
  K = permeability (gpd/ft^2)
  H = total head (ft)
  h_w = head in well (ft)
  Q = pumping rate (gpm)
  r_w = well radius (ft)
  T = transmissivity (gpd/ft)
  t = time (min)
  S = storage coefficient (dimensionless)

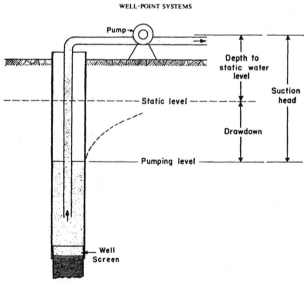

WELL-POINT SYSTEMS

Figure 2.1   When pumping wells by suction lift, the maximum drawdown obtainable is the difference between the suction limit of the pump and the distance from the static water level to the center of the pump. (Johnson Division, UOP Inc, 1975).

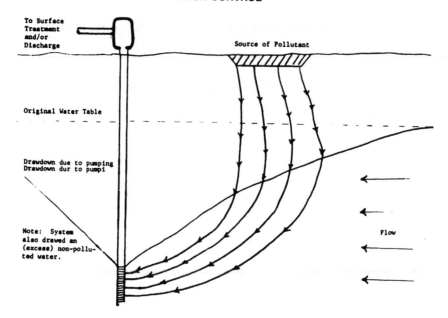

Figure 2.2. Principle of withdrawal wells.

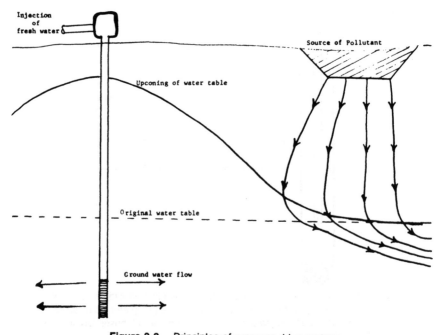

Figure 2.3. Principles of pressure ridge system.

Through the use of data obtained from the initial hydrogeologic study, and the principles of well hydraulics, the number, depth, spacing, and pumping or injection rate of the wells can be specified. In addition, estimates of the time required for the remedial actions can be made.

Application in Hydrocarbon Recovery

The discussion of the three types of well systems above has been general so as to be applicable to a wide variety of situations. However, there is one situation in which the purpose of the well system is very specific. That situation is when the contaminant is not water soluble and tends to float on top of the water table; this most often involves hydrocarbons. A unique aspect of this type of problem is that the contaminant provides a strong economic incentive for ground water cleanup since the product recovered from the water table can be utilized. Because of this economic incentive and the large number of instances of hydrocarbon leakage from storage tanks, considerable work has been directed toward developing hydrocarbon recovery systems.

Blake and Lewis (1982) have outlined four design options available for hydrocarbon recovery systems: (1) single-pump systems utilizing one recovery well, (2) single-pump systems utilizing multiple wells, (3) two-pump systems utilizing two recovery wells; and (4) two-pump systems utilizing one recovery well. The single-pump, single-well system is depicted in Figure 2.4. The advantages of this system include low equipment and drilling costs. However, this system produces an oil-water mix requiring separation of the surface. One-pump systems utilizing multiple well points are plagued by the same problems as the single-well arrangements, but they may be the most feasible option in low-permeability formations (Blake and Lewis, 1982).

The two-pump, two-well system is depicted in Figure 2.5. The deeper well serves the purpose of producing a gradient toward the shallower (product recovery) well. The advantage of this system is the ability to keep the product and the produced water separate; one disadvantage is the increased cost of multiple wells. Additionally, this type of system can require fairly sophisticated surface monitoring equipment to insure proper operation.

The most desirable product recovery system is that involving two pumps in one well as depicted in Figure 2.6. The basis of this system is similar to the previous system, i.e., the lower pump produces a gradient toward the well and the upper pump removes the accumulated product. This system separates the product from the produced water, but only

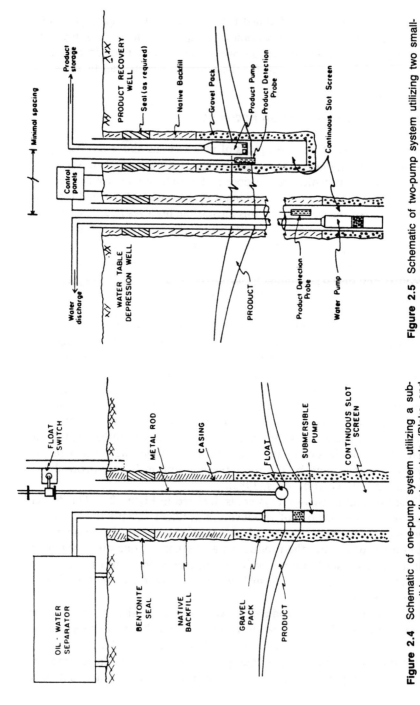

Figure 2.5 Schematic of two-pump system utilizing two small-diameter wells (Blake and Lewis, 1982).

Figure 2.4 Schematic of one-pump system utilizing a submersible pump and float controls (Blake and Lewis, 1982).

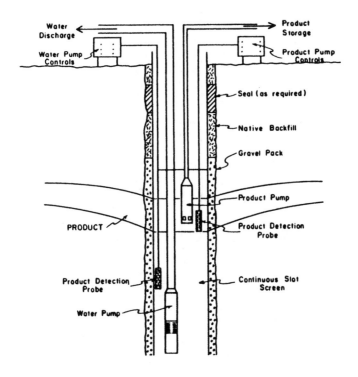

Figure 2.6.   Schematic of two-pump system utilizing one recovery well (Blake and Lewis, 1982).

involves a single well. This system can also be fully automated. The disadvantages of this system include: increased well diameter to house two pumps; expensive operational monitoring equipment; need for careful startup operations and calibration; and need for frequent inspections.

Costs

The total cost of a well system is dominated by two individual cost items: (1) the cost of the installation, and (2) operation and maintenance

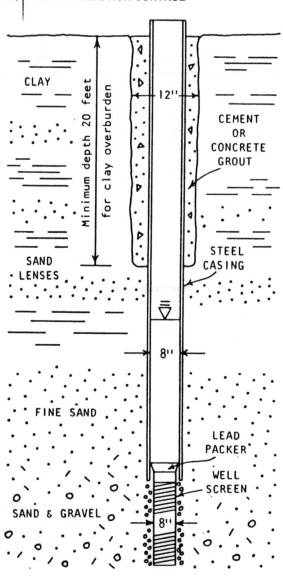

Figure 2.7. Well completed in unconsolidated sediments (sand and gravel) (Campbell and Lehr, 1977).

costs. In this time of rising energy cost, the power required to pump the well system will probably be the single most dominating cost factor. A cost analysis study for a variety of well systems was conducted by Gibb

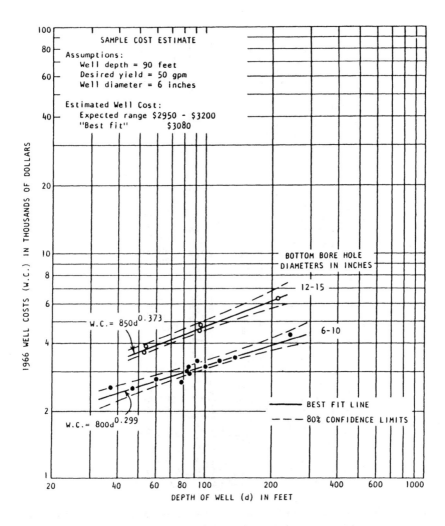

Figure 2.8.   Cost of cased wells finished in sand and gravel (Campbell and Lehr, 1977).

(1971) and included in Campbell and Lehr (1977). The actual cost figures are somewhat dated, but the principles employed would be applicable today. A schematic of a well completed in unconsolidated sediments (sand and gravel) is shown in Figure 2.7, with corresponding cost relationships shown in Figure 2.8. Similar information for a gravel-packed

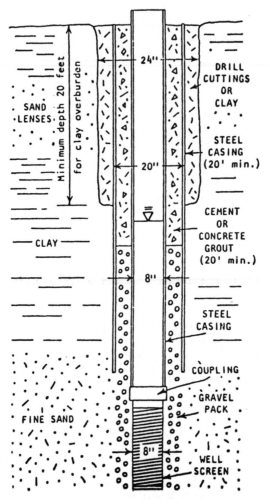

Figure 2.9. Gravel-packed well completed in sand and gravel—Unconsolidated to partly consolidated sediments (Campbell and Lehr, 1977).

well in consolidated to partly consolidated sediments (sand and gravel) is shown in Figures 2.9 and 2.10; for a shallow well in consolidated sediments (sandstone, limestone, or dolomite) in Figures 2.11 and 2.12; and for a deep well in consolidated sediments in Figures 2.13 and 2.14 (Campbell and Lehr, 1977).

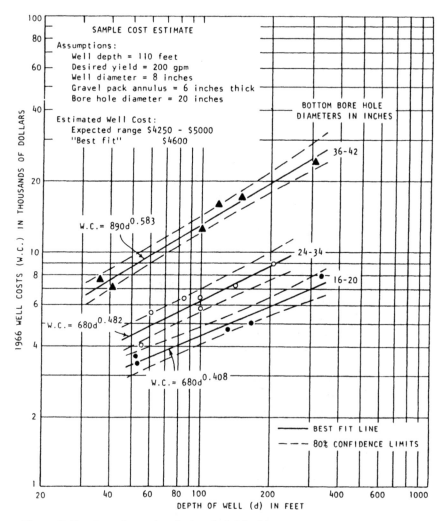

Figure 2.10. Cost of gravel-packed wells finished in sand and gravel (Campbell and Lehr, 1977).

Advantages and Disadvantages

The use of well systems is presently, and probably will continue to be, the most utilized method of ground water pollution control. This is not without good reason. Well systems represent the most assured means of controlling subsurface flows contaminants. Engineers and hydrogeolo-

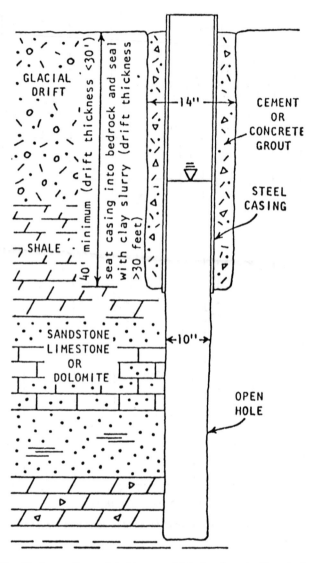

Figure 2.11.   Shallow well completed in consolidated sediments (Campbell and Lehr, 1977).

gists have a somewhat firmer understanding of the mechanics of well hydraulics than they do of some potential aquifer restoration strategies. However, well systems are not without their limitations. A comparison of some of the advantages and disadvantages of well systems is in Table 2.6.

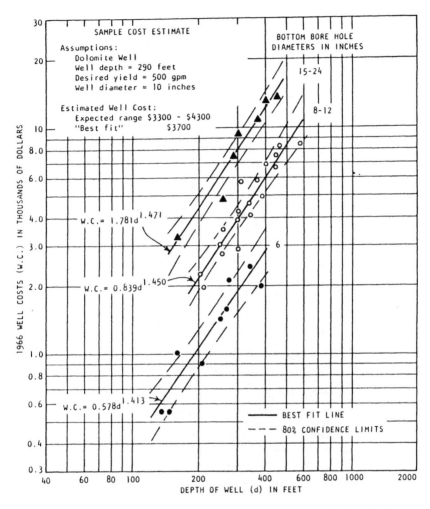

Figure 2.12. Cost of shallow sandstone, limestone, or dolomite bedrock wells (Campbell and Lehr, 1977).

INTERCEPTOR SYSTEMS

Interceptor systems involve excavation of a trench below the water table, and possibly the placement of a pipe in the trench. Subsurface drains or trenches function similarly to an infinite line of extraction wells, that is, they effect a continuous zone of depression which runs the

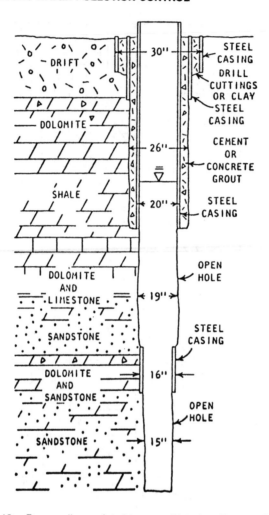

Figure 2.13. Deep well completed in consolidated sediments (Campbell and Lehr, 1977).

length of the drainage trench (Kufs et al., 1983). Figure 2.15 depicts an interceptor trench. Usually the trench receives perforated pipe and coarse backfill material for efficiency of collection and transport of incoming water. There are several names applied to interceptor systems; examples include their classification as preventive measures (leachate collection systems) or abatement measures (interceptor drains).

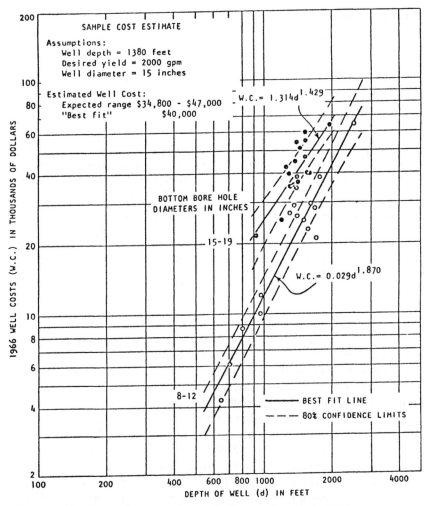

Figure 2.14. Cost of deep sandstone wells (Campbell and Lehr, 1977).

Construction

Collector Drains

Construction of interceptor systems is relatively simple and involves digging a system of trenches, placing perforated pipes in the trenches, and backfilling with a coarse material such as gravel. Usually collector

Table 2.6   Advantages/Disadvantages of Well Systems

Advantages	Disadvantages
1. Efficient and effective means of assuring ground water pollution control.	1. Operation and maintenance costs are high.
2. Can be installed readily.	2. Require continued monitoring after installation.
3. Previously installed monitoring wells can sometimes be employed as part of well system.	3. Withdrawal systems necessarily remove clean (excess) water along with polluted water.
4. Can sometimes include recharge of aquifer as part of the strategy.	4. Some systems may require the use of sophisticated mathematical models to evaluate their effectiveness.
5. High design flexibility.	5. Withdrawal systems will usually require surface treatment prior to discharge.
6. Construction costs can be lower than artificial barriers.	6. Application to fine soils is limited.

systems will consist of a series of feeder pipes (laterals) flowing into a main collecting pipe (header) as shown in Figure 2.16. From the header, the collected water will flow to a sump, where it is then pumped to the surface for treatment and/or discharge. Collector systems used in a preventive mode, such as a leachate collection system, usually include a series of 4- to 6-in. laterals draining into 8- to 12-in. headers, all arranged in either a herringbone or gridiron design as shown in Figure 2.17.

Interceptor systems used in an abatement mode can be classified as relief drains or interceptor drains. Relief drains are installed in areas where the hydraulic gradient is relatively flat. They are generally used to lower the water table beneath a site, or to prevent contamination from reaching a deeper, underlying aquifer. Relief drains are installed in parallel on either side of the site such that their areas of influence overlap and contaminated ground water does not flow between the drain lines. They can also be installed completely around the perimeter of the site (Kufs et al., 1983).

Interceptor drains are used to collect ground water from an upgradient source in order to prevent leachate from reaching wells and/or surface water located hydraulically downgradient from the site. They are installed perpendicular to ground water flow. A single interceptor drain located at the toe of the landfill, or two or more parallel interceptors may be needed, depending upon the circumstances (Kufs et al., 1983).

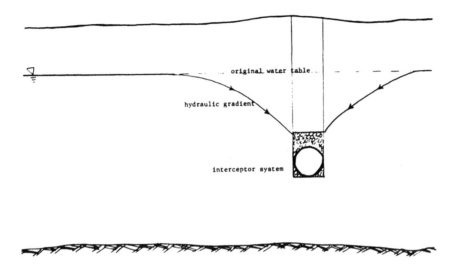

Figure 2.15. Hydraulic gradient toward interceptor system.

Interceptor Trenches

Interceptor trenches can be either active (pumped) or passive (gravity flow). Active systems will have intermediately spaced vertical removal wells or a perforated, horizontal removal pipe (collector drain) in the bottom of the trench. Active systems are usually backfilled with a coarse sand or gravel to promote wall stability. Passive systems are used almost exclusively for collection of pollutants that flow on the surface of water (e.g., oil and hydrocarbons). Passive systems are usually left open with the installation of a skimming pump for removal of the pollutant only.

All interceptor trenches require excavation to at least 3 or 4 ft below the water table to prevent the escape of inflowing pollutant and to speed up the inflow of further free pollutant (American Petroleum Institute, 1980). In active systems, the pumping capacity should be great enough to keep water drawn down to the bottom of the ditch. Additionally, pumping (active system) or skimming (passive system) must be continuous, or the collected pollutant will tend to seep into the trench walls and continue downgradient (American Petroleum Institute, 1980).

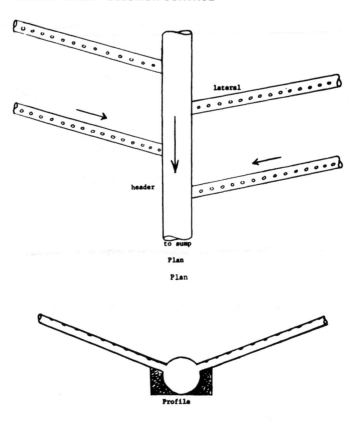

Figure 2.16.   Collector drain system.

The main construction activity associated with interceptor trenches is excavation of the trench. This can be accomplished with conventional equipment. The trench width should be wide enough to accommodate pipes or pumps if needed, yet it should be kept narrow to minimize the amounts of soil removed. Active (pumped) systems will require greater depths, once again to accommodate pipes or pumps.

It has been suggested (American Petroleum Institute, 1972, 1980) that the downgradient wall of the trench be lined with an impermeable material such as polyethylene film to prevent floating pollutants from passing through the trench and continuing downgradient. However, Pastrovich et al. (1979) indicate that the pollutant will tend to find its way around

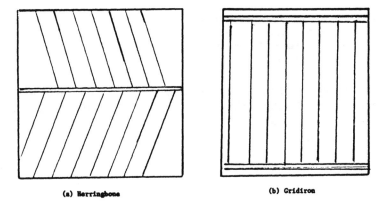

(a) Herringbone   (b) Gridiron

Figure 2.17. Collector system layouts.

the ends of the barrier and penetrate the ground beyond, while equalization of the hydrostatic pressure on both sides of the film tends to cause the pollutant to float away from the wall. As noted above, continuous removal of the pollutant and water from the trench will cause drawdown on both sides of the trench, thus preventing further migration of the pollutant.

After excavation of the trench, the passive system will usually require installation of a skimming pump. Active systems will require the installation of removal wells or collector drains, with subsequent backfilling of the trench with coarse sand or gravel.

Design

Collector Drains

The primary design parameters for a collector drain system are (1) the size of the pipe; and (2) the spacing of the pipe. To determine the size of the pipe required, the carrying capacity of the pipe is assumed to be equal to the design seepage, and the pipe velocity is described by Manning's equation. The resulting equation in a rearranged form is (Luthin, 1957):

$$d = 0.892 \ (SC)^{0.375} \ (A)^{0.375}(s)^{-0.1875}$$

where   d = inside diameter of pipe (in.)
            A = drainage area (acres)
         SC = seepage coefficient (in./day)
           s = hydraulic gradient

For purposes of design of drain spacing, from the Dupuit-Fochheimer theory the ground water surface between two tile lines may be taken as a portion of an ellipse as shown in Figure 2.18 and described as follows (Spangler and Handy, 1973):

$$\frac{S^2}{(b^2 - a^2)} = \frac{4k}{R}$$

where   S = tile line spacing (m)
         a = depth to impermeable barrier under tile (m)
         b = depth to impermeable barrier halfway between tiles (m)
         k = hydraulic conductivity of soil (m/day)
         R = infiltrating rainfall (m/day)

A wide variety of equations for parallel subsurface drains have been developed for varying aquifer conditions. These equations are outlined in Tables 2.3 and 2.4.

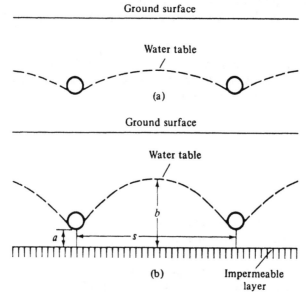

Figure 2.18. Shape of water table adjacent to tile drains: (a) relatively permeable soil; (b) less permeable soil (Spangler and Handy, 1973).

Interceptor Trenches

Design of interceptor trenches is different from that of relief drains. In order to decide where to position the trench to intercept the desired flow, the relationship between depth and flow and the upgradient and downgradient influence of the trench must be known. The upgradient influence can be determined from the following (Kufs et al., 1983):

$$D_u = \text{4/3 } H \tan \phi$$

where D_u = effective distance of drawdown upgradient (m)
   H = saturated thickness of the water-bearing strata not affected by drainage (m)
   ϕ = angle between the initial water table or ground surface and the horizontal plane

The theoretical expression for the downgradient influence is as follows:

$$D_d = \frac{K \tan \phi}{q}(h_1 - h_2 + D_2)$$

where D_d = distance downgradient from the drain where the water table is lowered to the desired depth (m)
   K = hydraulic conductivity (m/day)
   q = drainage coefficient (m/day)
   h_1 = effective depth of the drain (m)
   h_2 = desired depth to the water table after drainage (m)
   D_2 = distance from the ground surface to the water table before drainage at the distance D_d downgradient from the drain (m)

Costs

Shallow drainage by trenches or collector drains has been estimated to cost on the order of $1235 to $1730 capital cost per hectare drained ($500 to $700 per acre), depending on the depth of placement and materials used. Labor and maintenance costs are about $1500 per year (Tolman et al., 1978).

Advantages and Disadvantages

Listed in Table 2.7 are some of the advantages and disadvantages particular to subsurface drains (interceptor systems). The most obvious advantage of interceptor systems is their relatively simple construction methods. Just as obvious is the disadvantage of requiring continuous monitoring and maintenance. Other advantages of interceptor systems are:

(1) not only easy but also inexpensive to install;
(2) useful for intercepting landfill side seepage and runoff (U.S. Environmental Protection Agency, 1982);
(3) useful for collecting leachate in poorly permeable soils (U.S. Environmental Protection Agency, 1982);
(4) large wetted perimeter allows for high rates of flow (U.S. Environmental Protection Agency, 1982);
(5) possible to monitor and recover pollutants; and
(6) produces much less fluid to be handled than well-point systems.

Disadvantages of interceptor systems include:

(1) when dissolved constituents are involved, it may be necessary to monitor ground water downgradient of the recovery line (Quince and Gardner, 1982);
(2) open systems require safety precautions to prevent fires or explosions;

Table 2.7   Advantages/Disadvantages of Subsurface Drains (U.S. Environmental Protection Agency, 1982)

Advantages	Disadvantages
1. Operation costs are relatively cheap since flow to underdrains is by gravity.	1. Not well suited to poorly permeable soils.
2. Provides a means of collecting leachate without the use of impervious liners.	2. In most instances it will not be feasible to situate underdrains beneath an existing site.
3. Considerable flexibility is available for design of underdrains; spacing can be altered to some extent by adjusting depth or modifying envelope material.	3. System requires continuous and careful monitoring to assure adequate leachate collection.
4. System is fairly reliable, providing there is continuous monitoring.	
5. Construction methods are simple.	

(3) interceptor trenches are less efficient than well-point systems;

(4) operation and maintenance costs are high;

(5) not useful for deep disposal sites (U.S. Environmental Protection Agency, 1982); and

(6) may interfere with use of land.

SURFACE WATER CONTROL, CAPPING AND LINERS

Surface water control, capping and liners are three different technologies that are usually used in conjunction with each other. The three technologies serve different purposes in preventing ground water pollution. Surface water control measures are designed to minimize the amount of surface water flowing onto a site, thus reducing the amount of potential infiltration. Capping of a site is designed to minimize the infiltration of any surface water or direct precipitation that does come onto a site. Impermeable liners provide ground water protection by inhibiting downward flow of low-quality leachate and/or attenuating pollutants by adsorption processes. These technologies are most often used as preventive measures, although surface water control and capping have also been used successfully to abate ground water pollution problems (Emrich, Beck and Tolman, no date). Technologies also exist for placement of liners after a polluting activity has been initiated; these usually involve the use of grouts, which are discussed in a subsequent section.

Surface water control measures represent a relatively inexpensive means of minimizing future ground water pollution problems, and as such, should be part of any current site design. Surface water run-on is easily minimized by using standard civil engineering techniques such as diversion berms and drainage ditches (Smith, 1983). The design of a site utilizing surface caps or impermeable liners, or both, is based on a water balance for the site. Depicted in Figure 2.19 are the individual components of moisture at a site.

There are four combinations of capping and liners that could be used (in conjunction with surface water control measures) at a waste site, with these combinations depicted in Figure 2.20. The first type is called a natural attenuation site and utilizes neither capping or lining; it relies entirely on the pollutant attenuating capacity of the native soil. The second type of system utilizes an engineered liner (impermeable liner) to maximize the amount of leachate collected and minimize the amount escaping. The engineered cover (impermeable cap) sites are designed to minimize infiltration into the waste source by maximizing surface runoff

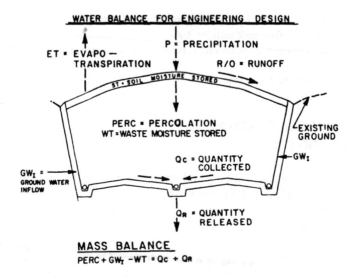

MASS BALANCE

$PERC + GW_I - WT = Qc + Q_R$

Figure 2.19.   Moisture components of landfills (Smith, 1983).

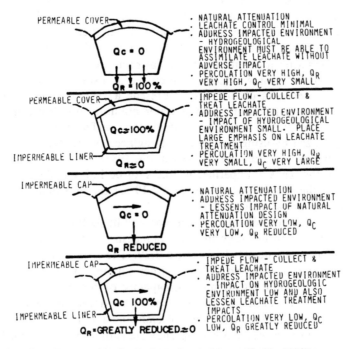

Figure 2.20.   Four combinations of capping and liners (Smith, 1983).

and evapotranspiration. The combination engineered cover and liner (total containment) sites are designed to minimize infiltration and maximize the leachate collected, thus all but eliminating the escape of any leachate. The advantages and disadvantages of each type of system are outlined in Table. 2.8.

Emrich, Beck and Tolman (no date) summarize a study of top-sealing a landfill to minimize leachate generation. Applications of liner technology are numerous and grow every year. Shultz and Miklas (1982) give an excellent discussion of the activities at fourteen lined facilities. Due to their potential for use in aquifer restoration projects, the emphasis herein will be on surface caps.

Construction—Surface Capping

In general, the steps involved in the construction of an engineered cover site include:

(1) closure of facility for future receipt of waste materials;
(2) compaction of waste material and laying and working of suitable subbase;
(3) placement of impermeable membrane (surface cap);
(4) placement and working of suitable cover material over cap;
(5) revegetation; and
(6) regrading.

The key steps involved with surface capping are the placement of the impermeable membrane over the source of contamination, and the revegetation and regrading of the site. The regrading of the site should be done so as to maximize surface runoff and channel it away from the site. Revegetation allows for evapotranspiration of water that soaks into the cover soil, with surface capping being designed to prevent any remaining water from infiltrating into the waste and moving down to the ground water. Additional measures that should be included at most sites include installation of gas vents and monitoring facilities for both surface water runoff and ground water.

Fung (1980) provides some general recommendations for layering the surface cap, assuming it is some type of soil or workable material, from the bottom up. These recommendations are:

(1) make buffer layer below barrier (cap) thick and dense enough to provide a smooth, stable base for compacting;
(2) compact all layers except topsoil and top lift of upper buffer;
(3) in barrier layer, strive for 90% of maximum density according to 5- or 15-blow compaction test;

Table 2.8 Advantages/Disadvantages of Capping and Liners (Smith, 1983)

System	Advantages	Disadvantages
Natural Attenuation	No leachate collection, transport and treatment costs. Reduced construction costs	Requires unique hydrogeologic setting. Regulatory acceptance difficult to obtain. Long-term liabilities.
Engineered Liner	Lessens hydrogeologic impact. Allows waste to stabilize quickly.	"Clay-bowl" effect. Increased construction costs. Chance for surface discharge.
Engineered Cover	Lessens hydrogeologic impact after closure. Reduces construction costs relative to liners.	Increases closure costs. No leachate control during site operations. Long-term monitoring and land surface care.
Engineered Cover and Liner	Lessens environmental impacts. Minimizes post-closure leachate collection, transport and treatment costs. Politically/socially acceptable.	High cost for engineering and construction. Need high-quality clay or synthetic material. Lengthens time for waste stabilization.

(4) cover barrier layer soon enough to prevent extensive drying;

(5) provide sufficient design thickness to assure performance of layer function; specifying a 6-in. to 12-in. minimum should prevent excessively thin spots resulting from poor spreading techniques;

(6) construct in units small enough to allow rapid completion; and

(7) consider seeding topsoil at time of spreading.

Surface Capping—
Selection and Design Considerations

The materials used for covering or capping are many and varied. Fung (1980) groups these materials into natural soils, commercial cover materials, and waste materials. Soil has traditionally been used as a cover material at waste disposal sites. The cover has to fulfill several functions, and Table 2.9 summarizes the types of soil classified according to the Unified Soil Classification System (USCS) in terms of several functions (Fung, 1980). The rating of soils accomplished in Table 2.9 is based on the following points, with the functional ratings from I (best) to as many as XIII (poorest). Quantitative parameters are also provided wherever available, so that the absolute position of a particular soil rating can be seen (Fung, 1980).

(1) The first column in Table 2.9 concerns the Go-No Go aspect of trafficability. The ratings of I to XII from coarse granular to fine peaty soils are based on the rating cone index values (RCI). The RCI expresses the probable strength of the specific site under repeated traffic. In column 2, stickiness trafficability is based on the weight percent of clay in the soil. Ranking for slipperiness trafficability in column 3 is similar to ranking in the previous column though based on a different parameter—sand and gravel percentage. Sand and gravel components decrease the slipperiness, i.e., improve the quality of soil from that functional point of view, by increasing frictional resistance. Attention is called to the fact that all three trafficability functions provide approximately the same soil ranking.

(2) Water percolation is addressed in columns 4 and 5. The best parameter for judging relative effectiveness for water percolation is the coefficient of permeability k, although it specifically reflects water movement in a saturated condition. The rankings in columns 4 and 5 are diametrically opposed, since one reflects impeded percolation and the other assisted percolation.

(3) Gas migration is addressed in columns 6 and 7 from contrasting viewpoints of impeded and assisted flow. The parameter on which the ranking is partly based is the capillary head (H_c) of the soil. This parameter reflects the tenacity with which the soil holds water.

(4) Two types of erosion control are considered in columns 8 and 9. Soils are rated for resistance to water erosion on the basis of values of the USDA soil erodibility K factor. (K is the average soil loss in tons per acre per unit of rainfall and runoff

Table 2.9 Ranking of USCS Soil Types According to Performance of Cover Functions (Fung, 1980)

USCS Symbol	Typical Soils	(1) Go-No Go (RCI Value)[a]	(2) Trafficability — Stickiness (Clay, %)	(3) Trafficability — Slipperiness (Sand-Gravel, %)	(4) Water Percolation — Impede (k, cm/s)[a]	(5) Water Percolation — Assist (k, cm/s)[a]	(6) Gas Migration — Impede (H_c, cm)[a]	(7) Gas Migration — Assist (H_c, cm)[a]
GW	Well-graded gravels, gravel-sand mixtures, little or no fines	I (>200)	I (0–5)	I (95–100)	X (10^{-2})	III (10^{-2})	XI (6)	I (6)
GP	Poorly graded gravels, gravel-sand mixtures, little or no fines	I (>200)	II (0–5)	II (95–100)	XII (10^{-1})	I (10^{-1})	X (6)	II (6)
GM	Silty gravels, gravel-sand-silt mixtures	III (177)	V (0–20)	V (60–95)	VII (5×10^{-4})	VI (5×10^{-4})	VII (68)	V (68)
GC	Clayey gravels, gravel-sand-clay mixtures	V (150)	IX (10–50)	VIII (50–90)	V (10^{-4})	VIII (10^{-4})	VI (68)	VI (68)
SW	Well-graded sands, gravelly sands, little or no fines	I (>200)	III (0–10)	III (95–100)	VIII (10^{-3})	IV (10^{-3})	IX (60)	III (60)
SP	Poorly graded sands, gravelly sands, little or no fines	I (>200)	IV (0–10)	IV (95–100)	XI (5×10^{-2})	II (5×10^{-2})	VIII (60)	IV (60)
SM	Silty sands, sand-silt mixtures	II (179)	VI (0–20)	VI (60–95)	IX (10^{-3})	V (10^{-3})	V (112)	VII (112)
SC	Clayey sands, sand-clay mixtures	IV (157)	X (10–50)	VII (60–95)	VI (2×10^{-4})	VII (2×10^{-4})	IV (112)	VIII (112)
ML	Inorganic silts and very fine sands, rock flour, silty or clayey fine sands, or clayey silts with slight plasticity	IX (104)	VII (0–20)	IX (0–60)	IV (10^{-5})	IX (10^{-5})	III (180)	IX (180)
CL	Inorganic clays of low to medium plasticity, gravelly clays, sandy clays, silty clays, lean clays	VII (111)	XI (10–50)	XI (0–55)	III (3×10^{-6})	X (3×10^{-6})	II (180)	X (180)
OL	Organic silts and organic silty clays of low plasticity	X (64)	VIII (0–20)	X (0–60)	—	—	—	—
MH	Inorganic silts, micaceous or diatomaceous fine sandy or silty soils, elastic silts	VIII (107)	XII (50–100)	XII (0–50)	II (10^{-7})	XI (10^{-7})	—	—
CH	Inorganic clays of high plasticity, fat clays	VI (145)	XIII (50–100)	XIII (0–50)	I (10^{-9})	XII (10^{-9})	I (200–400+)	XI (200–400+)
OH	Organic clays of medium to high plasticity, organic silts	XI (62)	—	—	—	—	—	—
PT	Peat and other highly organic soils	XII (46)	—	—	—	—	—	—

Table 2.9, continued

USCS Symbol	(8) Water (K-Factor)[a]	(9) Wind (Sand-Gravel, %)	(10) Fast Freeze (H_c, cm)[a]	(11) Saturation (Heave, mm/day)	(12) Crack Resistance (Expansion, %)	(13) Impede Vector Emergence	(14) Support Vegetation
	Erosion Control		Reduce Freeze Action				
GW	(<0.05) I	I	X	I	(0) I	X	X
GP	I	(95–100) I	IX	(0.1–3) III	(0) I	X	X
GM	IV	(95–100) III	VII	(0.1–3) IV	(0) III	VIII	VI
GC	III	(60–95) V	IV	(0.4–4) VII	II	V	V
SW	II	(50–90) II	VIII	(1–8) II	I	IX	IX
SP	(0.05) II	(95–100) II	VII	(0.2–2) II	(0) II	IX	IX
SM	VI	(95–100) IV	VI	(0.2–2) V	IV	VII	II
SC	(0.12–0.27) VII	(60–95) VI	V	(0.2–7) VI	VI	IV	I
ML	(0.14–0.27) XIII	(50–90) VII	III	(1–7) X	VIII	VI	III
CL	(0.60) XII	(0–60) VIII	II	(2–27) VIII	(1–10) VII	III	VII
OL	(0.28–0.48) XI	(0–55) VII	I	(1–6) VII	IX	VI	IV
MH	(0.21–0.29) X	(0–60) IX	—	(1–6) IX	X	II	IV
CH	(0.25) IX	(0.50) X	—	(0.8) III	(>10) IX	I	VIII
OH	(0.13–0.29) VIII	(0–50) VIII	—	—	—	—	VIII
PT	(0.13) V	—	—	—	—	—	III

[a] RCI is rating cone index, k is coefficient of permeability, H_c is capillary head, and K-Factor is the soil erodibility factor. The ratings I to XIII are for best through poorest in performing the specified cover function.

Note: Additional parameters for Table 2.9 are as follows:

Parameter	Description
Fire resistance	Same as Impede Gas Migration
Dust control	Same ranking and values as for Wind Erosion Control
Side slope stability	Determine on basis of laboratory testing
Side slope seepage	Same ranking and values as for Impede Water Percolation
Side slope drainage	Same ranking and values as for Assist Water Percolation
Discourage burrowing	Same ranking and values as for Slipperiness Trafficability
Discourage birds	All soils are suitable
Future use as natural settings	Same ranking as for Support Vegetation
Future use as foundation	Same ranking and values as for Go-No Go Trafficability

erosivity factor R, where R usually equals the pertinent rainfall index which is predictable from meteorological data.) For wind erosion resistance, soils are rated according to the sand and gravel content.

(5) Columns 10 and 11 concern cover functions of reducing freeze action. The fast freezing/fast thawing aspects, which also involve the tendency to freeze to greater depths, give soil ratings of I to X based on the capillary head. Ranking for crack resistance involves considerable judgment. It is, however, well-known that the extreme cracking is almost restricted to soils with high clay content, and only those soils rated VIII or worse should cause much concern.

(6) In order to impede vector emergence, a soil with a high clay content or well-graded combination of clay and silt, with or without sand, is preferred. Loamy soils are somewhat more suitable for supporting vegetation than soils at the sand, silt, and clay extremes of grain-size distribution. Thus, a ranking of cover soil of supporting vegetation in column 14 gives the high positions, somewhat subjectively, to loam and other well-graded mixtures. The fire resistance ranking is the same as that for impede gas migration in column 6, since fire resistance is increased if gas movement is impeded.

(7) With dust control, fine-grained particles are essential for dust, so that the appropriate quantitative rating parameter is considered to be the sand and gravel content. The ranking, therefore, is the same as for wind erosion control.

(8) Important aspects of the side slope cover are stability, seepage, and drainage. The seepage and drainage aspects have rankings and rating values identical to those for impede water percolation and assist water percolation, respectively. The stability of side slopes may in some cases be a very important function of the cover. In such cases, the choice of soil should be carefully considered and based on laboratory strength tests.

(9) The effectiveness of soils for discouraging burrowing animals is believed to increase with the percentage of sand content. Accordingly, the same ranking is used as was used for slipperiness trafficability, since in both cases the important parameter is sand and gravel content. Soil type is apparently not a pertinent consideration in discouraging birds, so no preference is given in the table.

(10) For the long-range future use of a landfill site for natural or parklike settings, the ranking should be the same as for supporting vegetation. Where the future use of the landfill will be to support pavement or light structures, a major concern is for high modulus. This aspect of cover soil is adequately reflected in RCI values as for go-no go trafficability. It should be remembered, however, that the supportive capability of the underlying solid waste is commonly the critical factor.

After selection of the soil for covering the waste, efforts should be directed to the most effective placement and treatment. Soil cover can be improved in several ways as it is constructed. Materials may be added for better gradation, hauling and spreading equipment can be operated beneficially, and the base or internal structure of the cover system may be improved. Soil blending increases the expense of surface capping operations and, therefore, has usually not been duly considered. The benefits derived from blending, however, are sometimes dramatic through the alteration of grain-size distribution or average grain size. Blending may be done in place using a blade or harrow. Examples of materials used for blending include gravel, sand, silt, and clay (Fung, 1980).

Soils to be used for surface capping may also be improved through blending with additives and/or cements. Additives and cements are defined as synthetic materials added in relatively small amounts to soil to achieve beneficial effects. Classifications include strengthening cements or stabilizers, dispersants, freeze-point suppressants, water repellants, and dust palliatives. Factors that enter into determining the cost effectiveness of additives and cements are their relatively high unit cost, manner of addition or incorporation, and duration of effect. A major concern in using any cover system with increased strength is the remaining susceptibility to cracking as a result of differential settlement or environmental deterioration, such as freeze-thaw; plans for patching repairs are essential. Examples of additives and cements which could be used are as follows (Fung, 1980):

(1) Soil-Cement—Soil cement is composed of sandy soil, portland cement, and water that have reacted and hardened. About 8% of dry cement is incorporated in place and compacted after water is added to about optimum content. Premixing and placement wet is another method that seems less amenable to covering waste because of extra costs, except possibly on steep slopes. Somewhat higher cement content is needed in poorly graded soils or where fines are lacking, but correspondingly less for well-graded mixtures.

(2) Soil-Bitumen—Bitumen stabilizes sandy soils by cementing and waterproofing. An optimum content increases the cohesion of the soil, yet does not separate grains enough to reduce frictional strength. From 4 to 8% bitumen (soil dry weight basis) is recommended as a first cut for sandy loam to be refined subsequently by testing mixtures of various contents. Sand should require somewhat less bitumen.

(3) Cement-Treated Soil—Incorporation of as little as 1% of portland cement has been shown to have stabilizing effects on granular soil.

(4) Lime-Treated Soil—Lime (calcium oxide or hydroxide) is principally mixed into clayey soils for its effects as a flocculant or base exchanger. However, lime also promotes some beneficial pozzolanic reaction in fine, cohesive soils, and strengthening may continue for many weeks of curing.

(5) Fly Ash-Lime (or Cement-)-Treated Soil—Fly ash is attractive as an additive to soil because of its pozzolanic properties (i.e., although not a cement itself, fly ash develops cementitious properties with lime or cement and water). Accordingly, a small amount of lime or portland cement should be included unless self-hardening has been confirmed by test or previous experience with the particular fly ash.

(6) Fly Ash-Treated Soil—Some fly ashes possess moderate self-hardening characteristics when moistened and compacted and, therefore, offer promise alone as major additions to cover waste. This phenomenon is partly due to the free lime content of the ash itself. Laboratory tests or field trials will be necessary to establish such useful peculiarities of specific fly ash sources where previous experience is lacking.

(7) Fly Ash-Lime (or Cement)-Sulfate-Treated Soil—This mixture is proposed as a locally viable option for covering waste following research on the uses of fly ash-lime-sulfate (or sulfite).

(8) Dispersants—Several soluble salts act as dispersing agents and are recommended for improving soil compaction and reducing K. The treatment is primarily for fine-grained soils containing clay minerals. Better compaction can be achieved on coarse-grained soils by other means, such as blending in fine material. Dispersion (deflocculation) involves the breakdown of clayey aggregates into the individual mineral particles. The effect is to increase dry unit weight, to lower permeability, and incidently, to facilitate compaction. Three dispersing agents have been found notably effective. Common sodium chloride produces satisfactory results in many cases and is inexpensive. Tetra-sodium pyrophosphate and sodium polyphosphate appear to have slight advantages in many cases over sodium chloride.

(9) Swell Reducers—Lime is the principal additive for reducing shrink/swell behavior in clay-rich soils. About 5% is optimum for effecting ion exchange (calcium for sodium), flocculation, and cementation. Quicklime contains more calcium and, therefore, is preferred to hydrated lime by some users provided a mellowing period is allowed for homogenization.

(10) Freeze-Point Suppressant—Poor compaction can result when soil pore water is frozen during cold weather operations. Ordinarily soils construction for roads, etc., is curtailed in the winter to avoid the poor results. Some improvements can be achieved by adding a freeze-point suppressant and may be advisable where covering operations are expected to continue during cold spells. Calcium chloride can be effective, either in solution or in dry, flaked form.

Layering is perhaps the most promising, yet presently underutilized, technique for designing final solid waste cover. By combining two or three distinct materials in layers, the designer may mobilize favorable characteristics of each together at little extra expense. The systematics are briefly set forth there. Figure 2.21 schematically illustrates a layered system (Fung, 1980). A layered system may include topsoil, barrier layer, buffer layer, water drainage layer, filter and gas drainage layer. Comments on each of these layers are as follows (Fung, 1980).

(1) Topsoil—A topsoil or a subsoil made amenable to supporting vegetation frequently forms the top of a layered cover system. Untreated subsoils are seldom suitable directly, so it has been often necessary to supplement subsoil with fertilizers, conditioners, etc., to obtain the desired result.

(2) Barrier Layer or Membrane—The primary feature in many layered systems is the barrier. This layer functions to restrict passage of water or gas. Barrier layers are almost always composed of clayey soil that has inherently low permeability; USCS types CH, CL and SC are examples of soil types. Soil barriers are susceptible to deterioration by cracking when exposed at the surface, so that a buffer layer is recommended to protect the clay from excessive drying. Synthetic membranes may be used in place of soil barriers. Cost are generally high in comparison to available soils, and problems in placement or with deterioration may arise. Nevertheless, membranes are a viable option.

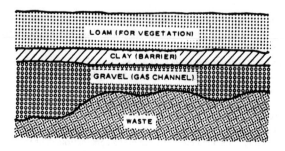

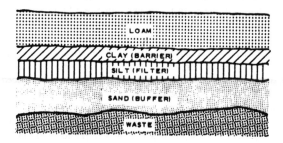

Source: EPA-600/2-79-165

Figure 2.21. Typical layered cover systems (Fung, 1980).

(3) Buffer Layer—A buffer layer may be described as a random layer having a
subordinate covering function and characteristics in comparison with the adjacent layer. The principal service of a buffer is to protect a barrier layer or
membrane located above or below. The buffer shields a vulnerable, thinner
barrier or sheet from tears, cracks, offsets, and punctures. Below a barrier, the
buffer layer also provides a smooth, regular base. Any soil type will serve as a
buffer ordinarily, but it should be free of clods.

(4) Water Drainage Layer or Channel—A water drainage layer, blanket, or channel
may be designed into the cover in numerous ways to provide a path for water to
exit rapidly. Poorly graded sand and gravel are possibilities as effective drainage
materials, i.e., soils classified GP and SP.

(5) Filter—Where layers with grossly discordant grain sizes are joined, there may be
a tendency for fine particles to penetrate the coarser layer. As a result, the
effectiveness of the coarse layer for water drainage may be reduced by clogging
of pores. Removals from the fine layer may promote additional bad effects,
such as internal erosion and settlement. A filter layer with a material such as silt
may be desirable.

(6) Gas Drainage Layer and Vents—A gas drainage layer has consistency and configuration similar to those of the water drainage layer or channel. Both layer
types function to transmit preferentially. The position in the cover system is the
main distinction. The gas drainage layer is placed on the lower side to intercept

gases rising from waste cells, whereas the drain for water is positioned on the upper side to intercept water percolating from the surface. Gas vents are frequently used in the present state-of-the-art in municipal waste landfilling. It has been found that dangerous concentrations of flammable gases can accumulate if not vented properly to the atmosphere. Figure 2.22 illustrates two recommended gas vents.

Commercially available capping materials or membranes may be preferable to soil-based systems over wastes that pose a high risk of serious pollution (where not properly shielded). Examples of commercial materials include bituminous and portland cement concrete barriers, and membranes of various types functioning alone or in conjunction with soil or granular waste layers to exclude water. A summary of commercial materials and their characteristics is presented elsewhere (Fung 1980).

A number of factors should be considered in the selection of a commercial cover material. Important characteristics of flexible membranes and rigid barriers include vulnerability to tearing or cracking, respectively. A smooth buffer of sand should reduce membrane puncture during construction, but extra long-term planning will be necessary against stressing effects of differential settlement. The rigid barriers sometimes have an advantage because cracks can be exposed, cleaned, and repaired (sealed with tar) with relative ease. Deterioration is another major problem for serious consideration. Concrete barriers are susceptible to chemical deterioration in harsh environments such as those rich in sulfates.

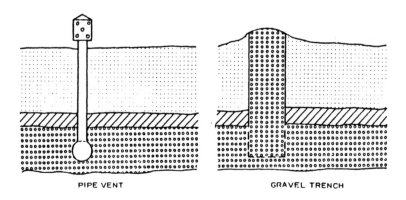

PIPE VENT   GRAVEL TRENCH

Figure 2.22. Two types of gas vents for layered cover systems (Fung, 1980).

Sunlight, burrowing animals, and plant roots increase the deterioration of membranes. Twenty years is about the extent of experience with membranes. Longer design service life may be somewhat risky in adverse environments and conditions. The cost of commercial coverings will often be the overriding consideration (Fung, 1980).

The other category of nonsoil materials that is available for covering waste ranges through a long and diverse list of granular wastes (some are in fact soils). An economic advantage may accrue from using these inexpensive, locally generated wastes in place of scarce, indigenous soils. Examples of waste materials used for covers include fly ash, fly ash modified with lime to enhance cementation, bottom ash and slag, furnace slag (from iron steel), incinerator residue, foundry sand, mine and pit wastes, mine mill tailings, plant sludges, reservoir and channel silt, dredged material, and composted sewage sludge (Fung, 1980).

Costs

The general range for costs of surface caps is $20,000 to $30,000 per acre covered, while the cost for installing a liner is more expensive, being in the range of $40,000 to $50,000 per acre. Cost estimates for various liners are outlined in Table 2.10.

Table 2.10   Costs for Various Sanitary Landfill Liner Materials (Haxo, 1973)

Material	Installed Cost[a] ($/sq yd)
Polyethylene (10 – 20[b] mil[c])	0.90 – 1.44
Polyvinyl chloride (10 – 30[b] mil)	1.17 – 2.16
Butyl rubber (31.3 – 62.5[b] mil)	3.25 – 4.00
Hypalon (20 – 45[b] mil)	2.88 – 3.06
Ethylene propylene diene monomer (31.3 – 62.5[b] mil)	2.43 – 3.42
Chlorinated polyethylene (20 – 30[b] mil)	2.43 – 3.24
Paving asphalt with sealer coat (2 in.)	1.20 – 1.70
Paving asphalt with sealer coat (4 in.)	2.35 – 3.25
Hot sprayed asphalt (1 gal/yd^2)	1.50 – 2.00 (includes earth cover)
Asphalt sprayed on polypropylene fabric (100 mil)	1.26 – 1.87
Soil-bentonite (9.1 lb/yd^2)	0.72
Soil-bentonite (18.1 lb/yd^2)	1.17
Soil-cement with sealer coat (6 in.)	1.25

[a]Cost does not include construction of subgrade nor the cost of earth cover. These can range from $0.10 to $0.50/yd^2/ft of depth.
[b]Material costs are the same for this range of thickness.
[c]One mil = 0.001 in.

SHEET PILING

Sheet piling involves driving lengths of steel that connect together into the ground to form a thin impermeable permanent barrier to flow. Sheet piling materials also include timber and concrete; however, their application to polluted ground water situations would be doubtful due to corrosive actions and costs, respectively.

Construction

Sheet piling requires that the steel sections be assembled prior to being driven into the ground. The lengths of steel have connections along both edges. The connections may be either slotted or ball- and socket-types. The sections are then driven individually into the ground by use of a pile hammer. The types of pile hammers include: drop, single-acting steam, double-acting steam, diesel, vibratory, and hydraulic. For each type of hammer listed the driving energy is supplied by a falling mass, which strikes the top of the pile (Bowles, 1977). After the piles have been driven to their desired depth, the remaining aboveground portions are cut off. Initially, sheet piles are not totally impermeable because of small gaps in the connections. As time passes, these gaps are closed as ground water flow carries fine particles into the gaps and closes them by clogging.

Steel sheet piling can be considered permanent because experience has shown that corrosion is not a factor in causing failures (Bowles, 1977). One strong advantage of sheet piling is that the sections are reusable and do not have to be left in place permanently. If conditions permit, steel piles may be removed and reused.

Costs

In situations where the function of the sheet pile is just to restrict ground water flow, i.e., no significant load resistance is required, a light-weight steel will be adequate. Construction costs of a 1700-ft-long and 60-ft-deep light-weight steel cut-off wall were reported to range from $650,500 to $956,000 (Tolman et al., 1978). To date, sheet piling has been proposed as a means of ground water pollution control, but specific applications have been minimal, if any.

Advantages and Disadvantages

Construction of steel sheet piles as a means of ground water control can potentially be effective and economical in specific cases. In general, however, this is probably an overelaborate technique to achieve a relatively simple result. As the size of a project increases, sheet piling will become uneconomical because of high material and shipping costs. In addition, pile driving requires a relatively uniform, loose, boulder-free soil for ease of construction. Other advantages/disadvantages are listed in Table 2.11 (Tolman et al., 1978).

GROUTING

Grouting is the process of injecting a liquid, slurry, or emulsion under pressure into the soil. The fluid injected will move away from the point of injection to occupy the available pore spaces. As time passes, the injected fluid will solidify, thus resulting in a decrease in the original soil permeability and an increase in the soil-bearing capacity. Grouts are usually classified as particulate or chemical. Particulate grouts consist of water plus particulate material which will solidify within the soil matrix. Chemical grouts usually consist of two or more liquids which will gel when they come in contact with each other. Listed in Table 2.12 are materials commonly used for grouts. Table 2.13 identifies the properties of cement grout additives.

Construction

Two of the more popular methods of grout installation are the stage and packer methods. In stage grouting, holes are drilled to the geologic

Table 2.11 Advantages/Disadvantages of Steel Sheetpiles (Tolman et al., 1978)

Advantages	Disadvantages
1. Construction is not difficult; no excavation is necessary.	1. The steel sheet piling initially is not watertight.
2. Contractors, equipment, and materials are available throughout the United States.	2. Driving piles through ground containing boulders is difficult.
3. Construction can be economical.	3. Certain chemicals may attack the steel.
4. No maintenance required after construction.	
5. Steel can be coated for protection from corrosion to extend its service life.	

Table 2.12   Materials Commonly Used for Grout

cement, water	asphalt
cement, rock flour, water	clay, water
cement, clay, water	chemical
cement, clay, sand, water	

seam closest to the land surface and the grouting fluid is injected. The holes are then cleaned, drilling continues down to the next seam and grouting continues. The process is repeated until a sufficient depth has been obtained. In general, the stage method proceeds downward, utilizing increasing injection pressures. In the packer method, holes are drilled to the maximum planned depth. A zone of specified thickness is then partitioned off by placing packers at the top and bottom of the zone. Grout is then injected into the zone between the two packers. The mechanism is then moved up to the next zone to be grouted and the process is repeated. The packer method moves upward from the bottom utilizing decreasing injection pressures. Advantages of the packer method include: grouting pressures can be adjusted specifically to a particular foundation depth; walls of the borehole remain smooth and an excellent seal with packers can be achieved; and high pressures used in the stage method which may cause fracturation are avoided. However, these advantages can be offset by increased equipment needs and time for installation (Bowen, 1981).

Table 2.13   Properties of Admixtures Used with Cement Grouts

Admixture	Property
Calcium chloride Sodium hydroxide Sodium silicate	Accelerates setting time
Gypsum Lime sugar Sodium tannate	Retards setting time
Finely ground bentonite	Increases plasticity Reduces grout shrinkage
Clay Ground shale Rock flour	Reduces cost of grout Reduces strength of grout

Another method of grout injection is the driven-rod method. In this method, a perforated rod is driven to a desired depth and grout is injected as the rod is slowly withdrawn. This method is limited to shallow depths and relatively boulder-free soils.

Problems that might arise during construction of grout systems include: leakage of grout around the injection pipe of the hole being grouted; loss of pressure below the water table resulting in sand and water being forced into the pipe; and grout surfacing in outlying areas due to lateral migration (Bowen, 1981). It is also desirable to deposit cement grouts in clean seams from which any clay or unconsolidated materials have been removed, thus adding the burden of washing bore-holes (Peurifoy, 1979).

Construction of grout cutoff walls requires certain equipment in addition to normal borehole or drilling equipment. The list includes: one or more air compressors; one or two grout mixers; one agitator-type reservoir tank; one or more grout pumps; and grout discharge pipes or hoses, valves, and pressures gauges (Peurifoy, 1979).

Design

The first consideration in the design of a grout cutoff system is the actual composition of the grout. The composition will be a function of several variables including: the soil type to be injected into; the pollutant to be inhibited; the time since pollution started; and the time for installation. In general, chemical grouts must be used in fine-grained soils. However, chemical grouts (usually silicate) are not suitable for highly acidic or alkaline environments because their gel formation is an acid-base reaction. For coarse or gravel soils, particulate grouts are suitable. The amount of cement or bentonite in a particulate grout varies widely. The quantity of bentonite which can be incorporated in a grout is dependent upon the following considerations: the workability of the mixture — increased bentonite concentrations increase the stiffness of the slurry until it may become unpumpable; adding bentonite to cement slurries decreases their compressive strength; increased bentonite concentrations yield lower-specific-gravity slurries, showing a reduced tendency to migrate through the soil after placement; and increased bentonite concentrations reduce settlement or sedimentation of the slurries before injection (Jones, 1963).

A general guide for the selection of grouts is shown in Figure 2.23. Knowing the soil type in the area to be protected, Figure 2.23 can be entered from the top to determine both the applicable types of grouts and

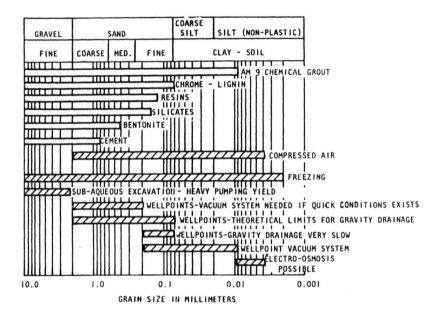

Figure 2.23. Soil gradation limits for grout injection (from American Cyanamid Company).

the available grouting procedures. The range of soil types to which a particular type of grout is applicable is covered by the solid white horizontal bars. The range of soil types to which a particular grouting procedure is applicable is indicated by the cross-hatched bars. Tables 2.14 to 2.16 summarize the compatibility of grouts with several contaminant types.

The second design consideration is the pressure at which the grout is to be injected. The use of excess pressures may weaken the strata by fissuring the rock or by opening fissures in otherwise closely jointed rock. Pressure-induced fissures will result in the wastage of grout. In contrast, the pressure should be kept sufficiently high to ensure penetration of the grout and decrease the time required for grouting. The allowable grouting pressure is best determined by conducting hydraulic fracture tests in the strata to be grouted (Morgenstern and Vaughan, 1963).

A number of chemical grouting agents have been banned or their use discontinued because of their toxicity. Huibregtse and Kastman (1981) note that one of the most successful grouts, AM-9, is highly toxic and has

Table 2.14   Interactions Between Grouts and Generic Chemical Classes (Hunt et al., no date)

| Chemical Group | Portland Cement | | Clay (Bentonite) | Clay-Cement | Silicate | Acrylamide | Polymers | | | | | Bitumen |
	Type I	Type II and V					Phenolic	Urethane	Urea-formaldehyde	Epoxy	Polyester	
Acid	1d	1d	?c	?c	3a	2c	?a§	2c	1a	?a§	?d	?a
Base	1d	1a†	?c**	?d	2c	3d	?d	?d	2c	?a	?d	?a
Heavy Metals	1a	?	?d	?	3?	3?	?	?	?	?	?	?
Non-Polar Solvent	2c	2?	?d	?	?	?a	?d	?a	2?	?d	?d	?d
Polar Solvent	2c	2?	?d	?	?	?	?d	?a	2?	?d	?d	?d
Inorganic Salts	2c	2a	2d	?d*	3?	3d‡	3a	?d	?a	?a	?d	?d

KEY: Compatibility Index

Effect on Set Time
1 No significant effect
2 Increase in set time (lengthen or prevent from setting)
3 Decrease in set time

Effect on Durability
a No significant effect
b Increase durability
c Decrease durability (destructive action begins within a short time period)
d Decrease durability (destructive action occurs over a long time period)

* Except sulfates, which are ?c
† Except KOH and NaOH, which are 1d
‡ Except heavy metal salts which are 2
§ Non-oxidizing
** Modified bentonite is d
? Data unavailable

Table 2.15 Interactions Between Grouts and Specific Chemical Groups (Hunt et al., no date)

Chemical Group	Bitumen	Portland Cement		Clay (Bentonite)	Clay-Cement	Grout Type		Polymers				
		Type I	Type II and V			Silicate	Acrylamide	Phenolic	Urethane	Urea-formaldehyde	Epoxy	Polyester
Organic Compounds												
Alcohols and Glycols	?a	?d	?d	?d	?d	?	?d	?	3a	?	?a	?a
Aldehydes and Ketones	?d■	?	?	?d	?	?	?a	3a	?d	?	?i	?i
Aliphatic and Aromatic Hydrocarbons	?d	2a	2?	?d	?	?	?a	?d‡	?a	2a	?d	?d
Amides and Amines	?	?	?	?	?	?	?	?	3?	?	?	?
Chlorinated Hydrocarbons	?d	2d	2d	?	?	?	?a	?d	?a	2a	?d	?d
Ethers and Epoxides	?	?	?	?	?	?	?a	?	?a	?d	?	?
Heterocyclics	?	?	?	?d	?	?	?a	?	?	?	?	?
Nitrites	?	?	?	?	?	?	?	?	?	?	?	?
Organic Acids and Acid Chlorides	?a	1d	1d	?d	?d	?a	2a	?a	2a	1a	?d	?d
Organometallics	?	?	?	?	?	?	?	?	?	?	?	?
Phenols	?d	1d	?	?d	?d	?	?	2a	?c	?	?	?
Organic Esters	?	?	?	?	?	1a	?	?	?	?	?	?

Table 2.15, continued

Chemical Group	Bitumen	Portland Cement — Type I	Portland Cement — Type II and V	Clay (Bentonite)	Clay-Cement	Silicate	Acrylamide	Polymers — Phenolic	Polymers — Urethane	Polymers — Urea-formaldehyde	Polymers — Epoxy	Polymers — Polyester
Inorganic Compounds												
Heavy Metal Salts and Complexes	?d	2c	2a	?d	2c	3?	2?	?	?	?a	?	?
Inorganic Acids	?a§□	1d	1a	?c**	?c	3a	2c	?a§	2c	1d	?a§	?a•
Inorganic Bases	?a	1a	1a†	?c**	?d	2c	3d	?d	?d†	2c	?a	?d
Inorganic Salts	?d	2c	2a	2d	?d*	3?	3d	3a#	?d	2c	?a	?a

KEY: Compatibility Index

Effect on Set Time
1 No significant effect
2 Increase in set time (lengthen or prevent from setting)
3 Decrease in set time

Effect on Durability
a No significant effect
b Increase durability
c Decrease durability (destructive action begins within a short time period)
d Decrease durability (destructive action occurs over a long time period)

* Except sulfates, which are ?c
† Except KoH and NaOH, which are 1d
‡ Low-molecular-weight polymers only
§ Non-oxidizing
• Non-oxidizing, except HF
□ Except concentrated acids
■ Except aldehydes which are 1a
Except bleaches which are 3d
** For modified bentonites, ?d
? Data unavailable

Table 2.16   Predicted Grout Compatibilities (Hunt et al., no date)

Chemical Group	Silicate	Acrylamide	Phenolic	Urethane	Urea-formaldehyde	Epoxy	Polyester
Organic Compounds							
Alcohols and Glycols	1a	1 –	3b	–	1	1 –	1 –
Aldehydes and Ketones	1a	—	—	1a		1a	1a
Aliphatic and Aromatic Hydrocarbons	1d	1 –	—	1 –	—	1 –	1 –
Amides and Amines	3a	3d	3b	– a	1a	1a	3a
Chlorinated Hydrocarbons	1s	1 –	1a	1 –		1a	1a
Ethers and Epoxides	1a	1 –	1a	1 –		1a	1a
Heterocyclics	1d	1 –	1a	1a	1a	1a	1a
Nitrites	1a	3 –	1a	1a	1a	1a	1a
Organic Acids and Acid Chlorides	1 –	—	3 –	2 –	—	—	1 –
Organometallics	1a	3a	—	—	1a	1a	3†?
Phenols	1a	1a	—	2 –	1a	1a	1?
Organic Esters	—	?	?	?	?	?	1d
Inorganic Compounds							
Heavy Metal Salts and Complexes	– a	—	—	—	—	3 –	3?
Inorganic Acids	—	—	2 –	—	—	1 –	1 –
Inorganic Bases	—	—	3 –	—	—	—	1 –
Inorganic Salts	– d	—	—	—	—	—	3* –

KEY: Compatibility Index

Effect on Set Time
1 No significant effect
2 Increase in set time (lengthen or prevent from setting)
3 Decrease in set time

Effect on Durability
a No significant effect
b Increase durability
c Decrease durability (destructive action begins within a short time period)
d Decrease durability (destructive action occurs over a long time period)

* If metal salts that are accelerators
† If metal is capable of acting as an accelerator
? Data unavailable

been removed from most markets because of its potential undesirable effects on ground water. Other grouts considered to be toxic and presenting a potential for ground water pollution are the lignin- and formaldehyde-based grouts.

Costs

The costs for grout cutoff systems are high, hence they will be applicable only to small localized cases of ground water pollution. Costs have been reported to range from $142 to $357 per installed cubic foot (Lu, Morrison and Stearns, 1981). Table 2.17 lists the relative costs of different types of grout.

Advantages and Disadvantages

The technology of grouting has been used in the construction industry for years. To date, most applications of grouting technology have been for increasing a soil's bearing capacity (to aid in tunnel construction for example) or to decrease the permeability of a soil to inhibit water movement (such as a cutoff wall for a dam). The applications of this technology to ground water pollution control are very recent. For example, Huibregtse and Kastman (1981) have analyzed the feasibility of mobile grouting units for protecting ground water threatened by hazardous spills on land. A number of physical/chemical advantages and disadvantages for grouting are listed in Table 2.18.

Table 2.17 Relative Costs of Grout[a]

Type of Grout	Basic Cost Figure
Portland cement	1.0
Silicate base—15%	1.3
Lignin base	1.65
Silicate base—30%	2.2
Silicate base—40%	2.9
Urea formaldehyde resin	6.0
Acrylamide (AM-9)	7.0

[a]Base unit = 1.0. Under a given set of conditions, where portland cement grout costs 1.0 times $/unit, other types of grout will cost the given figure times $/unit. (Tolman et al., 1978).

Table 2.18 Advantages/Disadvantages of Grout Systems

Advantages[a]	Disadvantages
1. When designed on basis of thorough preliminary investigations, grouts can be very successful.	1. Grouting limited to granular types of solid that have a pore size large enough to accept grout fluids under pressure yet small enough to prevent significant pollutant migration before implementation of grout program.[b]
2. Grouts have been used for over 100 years in construction and soil stabilization projects.	2. Grouting in a highly layered soil profile may result in incomplete formation of a grout envelope.[b]
3. Many kinds of grout to suit a wide range of soil types available.	3. Presence of high water table and rapidly flowing ground water limits groutability through;
	a. extensive transport of contaminants.
	b. rapid dilution of grouts[b]
	4. Some grouting techniques are proprietary.[a]
	5. Procedure requires careful planning and pretesting. Methods of ensuring that all voids in the wall have been effectively grouted are not readily available.[a]

[a]Tolman et al., 1978
[b]Huibregtse and Kastman, 1981.

SLURRY WALLS

Slurry walls represent a technology for encapsulating an area to either prevent ground water pollution or restrict the movement of previously contaminated ground water. The technology involves digging a trench around an area and backfilling with an impermeable material. Slurry walls can either be placed upgradient from a waste site (Figure 2.24) to prevent flow of ground water into the site, or placed around a site (Figure 2.25) to prevent movement of polluted ground water away from a site. Usually, slurry walls will require a complementary technology such as surface capping or purge wells.

Construction

The most common type of slurry wall construction is the trench method. In this method a trench is excavated, in the presence of a

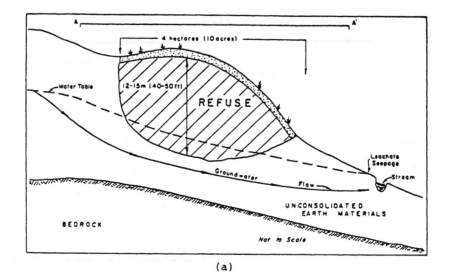

(a)

(b)

Figure 2.24. Cross section of landfill (a) before and (b) after slurry-trench cutoff wall installation (Tolman et al., 1978).

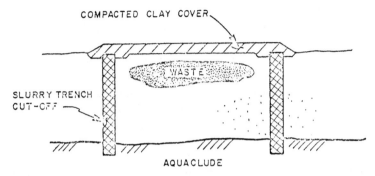

Figure 2.25.   Isolation of existing buried waste (Ryan, 1980).

bentonite-water slurry, to a desired depth. After excavation, the trench can be solidified by backfilling with a mixture of bentonite and the excavated material, or it can be allowed to solidify itself by incorporating cement in the original slurry. The backfilled trench is usually called a soil-bentonite (S-B) trench, and the other method is called a cement-bentonite (C-B) trench. Comparative advantages of each method are outlined in Table 2.19 (Ryan, 1980).

Another method of slurry wall construction is the vibrating beam method. In this approach, a beam with a pressure hose attached is driven into the ground, then slurry is injected under pressure as the beam is gradually withdrawn. This is similar to grouting and possesses similar advantages/disadvantages. This approach can be more economical and can proceed faster than the trench method. However the continuity of the wall cannot be guaranteed, the in-place thickness of the wall is considerably smaller, and the presence of boulders in the soil hinders this method more than the trench method.

Table 2.19   Comparison of C-B and S-B Slurry Trenches (Ryan, 1980)

Cement-Bentonite	Soil-Bentonite
1. Independent of availability or quality of soil for backfill.	1. Lower material costs
2. More suitable for limited access areas.	2. Can achieve higher permeability than C-B.
3. Cement sets quickly. Can cut trenches or allow traffic over wall in just a few days.	
4. Can be constructed in sections. S-B requires continuous trenching in one direction.	

Construction of slurry trenches is generally simple and consists only of excavating, recirculating the slurry, and backfilling. Excavation can be accomplished by any one or combinations of the following: backhoe, draglines, clamshells, bucket scrapers, or rotary drilling equipment. The choice of the specific type of excavation equipment is generally governed by the depth and width of excavation (Xanthakos, 1979). The backhoe is usually desirable when depths required are shallow. For depths of 30 m or more, draglines are required. Because dragline bucket widths usually exceed 2 m, they are not economical for C-B trenches due to high material costs. Clamshells can be cable-mounted or, like the bucket scraper, mounted to a rigid sliding bar and used for deep trench excavation (Ryan, 1980).

Recirculation of the slurry is important for maintaining the integrity of the slurry. During excavation the slurry will be subjected to losses through seepage and changes in density through addition of excavated material. Control of these changes is achieved by continuously recirculating the slurry through a central mixing unit which may have provisions for separating excavated materials from the original slurry. Backfilling operations may require mixing of different soil types prior to placement in the trench. Mixing can be accomplished by disking and blading. The mixed soil is then bulldozed into the trench, partially mixing with and displacing the slurry. Backfilling follows trenching after an interval of time sufficient to prevent interference between the two activities.

An important final aspect of slurry trench construction is keying into an underlying impervious zone. Trenches will require 2 or 3 ft for keying into clay materials, and will require a grout connection when keying into impervious bedrock (D'Appolonia, no date).

Design

The important design considerations of any slurry trench are the composition of the slurry and the ensuing impermeable wall. Design procedures for slurry walls range from general rules of thumb to detailed mathematical analyses of all aspects of the system. A general guideline approach is probably the most helpful.

D'Appolonia (no date) suggests that viscosities of both C-B and S-B slurries should be such that drain times from a Marsh Funnel range from 40 to 50 sec. It is also recommended that the specific gravity of the slurry be at least 15 lb/ft³ less than the unit weight of the backfill. The permeability of the S-B backfill will depend on the soil and the amount of bentonite blended in. These characteristics are depicted in Figures 2.26

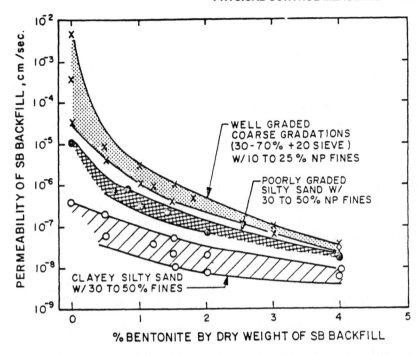

Figure 2.26. Relationship between permeability and quantity of bentonite added to S-B backfill (Ryan, 1980).

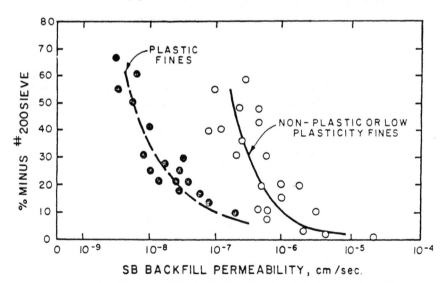

Figure 2.27 Permeability of soil-bentonite backfill related to fines content (Ryan, 1980).

and 2.27. The ideal consistency for backfill placement is a paste having a water content slightly above the liquid limit of the sand-clay-bentonite backfill mix. This usually corresponds to a slump cone reading of 2–6 in.

The durability of the slurry trench refers to its resistance to attack from contaminants. In the presence of clean ground water, both C-B and S-B trenches show little deterioration and can be considered permanent. However, C-B trenches show poor performance records where acids or sulfates are present. Similarly, exposure of an S-B trench to certain contaminants can lead to increased permeability through pore fluid substitution or the increased solubility of barrier minerals in the contaminant fluid. Pore fluid substitution can be the result of high concentrations of salts, which attract the waters of hydration, or it can result from ion substitution within the clay matrix. Permeabilities of S-B trenches have been shown to increase in the presence of certain organics, calcium, magnesium, heavy metals and solutions of high ionic strength (D'Appolonia, no date).

Xanthakos (1979) suggests that a slurry system should be designed based on the functions of the slurry. These functions are:

(1) support the face of the excavation, and also prevent the soil from sloughing and peeling off;

(2) seal the formation and form the filter cake, preventing slurry loss to the ground;

(3) suspend detritus, thereby preventing sludgy unconsolidated layers from accumulating on the bottom of the trench; and

(4) carry the cuttings in the slurry volume, thereby preventing sedimentation in the mud circuit.

All of the above functions can be controlled by manipulating the physical properties of the slurry, i.e., density, viscosity, etc. Xanthakos (1979) has outlined a series of simple steps and procedures for proportioning the materials that comprise a slurry. These are as follows:

(1) Determine the density required for trench stability. The density can be controlled by the presence of colloid and noncolloid solid materials. If the depth to an impermeable formation (H) is known, the required slurry density can be calculated by the following (see Figure 2.28).

$$\gamma_f = \frac{\gamma(1-m^2)\, K\alpha \; + \; \gamma^1 m^2 K\alpha \; + \; \gamma_w m^2}{n^2}$$

where γ_f = required slurry density (g/m³)
         γ = unit weight of soil (g/m³)
         $K\alpha$ = $\tan^2 (45\text{-}\theta^{1/2})$
         m = natural ground water level as a fraction of total depth of excavation (H)

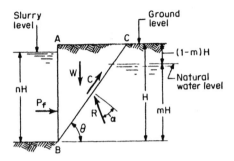

Figure 2.28. Stability of a trench for arbitrary slurry and natural water level (Xanthakos, 1979).

γ_w = unit weight of water (g/m³)
$\gamma 1$ = effective (buoyant) weight of soil (g/m³)
θ^1 = angle of shear resistance
n = slurry level as a fraction of total depth of excavation (H)

(2) Select the funnel viscosity by reference to Table 2.20.

(3) Establish any applicable control limits from Table 2.21.

(4) Determine whether control agents (peptizers, polyelectrolytes, fluid-loss-control materials, etc.) are necessary and economically justified.

(5) Proportion the constituent materials (water, bentonite, control agents, and noncolloid solids). This phase merely consists of a quantitative estimation. The proportioning may be empirical and depend on experience if the properties of the materials selected are known, or it may have a technical basis of tests and estimations.

Table 2.20   Funnel Viscosity for Common Types of Soil (Xanthakos, 1979)

Type of Soil	Funnel Viscosity, s/946 cm³	
	Excavation in Dry Soil	Excavation with Groundwater
Clay	27 – 32	
Silty sand, sandy clay	29 – 35	
Sand, with silt	32 – 37	38 – 43
Fine to coarse	38 – 43	41 – 47
and gravel	42 – 47	55 – 65
Gravel	46 – 52	60 – 70

Table 2.21   Control Limits[a] for the Properties of Slurries (Xanthakos, 1979)

Function	Property							
	Average bentonite concentration,[b] %	Density, lb/ft³	sp gr	Plastic viscosity, cP	Marsh cone viscosity	10-min gel strength (Fann), lb/100 ft²	pH	Sand content, %
Face support	>3–4	>64.3	>1.03	—	Limits	c	—	>1[d]
Sealing process	>3–4	—	—	—	established	—	—	1
Suspension of detritus	>3–4	—	—	—	by soil type	>12–15	—	
Displacement by concrete	<15	<78	<1.25	<20	—	—	<12	<25
Separation of noncolloids	—	—	—	—	—	—	—	<30
Physical cleaning	<15	<78	<1.25	—	—	—	—	<25
Pumping of slurry	—	—	—	—	—	Variable	—	
Limits	>3–4	>64.3	>1.03	<20		>12–15	<12	>1
	<15	<78	<1.25	—		—	—	<25

[a]Controls are not considered necessary for apparent viscosity and yield stress. Whereas fluid loss commonly is judged by standard filtration test and a maximum film thickness of 2 mm, better control limits are established by stagnation-gradient tests.
[b]Should be expected to vary widely because of different bentonite brands.
[c]The shear strength of filter cake is more applicable to peel-off control (also the time required for its formation).
[d]Optional.

Table 2.22   Approximate Slurry Wall Costs[a] as a Function of Medium and Depth (Spooner, Wetzel and Grube, 1982)

Soil Type	Soil-Bentonite Wall			Cement-Bentonite Wall		
	Depth ≤30 ft	Depth 30–75 ft	Depth 75–120 ft	Depth ≤60 ft	Depth 60–150 ft	Depth >150 ft
Soft to Medium Soil N≤40	2–4	4–8	8–10	15–20	20–30	30–75
Hard Soil N 40–200	4–7	5–10	10–20	25–30	30–40	40–95
Occasional Boulders	4–8	5–8	8–25	20–30	30–40	40–85
Soft to Medium Rock N≥200 Sandstone, Shale	6–12	10–20	20–50	50–60	60–85	85–175
Boulder Strata	15–25	15–25	50–80	30–40	40–95	95–210
Hard Rock Granite, Gneiss, Schist[b]				95–140	140–175	175–235

[a] In 1979 dollars/ft². For standard reinforcement in slurry walls add $8.99/ft². For construction in urban environment add 25–50% of price.
[b] Nominal penetration only.

In general, slurry trenches are attractive alternatives when an impervious natural barrier exists at a reasonable depth and the waste area is relatively large. A clayey sand or sandy clay containing 30–60% fines blended with the bentonite slurry is usually satisfactory for most waste isolation applications (D'Appolonia, no date).

Costs

Table 2.22 shows the comparative costs of S-B and C-B slurry walls.

Advantages and Disadvantages

There are many instances where slurry wall technologies have been employed. Slurry systems have been employed for their impermeability as cutoff and diaphragm walls, and slurries have been used to aid in the construction of load-bearing elements and foundations (Xanthakos, 1979).

To date, there has been very little information provided on the performance of slurry walls applied to ground water pollution control problems. This does not mean the technology has not been applied. Slurry walls have been constructed at the Rocky Mountain Arsenal in Colorado and the Gilson Road Hazardous Waste Dump in New Hampshire. A list of the advantages/disadvantages of slurry trenches compared to grouting, sheet piling, pumping or other techniques is shown in Table 2.23.

Table 2.23   Advantages/Disadvantages of Slurry Trenches

Advantages	Disadvantages[a]
1. Construction methods are simple.[a]	1. Cost of shipping bentonite from West.
2. Adjacent areas not affected by ground water drawdown.[a]	2. Some construction procedures are patented and will require a license.
3. Bentonite (mineral) will not deteriorate with age.[a]	3. In rocky ground, overexcavation is necessary because of boulders.
4. Leachate-resistant bentonites are available.[a]	4. Bentonite deteriorates when exposed to high-ionic-strength leachates.
5. Low maintenance requirements.[a]	
6. Eliminate risks due to pump breakdowns, or power failures.[b]	
7. Eliminate headers and other above ground obstructions.[b]	

[a]Tolman et al., 1978.
[b]Ryan, 1980.

SELECTED REFERENCES

American Petroleum Institute, "The Migration of Petroleum Products in Soil and Ground Water—Principles and Counter Measures", Publication No. 4149, Dec. 1972, Washington, D.C.

American Petroleum Institute, "Underground Spill Cleanup Manual", API Publications 1628, June 1980, Washington, D.C.

Blake, S. B. and Lewis, R. W., "Underground Oil Recovery", *Proceedings of the Second National Symposium on Aquifer Restoration and Ground Water Monitoring*, 1982, National Water Well Association, Worthington, Ohio, pp. 69-76.

Bowen, R., "Specifications for Grouting", *Grouting in Engineering Practice*, 2nd ed., Applied Science Publishers LTD, Great Britain, 1981, pp. 197-239.

Bowles, J. E., "Sheet-Pile Walls—Cantilevered and Anchored", *Foundation Analysis and Design*, 2nd ed., McGraw-Hill, 1977, pp. 412-444.

Campbell, M. D. and Lehr, J. H., "Well Cost Analysis", *Water Well Technology*, 4th ed., McGraw Hill, 1977.

Chapman, T. G., "Modeling Ground Water Over Sloping Beds", *Water Resources Research*, Vol. 16, No. 6, 1980, pp. 1114-1118.

Chauhan, H. S., Schwab, G. O. and Hamdy, M. Y., "Analytical and Computer Solutions of Transient Water Tables for Drainage of Sloping Land", *Water Resources Research*, Vol. 4, No. 3, 1968, pp. 573-579.

Childs, E. C., "Drainage of Ground Water Resting on a Sloping Bed", *Water Resources Research*, Vol. 7, No. 5, 1971, pp. 1256-1263.

Cohen, R. M. and Miller, W. J., "Use of Analytical Models for Evaluating Corrective Actions at Hazardous Waste Disposal Facilities", *Proceedings of the Third National Symposium on Aquifer Restoration and Ground Water Monitoring*, 1983, National Water Well Association, Worthington, Ohio, pp. 85-97.

D'Appolonia, D. J., "Slurry Trench Cutoff Walls for Hazardous Waste Isolation", Engineered Construction Internation, Inc., Pittsburgh, Pennsylvania.

Emrich, G. H., Beck, W. W. and Tolman, A. L., "Top-Sealing to Minimize Leachate Generation: Case Study of the Windham, Connecticut Landfill", SMC-Martin, King of Prussia, Pennsylvania.

Ferris, J. G., et al., "Theory of Aquifer Tests", USGS Water Supply Paper 1536E, 1962.

Fung, R., editor, *Protective Barriers for Containment of Toxic Materials*, 1980, Noyes Data Corporation, Park Ridge, New Jersey.

Gibb, J. P., "Cost of Domestic Wells and Water Treatment in Illinois", *Ground Water*, Vol. 9, No. 5, Sept.-Oct. 1971, pp. 40-49.

Hantush, M. S., "Growth and Decay of Ground-Water Mounds in Response to Uniform Percolation", *Water Resources Research*, Vol. 3, No. 1, 1967, pp. 227-234.

Haxo, H. E., Jr., "Evaluation of Liner Materials", Oct. 1973, U.S. Environmental Protection Agency, Cincinnati, Ohio.

Huibregtse, K. R. and Kastman, K. H., "Development of a System to Protect Groundwater Threatened by Hazardous Spills on Land", EPA-600/2-81-085, May 1981, National Technical Information Service, Springfield, Virginia.

Huisman, L., "Ground-Water Recovery", Winchester Press, New York, 1972, 336 pp.

Hunt, G. E., et al., "Collection of Information on the Compatibility of Grouts with Hazardous Wastes", no date, U.S. Environmental Protection Agency, Cincinnati, Ohio.

Johnson Division, UOP Inc., *Ground Water and Wells*, 4th ed., St. Paul, Minnesota, 1975.

Jones, G. K., "Chemistry and Flow Properties of Bentonite Grouts", *Symposium on Grouts and Drilling Muds in Engineering Practice*, International Society of Soil Mechanics and Foundation Engineering, 1963, pp. 22-28.

Kirkham, D., "Seepage of Steady Rainfall through Soil into Drains", *AGU Trans.*, Vol. 39, 1958, pp. 892-908.

Kirkham, D., "Explanation of Paradoxes in the Dupuit-Fochheimer Seepage Theory", *Water Resources Research*, Vol. 3, 1967, pp. 609-622.

Kufs, C., et al., "Procedures and Techniques for Controlling the Migration of Leachate Plumes", *Ninth Annual Research Symposium — Land Disposal, Incineration and Treatment of Hazardous Waste*, 1983.

Lu, J. C. S., Morrison, R. D. and Stearns, R. J., "Leachate Production and Management from Municipal Landfills: Summary and Assessment", *Land Disposal: Municipal Solid Waste — Proceedings of the Seventh Annual Research Symposium*, U.S. Environmental Protection Agency, 1981, pp. 1-17.

Lundy, D. A. and Mahan, J. S., "Conceptual Designs and Cost Sensitivities of Fluid Recovery Systems for Containment of Plumes of Contaminated Groundwater", *Proceedings of the National Conference on Management of Uncontrolled Hazardous Waste Sites*, 1982, Hazardous Materials Control Research Institute, Silver Spring, Maryland, pp. 136-140.

Luthin, J. N., *Drainage of Agricultural Lands*, 1st ed., American Society of Agronomy, Madison, Wisconsin, 1957.

Marei, S. M. and Towner, G. D., "A Hele-Shaw Analog Study of the Seepage of Ground Water Resting on a Sloping Bed", *Water Resources Research*, Vol. 11, No. 4, 1975, pp. 589-594.

McBean, E. A, et al., "Leachate Collection Design for Containment Landfills", *American Society of Civil Engineers Journal of Environmental Engineering Division*, Vol. 108, No. 1, 1982, pp. 204-209.

Morgenstern, N. R. and Vaughan, P. R., "Some Observations on Allowable Grouting Pressures", *Symposium on Grouts and Drilling Muds in Engineering Practice*, International Society of Soil Mechanics and Foundation Engineering, 1963, pp. 63-46.

Mualem, Y. and Bear, J., "Steady Phreatic Flow over a Sloping Semipervious Layer", *Water Resources Research*, Vol. 14, No. 3, 1978, pp. 403-408.

Pastrovich, T. L., et al., Protection of Groundwater From Oil Pollution", Report No. 3/79, CONCAWE, the Hague 1979.

Peurifoy, R. L., "Foundation Grouting", *Construction Planning, Equipment, and Methods*, 3rd Ed., McGraw-Hill, New York, 1979, pp. 491–505.

Pye, V. I., Patrick, R. and Quarles, J., *Groundwater Contamination in the United States*, University of Pennsylvania Press, Philadelphia, 1983, p. 57.

Quince, J. R. and Gardner, G. L., "Recovery and Treatment of Contaminated Ground Water", The Second National Symposium on Aquifer Restoration and Ground Water Monitoring", May 1982.

Ryan, C. R., "Slurry Cut-Off Walls Methods and Applications", Mar. 1980, Geo-Con, Inc., Pittsburgh, Pennsylvania.

Shultz, D. W., and Miklas, M. P., "Procedures for Installing Liner Systems", *Land Disposal of Hazardous Waste: Proceedings of the Eighth Annual Research Symposium*, U.S. Environmental Protection Agency, 1982, pp. 224–238.

Singh, S. R., and Jacob, C. M., "Transient Analysis of Phreatic Aquifers Lying Between Two Open Channels", *Water Resources Research*, Vol. 13, No. 2, 1977, pp. 411–419.

Smith, M. E., "Sanitary Landfills", *Conference on Reclamation of Difficult Sites*, 1983, University of Wisconsin Extension, Madison, Wisconsin.

Spangler, M. G., and Handy, R. L., "Soil Water", *Soil Engineering*, 3rd ed., Intext Educational Publishers, New York, 1973, pp. 191–209.

Spooner, P. A., Wetzel, R. S. and Grube, W. E., "Pollution of Migration Cutoff Using Slurry Trench Construction", *Management of Uncontrolled Hazardous Waste Sites*, Hazardous Materials Control Research Institute, 1982, pp. 191–197.

Tolman, A.L, et al., "Guidance Manual for Minimizing Pollution from Waste Disposal Sites", EPA-600/2-78-142, Aug. 1978, U.S. Environmental Protection Agency, Cincinnati, Ohio.

Towner, G. .D., "Drainage of Ground Water Resting on a Sloping Bed with Uniform Rainfall", *Water Resources Research*, Vol. 11, No. 1, 1975, pp. 144–147.

U.S. Environmental Protection Agency, *Survey of Solidification/Stabilization Technology for Hazardous Industrial Wastes*, EPA-600/2-79-056, July 1979.

U.S. Environmental Protection Agency, "Handbook for Remedial Action at Waste Disposal Sites", EPA-625/6-82-006, June 1982, Cincinnati, Ohio.

Xanthakos, P., *Slurry Walls*, 1st edition, McGraw-Hill 1979.

CHAPTER 3

Treatment of Ground Water

D.F. Kincannon and E.L. Stover, Oklahoma State University

Concepts for treatment of contaminated ground water following its pumping from the subsurface are well established. These concepts have been developed from years of experience treating industrial wastewaters. However, confusion may occur when these concepts are applied to site-specific contaminated ground waters. The type of treatment required depends primarily on the contaminants being removed. Treatment systems may be relatively simple when a single chemical is the only contaminant, or extremely complex for cases involving numerous contaminants. In most cases, treatability studies should be conducted with representative samples to determine appropriate treatment components. This chapter addresses air stripping, activated carbon, and biological treatment for organics in ground water, and chemical precipitation for inorganics in ground water. The orientation herein is toward describing the principles, design considerations and applications of the treatment processes. Theoretical information is found elsewhere in numerous books on wastewater treatment.

AIR STRIPPING

Air stripping is a mass transfer process in which a substance in solution in water is transferred to solution in a gas. The rate of mass transfer depends upon several factors according to the following equation.

$$M = K_L\, a(C_L - C_g)$$

where M = mass of substance transferred per unit time and volume (g/hr/m³)
K_L = coefficient of mass transfer (m/hr)
a = effective area (m²/m³)
$(C_L - C_g)$ = driving force (concentration difference between liquid phase and gas phase, g/m³)

The driving force is an important aspect in the success or failure of an air stripping process. The driving force is the difference between actual conditions in the air stripping unit and conditions associated with equilibrium between the gas and liquid phases. If equilibrium exists at the air-liquid interface, the liquid phase concentration is related to the gas phase concentration by Henry's law, which states:

$$C_{ig} = HC_{il}$$

where C_{ig} = equilibrium concentration in gas phase (g/m³)
       C_{il} = equilibrium concentration in liquid phase (g/m³)
       H = Henry's law constant

another form of Henry's law is

$$P_A = H'C$$

where P_A = partial pressure of substance in the air mixture in contact with the water at equilibrium (atm)
       H' = Henry's law constant (atm/g/m³)
       C = concentration of substance in the water at equilibrium (g/m³)

The Henry's law constant can be used to predict the strippability of a chemical. A compound with a high Henry's law constant generally is more easily stripped from water than one with a lower Henry's law constant.

The mass of a contaminant that can be transferred not only is a function of the driving force, but is also a function of the mass transfer. Due to the difficulty in separating these two parameters, they are usually evaluated as one coefficient, $K_L a$. The coefficient $K_L a$ is a function of the geometry and physical characteristics of the air stripping equipment, the loading rate (air-to-water ratio), and the temperature. These are factors that must be considered in the design of an air stripping process.

Description of Process

As shown in Figure 3.1, there are four basic equipment configurations used for air stripping, including diffused aeration, countercurrent packed columns, cross-flow towers, and coke tray aerators. Diffused aeration stripping uses aeration basins similar to standard wastewater treatment aeration basins. Water flows through the basin from top to

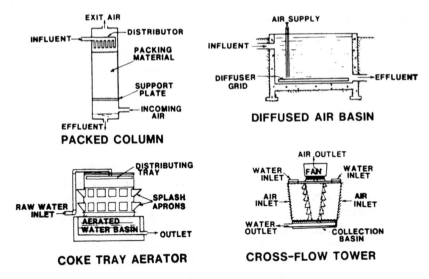

Figure 3.1 Air stripping equipment configurations.

bottom with the air dispersed through diffusers at the bottom of the basin. The air-to-water ratio is significantly lower than in either the packed column or the cross-flow tower.

In the countercurrent packed column, water containing one or more impurities is allowed to flow down through a column containing packing material with air flow countercurrently up through the column. In this way the contaminated water comes into intimate contact with clean air. Packing materials are used which provide high void volumes and high surface area. In the cross-flow tower, water flows down through the packing as in the countercurrent packed column, however, the air is pulled across the water flow path by a fan. The coke tray aerator is a simple, low-maintenance process. The water being treated is allowed to trickle through several layers of trays. This produces a large surface area for gas transfer.

The countercurrent packed tower appears to be the most appropriate equipment configuration for treating contaminated ground waters for the following reasons:

(1) It provides the most liquid interfacial area.
(2) High air-to-water volume ratios are possible due to the low air pressure drop through the tower.
(3) Emission of stripped organics to the atmosphere may be environmentally unacceptable; however, a countercurrent tower is relatively small and can be readily connected to vapor recovery equipment.

Design Parameters and Procedures

The design of an air stripping process for stripping volatile organics from contaminated ground water is accomplished in two steps. The cross-sectional area of the column is determined and then the height of the packing is determined. The cross-sectional area of the column is determined by using the physical properties of the air flowing through the column, the characteristics of the packing, and the air-to-water flow ratio. A key factor is the establishment of an acceptable air velocity. A general rule of thumb used for establishing the air velocity is that an acceptable air velocity is 60% of the air velocity at flooding. Flooding is the condition in which the air velocity is so high that it holds up the water in the column to the point where the water becomes the continuous phase rather than the air. If the air-to-water ratio is held constant, the air velocity determines the flooding condition. For a selected air-to-water ratio, the cross-sectional area is determined by dividing the air flow rate by the air velocity. The selection of the design air-to-water ratio must be based upon experience or pilot-scale treatability studies. Treatability studies are particularly important for developing design information for contaminated ground water.

The height of column packing may be determined by the following equation:

$$Z = \frac{\ln\left[\dfrac{(X_2 - Y_1/H)}{(X_1 - Y_1/H)} (1-A) + A \right] L}{K_L a \, C \, (1-A) \, (1-X) \, M}$$

where
$\quad Z$ = height of packing, ft
$\quad L$ = water velocity, lb-mole/hr/ft^2
$\quad X_2$ = influent concentration of pollutant in ground water, mole fraction
$\quad X_1$ = effluent concentration of pollutant in ground water, mole fraction
$\quad K_L a$ = mass transfer coefficient, 1/hr
$\quad C$ = molar density of water = 3.47 lb-mole/ft^3
$\quad H$ = Henry's law constant, mole fraction in air per mole fraction in water
$\quad G$ = air velocity, lb-mole/hr/ft^2
$\quad A$ = L/HG
$\quad (1-X)M$ = the average of one minus the equilibrium water concentration through the column
$\quad Y_1$ = influent concentration of pollutant in air, mole fraction

In most cases, the following assumptions can be made:

(1) $Y_1 = 0$, there should be no pollutants in the influent air.
(2) $(1-X)M = 1$, the influent concentrations should be too small when converted to mole fraction to shift this term significantly from 1.0.

The equation for determining the column packing height is then reduced to:

$$Z = \frac{\ln \left[\dfrac{X_2}{X_1} (1\text{-}A) + A \right] L}{K_L a \, C \,(1\text{-}A)}$$

All variables except the mass transfer coefficient are set by the design conditions. The selection of a suitable mass transfer coefficient must be based upon pilot-scale treatability studies and the judgment of the design engineer. The mass transfer coefficient is a function of:

(1) Type of volatile organic to be removed
(2) Air-to-water ratio
(3) Temperature of the ground water
(4) Type of packing
(5) Geometry of the tower

A plot of experimental data as shown in Figure 3.2 can be helpful in selecting the proper mass transfer coefficient.

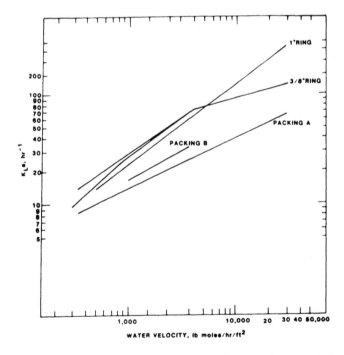

Figure 3.2 Mass transfer coefficients as a function of water velocity.

Applications to Ground Water

Air stripping has been successfully used for removing volatile organics from contaminated ground waters. These successful operations include an industrial park investigated by Stover (1982), a case study reported by Pekin and Moore (1982), and Rockaway Township, New Jersey, reported by McKinnon and Dyksen (1982). A summary of the removals achieved in these operations for various organic contaminants at various air-to-water ratios is given in Table 3.1.

It has been pointed out that temperature has an effect on the mass transfer coefficient. This becomes important when the contaminated ground water contains compounds that are very soluble. This makes their removal by ambient-temperature air stripping almost impossible. Lamarre, McGarry, and Stover (1983) have reported on such a case. It was found in methyl ethyl ketone removal, that at 54°F and an air-to-water ratio of 490, only 43% removal occurred; at 90°F and an air-to-water ratio of 513, removal was 92%; and at 136°F and an air-to-water ratio of 469, the removal was 99%. Removal efficiency increased dra-

Table 3.1   Packed Column Air Stripping of Volatile Organics

Organic Contaminant	Air-Water Ratio	Influent, µg/l	Effluent, µg/l
1,1,2-Trichloroethylene	9.3	80	16
	96.3	80	3
	27.0	75	16
	156.0	813	52
	44.0	218	40
	75.0	204	36
	125.0	204	27
1,1,1-Trichloroethane	9.3	1200	460
	96.3	1200	49
	27.0	90	31
	156.0	1332	143
1,1-Dichloroethane	9.3	35	9
	96.3	35	1
1,2-Dichloropropane	27.0	50	<5
	146.0	70	<5
	156.0	377	52
Chloroform	27.0	50	<2
	146.0	57	<2
Diisopropyl ether	44.0	15	7
	75.0	14	6
	125.0	4	4

matically with temperature and less sharply with the air-to-water ratio. Therefore, high-temperature air stripping, or steam stripping, should be investigated for each case encountered.

Costs

Although the variation in the design of packed column air stripping systems results in various costs, the major components of an air stripping system for removing organic contaminants from a ground water would include the packed column, the air supply equipment, and repumping. Annual costs per 1000 gal water treated have been presented by Dyksen et al. (1982), and Figure 3.3 provides a summary. The costs are 1982 costs and are based upon preliminary designs for achieving 90% removal of trichloroethylene (TCE).

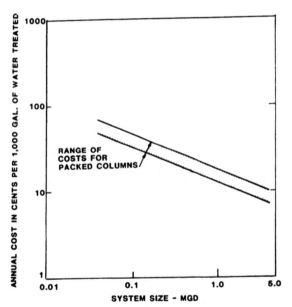

NOTES:

1. ANNUAL COSTS INCLUDES AMORTIZED CAPITAL COSTS AND ANNUAL OPERATING COSTS

2. SYSTEM SIZE REPRESENTS AVERAGE PLANT CAPACITY.

3. RANGE OF COSTS FOR PACKED COLUMN ACCOUNTS FOR DIFFERENCES IN MATERIALS OF CONSTRUCTION AS DISCRIBED IN THE TEXT

Figure 3.3 Comparison of costs for packed column aeration (Dyksen et al., 1982).

CARBON ADSORPTION

Adsorption occurs when an organic molecule is brought to the activated carbon surface and held there by physical and/or chemical forces. The quantity of a compound or group of compounds that can be adsorbed by activated carbon is determined by a balance between the forces that keep the compound in solution and the forces that attract the compound to the carbon surface. Factors that affect this balance include:

(1) Adsorptivity increases with decreasing solubility.
(2) The pH of the water can affect the adsorptive capacity. Organic acids adsorb better under acidic conditions, whereas amino compounds favor alkaline conditions.
(3) Aromatic and halogenated compounds adsorb better than aliphatic compounds.
(4) Adsorption capacity decreases with increasing temperature although the rate of adsorption may increase.
(5) The character of the adsorbent surface has a major affect on adsorption capacity and rate. The raw materials and the process used to activate the carbon determine its capacity.

When activated carbon particles are placed in water containing organic chemicals and mixed to give adequate contact, the adsorption of the organic chemicals occurs. The organic chemical concentration will decrease from an initial concentration, C_o, to an equilibrium concentration, C_e. By conducting a series of adsorption tests, it is usually possible to obtain a relationship between the equilibrium concentration and the amount of organics adsorbed per unit mass of activated carbon.

The Freundlich isotherm and the Langmuir isotherm are most often used to represent the adsorption equilibrium. One form of the Freundlich isotherm is:

$$\frac{X}{m} = KC_e^{1/n}$$

where
$$X = \text{mass of organic adsorbed (g)}$$
$$m = \text{mass of activated carbon (g)}$$
$$C_e = \text{equilibrium concentration of organics (g/m}^3)$$
$$K, n = \text{experimental constants}$$

The Langmuir isotherm has the following form:

$$\frac{X}{m} = \frac{a K C_e}{1 + K C_e}$$

where   a = mass of adsorbed organic required to completely saturate a unit mass of
            carbon (g)
         K = experimental constant

From an isotherm test it can be determined whether or not a particular organic material can be removed effectively. It will also show the approximate capacity of the carbon for the application, and provide a rough estimate of the carbon dosage required. Isotherm tests also afford a convenient means of studying the effects of pH and temperature on adsorption. Isotherms put a large amount of data into concise form for ready evaluation and interpretation. Isotherms obtained under identical conditions using the same contaminated ground water for two or more carbons can be quickly and conveniently compared to determine the relative merits of the carbons.

Figure 3.4 represents a typical Freundlich isotherm for the comparison of two carbons removing the organics from a contaminated ground water. The isotherm for Carbon A is at a high level and has only a slight slope. This means that adsorption is large over the entire range studied. The isotherm for carbon B is at a lower range and with a steeper slope. This means lower adsorption than carbon A. However, adsorption improves at higher concentrations. In general, the steeper the slope of its isotherm, the greater the efficiency of a carbon in column operation.

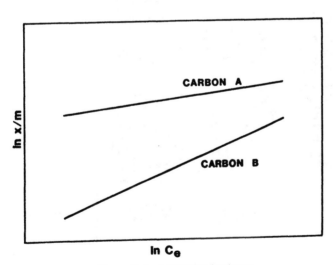

Figure 3.4 Adsorption isotherm.

Description of Process

Activated carbon adsorption may be accomplished by batch, column, or fluidized-bed operations. The usual contacting systems are fixed bed or countercurrent moving beds. The fixed beds may employ downflow or upflow of water. The countercurrent moving beds employ upflow of the water and downflow of the carbon since the carbon can be moved by the force of gravity. Both fixed beds and moving beds may use gravity or pressure flow.

Figure 3.5 shows a typical fixed-bed carbon column employing a single column with downflow of the water. The column is similar to a pressure filter and has an inlet distributor, an underdrain system, and a surface wash. During the adsorption cycle the influent flow enters through the inlet distributor at the top of the column, and the ground water flows downward through the bed and leaves through the underdrain system. The unit hydraulic flow rate is usually from 2 to 5 gpm/ft². When the head loss becomes excessive due to the accumulated suspended solids, the column is taken off-line and backwashed.

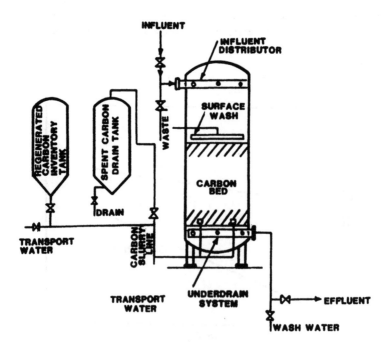

Figure 3.5 Fixed-bed adsorption system.

Figure 3.6 shows a typical countercurrent moving-bed carbon column employing upflow of the water. Two or more columns are usually provided and are operated in series. The influent contaminated ground water enters the bottom of the first column by means of a manifold system which uniformly distributes the flow across the bottom. The ground water flows upward through the column. The unit hydraulic flow rate is usually 2 to 10 gpm/ft². The effluent is collected by a screen and manifold system at the top of the column and flows to the bottom manifold of the second column. The carbon flow is not continuous but instead is pulsewise.

The fluidized bed consists of a bed of activated carbon. The water flows upward through the bed in the vertical direction. The upward liquid velocity is sufficient to suspend the activated carbon so that the carbon does not have constant interparticle contact. At the top of the carbon there is a distinct interface between the carbon and the effluent water. The principal advantage of the fluidized bed is that waters with appreciable suspended solids content may be given adsorption treatment without clogging the bed, since the suspended solids pass through the bed and leave with the effluent. This should not be a concern with ground waters.

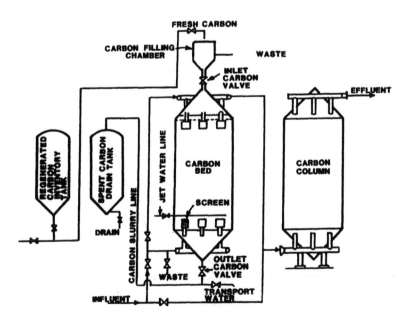

Figure 3.6 Moving-bed adsorption system.

Design Parameters and Procedures

Although the treatability of a particular wastewater by carbon and the relative capacity of different types of carbon for treatment may be estimated from adsorption isotherms, carbon performance and design criteria are best determined by pilot column tests. Design-related information which can be obtained from pilot tests includes:

(1) Contact time
(2) Bed depth
(3) Pretreatment requirements
(4) Breakthrough characteristics
(5) Headloss characteristics
(6) Carbon dosage in pounds of pollutants removed per pound of carbon.

The design of an activated carbon adsorption column can be accomplished by using a kinetic equation which is based on a derivation by Thomas (1948). The method requires data obtained from a breakthrough curve. A typical breakthrough curve obtained from a pilot test is shown in Figure 3.7. The expression by Thomas for an adsorption column is as follows:

$$\frac{C_o}{C} = 1 + \exp\frac{K_1}{Q}(A_oM - C_oV)$$

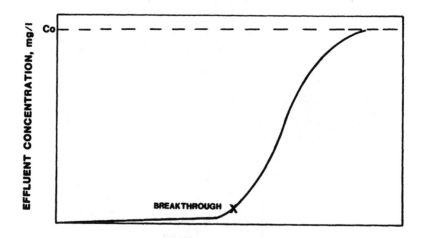

Figure 3.7 Typical breakthrough curve.

where C = effluent pollutant concentration (g/m^3)
           C_o = influent pollutant concentration (g/m^3)
           K_1 = rate constant $m^3/day/g$
           Q = flow rate, m^3/day
           A_o = adsorption capacity, g/g
           M = mass of carbon, g
           V = throughput volume, m^3

Rearranging and taking the logarithms of both sides yields

$$\ln \left(\frac{C_o}{C} - 1\right) = \frac{K_1 A_o M}{Q} - \frac{K_1 C_o V}{Q}$$

This is the equation of a straight line in which $K_1 A_o M/Q$ is the y-intercept and $K_1 C_o/Q$ is the slope of the line. The pilot column tests provides all parameters except K_1 and A_o. If $\ln (C_o/C - 1)$, is plotted versus V, the values of K_1 and A_o can be determined by a graphical solution. Then the mass of carbon needed for a selected breakthrough volume may be determined.

Applications to Ground Water

Activated carbon adsorption has been successfully used for removing organics from contaminated ground waters. Kaufman (1982) reported on 17 separate cases. The aquifers had been contaminated by leachate from lagoons or dumpsites, industrial accidents, and chemical spills due to railroad or truck accidents. Activated carbon successfully treated these contaminated aquifers. McDougall, Fusco, and O'Brien (1980) reported on the successful treatment of the Love Canal landfill leachate by activated carbon adsorption. O'Brian and Fisher (1983) have reported on 31 separate cases where activated carbon adsorption has been successful in treating contaminated ground waters. The reasons for treatment were to prevent the spread of contamination through an aquifer, for process water reuse, and for decontamination of potable water. Table 3.2 provides some results obtained in these operations. In all cases, the activated carbon was very successful in removing priority pollutants.

Costs

The cost of treating ground water by activated carbon adsorption is dependent on a number of factors, such as: flow rates, concentrations,

Table 3.2   Activated Carbon Adsorption of Organics

Contaminants	Influent Concentration, $\mu g/l$	Effluent Concentration, $\mu g/l$
Phenol	63,000	<100
	2,400	< 10
	40,000	< 10
Carbon tetrachloride	61,000	< 10
	130,000	< 1
	73,000	< 1
1,1,2- Tetrachloroethane	80,000	< 10
Tetrachloroethylene	44,000	12
	70,000	< 1
1,1,1-Trichloroethane	23,000	ND[a]
	1,000	< 1
	3,300	< 1
	12,000	< 5
	143,000	< 1
	115	< 1
Benzene	2,800	< 10
	400	< 1
	11,000	<100
2,4- Dichlorophenol	5,100	ND

[a]ND—Nondetectable

type of contaminant, type of application, and site requirements. These factors determine equipment and carbon usage; thus it is difficult to be specific on costs. However, Kaufmann (1982) reported that for municipal potable treatment the costs have been in the $0.40 to $1.00 per person per month range. O'Brien and Fisher (1983) have reported that if the influent concentration is in the mg/l range the costs have ranged from $0.48 to $2.52 per 1000 gal treated. However, if the influent concentrations are in the $\mu g/l$ range, the costs have varied from $0.22 to $0.55 per 1000 gal treated.

BIOLOGICAL TREATMENT

In biological treatment of contaminated ground water, the objective is to remove or reduce the concentration of organic and inorganic compounds. Because many of the compounds that may be present in contaminated ground water are toxic to microorganisms, pretreatment of the ground water may be required. When a ground water containing organic compounds is contacted with microorganisms, the organic material is removed by the microorganisms through metabolic processes. The organic compounds may be used by the microorganisms to form new cellular material or to produce energy that is required by the microorganisms for their life systems. It has been found by Tabak et al. (1981); Kincannon and Stover (1981a, b, c, no date); Kincannon, Stover and Chung (1981); and Kincannon et al. (1981; 1982a,b) that many organic compounds that are considered to be toxic are biodegraded by microorganisms when the proper environment is provided.

Heterotrophic microorganisms are the most common group of microorganisms providing the metabolic process for removing organic compounds from contaminated ground water. Heterotrophs use the same substances as sources of carbon and energy. A portion of the organic material is oxidized to provide energy while the remaining portion is used as building blocks for cellular synthesis. Three general methods exist by which heterotrophic microorganisms can obtain energy. These are fermentation, aerobic respiration, and anaerobic respiration.

In the case of fermentation, the carbon and energy source is broken down by a series of enzyme-mediated reactions which do not involve an electron transport chain. In aerobic respiration, the carbon and energy source is broken down by a series of enzyme-mediated reactions in which oxygen serves as an external electron acceptor. In anaerobic respiration, the carbon and energy source is broken down by a series of enzyme-mediated reactions in which sulfates, nitrates, and carbon dioxide serve as the external electron acceptors. These three processes of obtaining energy form the basis for the various biological waste water treatment processes.

Description of Process

Biological treatment processes are typically divided into two categories: suspended growth systems and fixed-film systems. Suspended growth systems are more commonly referred to as activated sludge processes, of which several variations and modifications exist. The basic system consists of a large basin into which the contaminated water is introduced, and air or oxygen is introduced by either diffused aeration or mechanical aeration devices. The microorganisms are present in the aeration basin as suspended material. After the microorganisms remove the organic material from the contaminated water they must be separated from the liquid stream. This is accomplished by gravity settling. After separating the biomass from the liquid, the biomass increase resulting from synthesis is wasted and the remainder returned to the aeration tank.

Thus, a relatively constant mass of microorganisms is maintained in the system. The performance of the process depends on the recycle of sufficient biomass. If biomass separation and concentration fail the entire process fails. The process requires the skills of well-trained operators.

Fixed-film biological processes differ from suspended growth systems in that microorganisms attach themselves to a medium which provides an inert support. Biological towers (trickling filters) and rotating biological contactors are the most common forms of fixed-film processes. The original trickling filter consisted of a bed of rocks over which the contaminated water was sprayed. The microbes, forming a slime layer on the rocks, would metabolize the organics, with oxygen being provided as air moved countercurrent from the water flow. A major drawback to the rock-bed trickling filter was the inefficient use of space and poor oxygen transfer.

Biological towers are a modification of the trickling filter. The medium, which is comprised of polyvinyl chloride (PVC), polyethylene, polystyrene, or redwood, is stacked into towers which typically reach 16–20 ft. The contaminated water is sprayed across the top and, as it moves downward, air is pulled upward through the tower. A slime layer of microorganisms forms on the media and removes the organic contaminants as the water flows over the slime layer.

A rotating biological contactor (RBC) consists of a series of rotating discs, connected by a shaft, set in a basin or trough. The contaminated water passes through the basin where the microorganisms, attached to the discs, metabolize the organics present in the water. Approximately 40% of the disc's surface area is submerged. This allows the slime layer to alternately come in contact with the contaminated water and the air where oxygen is provided to the microorganisms.

Removal efficiencies are generally the same for fixed-film and suspended growth processes. However, fixed-film processes have the potential to be lower in cost, due to the absence of aeration equipment, and are easier to operate. Both systems may be operated under anaerobic conditions, which may offer advantages for certain contaminated waters.

Design Parameters and Procedures

The design of biological treatment processes is usually accomplished by using some type of kinetic model. The most widely used models have been developed by Eckenfelder, McKinney, Lawrence and McCarty, and Gaudy. However, Kincannon and Stover (no date) have found that a great amount of variability exists in the biokinetic constants for these

models, and concluded that these models were not ideal for waters containing priority pollutants; therefore, they have developed the following models that are reliable for these types of waters:
Activated Sludge

$$V = \frac{FS_i/X}{\dfrac{U_{max} S_i}{S_i - S_e} - K_B}$$

Biological Tower and Rotating Biological Contactor

$$A = \frac{FS_i}{\dfrac{U_{max} S_i}{S_i - S_e} - K_B}$$

where
V = volume of aeration tank (m³)
F = flow rate (m³/day)
X = mixed liquor volatile solids (mg/l)
S_i = influent BOD, COD, TOC, or specific organics (mg/l)
S_e = effluent BOD, COD, TOC, or specific organics (mg/l)
U_{max} and K_B = biokinetic constants (day^{-1})
A = surface area of biological tower or rotating biological contactor (m²)

The biokinetic constants are determined by conducting laboratory or pilot plant studies. After the biokinetic constants are determined, the required volume of aeration tank or the required surface area for a biological tower or rotating biological contactor can be determined for any flow rate; influent concentration of BOD, COD, TOC, or specific organic, and a required effluent concentration of BOD, COD, TOC, or specific organic.

Applications to Ground Water

Biological treatment of contaminated ground water has not been actively employed in the past. However, pilot plant studies by Kincannon and Stover (1983) showed that biological treatment of a contaminated ground water was feasible. In addition, Kincannon and Stover (no date) have conducted extensive research on the fate of priority pollutants in activated sludge treatment systems. The removal mechanisms of 24 priority pollutants are shown in Table 3.3. It is seen that a majority of the 24 organic priority pollutants are removable by biological treatment. Figure 3.8 shows the effluent phenol concentration achieved by an RBC as a

Table 3.3   Removal Mechanisms of Toxic Organics

Compound	Percent Treatment Achieved		
	Stripping	Sorption	Biological
Nitrogen Compounds			
Acrylonitrile			99.9
Phenols			
Phenol			99.9
2,4-DNP			99.3
2,4-DCP			95.2
PCP		0.58	97.3
Aromatics			
1,2-DCB	21.7		78.2
1,3-DCB			
Nitrobenzene			97.8
Benzene	2.0		97.9
Toluene	5.1	0.02	94.9
Ethylbenzene	5.2	0.19	94.6
Halogenated Hydrocarbons			
Methylene Chloride	8.0		91.7
1,2-DCE	99.5	0.50	
1,1,1-TCE	100.0		
1,1,2,2-TCE	93.5		
1,2-DCP	99.9		
TCE	65.1	0.83	33.8
Chloroform	19.0	1.19	78.7
Carbon Tetrachloride	33.0	1.38	64.9
Oxygenated Compounds			
Acrolein			99.9
Polynuclear Aromatics			
Phenanthrene			98.2
Naphthalene			98.6
Phthalates			
Bis(2-Ethylhexyl)			76.9
Other			
Ethyl Acetate	1.0		98.8

function of the phenol loading. It is seen that very low levels of effluent phenol can be achieved by the RBC process.

Kincannon and Stover (1983) have suggested alternative designs for biological treatment of contaminated ground water in which treatability studies were conducted. In this case, the contaminated ground water undergoes metals removal and high-temperature air stripping before biological treatment. The quality of the water after metals removal and stripping is shown in Table 3.4. Activated sludge and RBC pilot plant

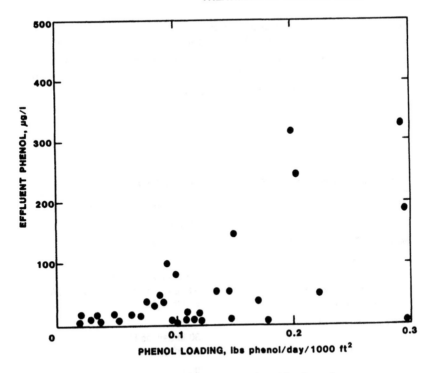

Figure 3.8 Effluent phenol versus phenol loading rate.

Table 3.4   High-Temperature Air Stripper Extract Water Qualities for Biological Studies

Parameter	Well #1	Well #2
pH	6.7	7.2
Specific Conductance, μmho$_s$	450	1300
TOC,mg/l	250	590
COD, mg/l	700	1960
BOD$_5$, mg/l	450	1025
Total Phenols, mg/l	2.5	4.0
SS, mg/l	25	30
VSS, mg/l	10	16
TDS, mg/l	380	1860
TVDS, mg/l	120	1220
Fe, mg/l	1.7	0.4
Mn, mg/l	0	2.3
NH$_3$-N, mg/l	5.5	9.0
Ortho-P, mg/l	0.2	0.2

studies were conducted and designs suggested for both systems. These designs are shown in Tables 3.5 and 3.6. The projected effluent qualities from the two systems are shown in Tables 3.7 and 3.8. It is seen that an excellent effluent is expected. Table 3.9 gives the expected priority pollutant levels for various biological treatment processes.

Table 3.5   Design Summary for Activated Sludge System

Reactor Volume	60,000 gal, or two 30,000 gal each
MLVSS	1500 mg/l
Maximum BOD Removed	600 lb BOD_5/day
Oxygen Requirements	
Energy Oxygen	0.4 lb O_2/lb BOD_5 removed
Endogenous Oxygen	6.3 lb O_2/hr/1,000 lb MLVSS UA
Total Oxygen	710 lb O_2/day
Nutrient Requirements	
Nitrogen	30 lb/day, as N
Phosphorus	6 lb/day, as P
Alpha (α)	0.61
Beta (β)	0.99
Water Temperature	90°F (32°C)
Elevation	300 ft
Oxygen Transfer Characteristics	
N_o	3.0 lb O_2/hp-hr
N	1.9 lb O_2/hp-hr
Diffused Air	500 scfm
Net Sludge Production	0.42 lb/lb BOD_5 removed
	250 lb/day
Secondary Clarifier	
Overflow Rate	500 gpd/ft²
Underflow Solids Concentration	10,000 mg/l
Recommended Recycle Pump Capacity	Variable up to 50 gpm
Sludge Dewatering Characteristics	
Vacuum Filter Yield	1.0 to 1.5 lb/hr/ft²
Cake Dry Solids Content	12 to 18%
High-Molecular-Weight Cationic	
Polymer	20 – 40 lb/ton
Ferric Chloride	70 – 100 lb/ton

Table 3.6   Design Summary for RBC System

Surface Area	300,000 ft²
	Three Standard Density RBC Shafts
	Two Stages
	Two Shafts/First Stage
Maximum BOD Removed	600 lb BOD$_5$/day
Nutrient Requirements	
Nitrogen	30 lb/day, as N
Phosphorus	6 lb/day, as P
Net Sludge Production	0.25 lb/lb BOD$_5$ removed
	150 lb/day
Secondary Clarifier	
Overflow Rate	500 gpd/ft²
Underflow Solids Concentration	10,000 mg/l
Sludge Dewatering Characteristics	Similar to Activated Sludge
	(Based on CST Comparisons)

Table 3.7   Projected Effluent Qualities from Activated Sludge Design

Soluble Effluent Qualities, mg/l	Flow = 25 gpm		Flow = 50 gpm	
	x = 1500[a]	x = 2000	x = 1500	x = 2000
Well #1 Quality				
V = 30,000 gal				
BOD$_5$	10	9	13	11
COD	53	52	57	55
TOC	28	26	33	30
Well #1 Quality				
V = 60,000 gal				
BOD$_5$	8	7	10	9
COD	51	50	53	52
TOC	25	24	28	26
Well #2 Quality				
V = 30,000 gal				
BOD$_5$	12	5	28	19
COD	148	125	231	191
TOC	88	74	137	114
Well #2 Quality				
V = 60,000 gal				
BOD$_5$	1	1	10	5
COD	104	95	145	126
TOC	60	55	85	75

[a]x denotes mixed liquor volatile suspended solids (mg/l).

Table 3.8 Projected Effluent Quality from RBC Design

Soluble Effluent Quality	Flow = 25 gpm	Flow = 50 gpm
Well #2 Quality		
3 RBC Shafts		
BOD_5	1	2
COD	94	96
TOC	60	65

Costs

There are no good data available at this time in regard to the cost of biological treatment of contaminated ground waters. However, the costs should be comparable to those for treating industrial wastewaters.

CHEMICAL PRECIPITATION

Chemical addition for the removal of inorganic compounds is a well-established technology. There are three common basic types of chemical addition systems which depend upon the low solubility of inorganics at a specific pH; (1) the carbonate system, (2) the hydroxide system, and (3) the sulfide system. In reviewing the basic solubility products for these systems, the sulfide system removes the most inorganics, with the exception of arsenic, because of the low solubility of sulfide compounds. This increased removal capability is offset by the difficulty in handling the chemicals and the fact that sulfide sludges are susceptible to oxidation to sulfate when exposed to air, resulting in resolubilization of the metals. The carbonate system is a method which relies on the use of soda ash and pH adjustment between 8.2 and 8.5. The carbonate system, although workable in theory, is difficult to control. The hydroxide system is the most widely used inorganics/metals removal system. The system responds directly to pH adjustment, and usually uses either lime (CaOH) or sodium hydroxide (NaOH) as the chemical to adjust the pH upwards. Sodium hydroxide has the advantage of ease in chemical handling and low volume of sludge. However, the hydroxide sludge is often gelatinous and difficult to dewater.

Table 3.9 Results of Specific Compound Studies[a]

Specific Compound	HTAS Extract	Well #1 AS #1	Well #2 AS #2	Well #2 Water AS #3	Well #2 Water AL	Well #2 Water RBC
Total Phenols	2500 (4000)	80 (130)	60 (115)	50 (100)	50 (70)	– (100)
Acid Extractables						
2-Chlorophenol	1080 (200)	≤10 (≤10)	≤10 (≤10)	≤10 (≤10)	≤10 (≤10)	– (≤10)
Phenol	675 (990)	≤10 (≤10)	≤10 (≤10)	≤10 (≤10)	≤10 (≤10)	– (≤10)
2, 4,-Dimethylphenol	≤10 (5300)	≤10 (≤10)	≤10 (≤10)	≤10 (≤10)	≤10 (≤10)	– (≤10)
Pentachlorophenol	ND (510)	ND (95)	ND (80)	ND (70)	ND (≤10)	ND (90)
Base Neutral Extractables						
1, 4-Dichlorobenzene	105 (85)	≤10 (≤10)	≤10 (≤10)	≤10 (≤10)	≤10 (≤10)	– (≤10)
1, 2-Dichlorobenzene	1920 (430)	335 (425)	335 (155)	110 (270)	100 (≤10)	– (≤10)
Naphthalene	715 (55)	70 (50)	80 (25)	25 (≤10)	25 (≤10)	– (≤10)
Volatile Organics	ND (ND)					

[a]All analyses in µg/l; ND—none detected; AS #1—high F/M activated sludge; AS #2—medium F/M activated sludge; AS #3—low F/M activated sludge; AL—aerated lagoon; RBC—rotating biological contactor.

Description of Process

Chemical precipitation can be accomplished by either batch or continuous flow operations. If the flow is less than 30,000 gpd, a batch treatment system would be the most economical. In the batch system, two tanks are provided, each with a capacity of one day's flow. One tank undergoes treatment while the other tank is being filled.

When the daily flow exceeds 30,000 gpd, batch treatment is usually not feasible because of the large tankage required. Continuous treatment may require a tank for acidification and reduction, then a mixing tank for chemical addition, and a settling tank.

Design Parameters and Procedures

The important design factors that must be determined for a particular water during treatability studies include:

(1) Best chemical addition system
(2) Optimum chemical dose
(3) Optimum pH conditions
(4) Rapid mix requirements
(5) Flocculation requirements
(6) Sludge production
(7) Sludge flocculation, settling, and dewatering characteristics

Laboratory-scale test procedures consisting of jar test studies have been used for years, and the test methodology developed is such that full-scale designs can be developed from these studies with a high degree of confidence. Some general design considerations include:

(1) The retention time in the reduction tank should be at least four times the theoretical time for complete reduction.
(2) Twenty minutes will usually be adequate for flocculation.
(3) Final settling should not be designed for an overflow rate in excess of 500 gal/day/ft^2.

Applications to Ground Water

Chemical precipitation has been successfully used for removing heavy metals from various waters. Kincannon and Stover (1983) reported the results of treating a contaminated ground water by various processes. The results of chemical precipitation to remove metals is given in Table

3.10. They found that all the metals found in the particular ground water were removed to acceptable levels.

Brantner and Cichon (1981) compared the three precipitation processes. The results of this work are shown in Table 3.11. Their work also shows that chemical precipitation is effective in removing metals. However, all methods are not equally as effective. In addition, they found that the carbonate system was the simplest process to operate, the hydroxide process the most reliable, and the sulfide precipitation process was the most complex to operate.

Costs

There is very little information in the literature regarding costs of chemical precipitation. However, Krause and Stover (1982) have presented an excellent cost comparison for removing barium from water supplies. Their cost evaluations for ion exchange, lime softening, and chemical precipitation–direct filtration are shown in Tables 3.12 through 3.14. All costs are April 1980 dollars. Table 3.15 shows a cost comparison of the three processes.

Table 3.10   Removal Chemical Precipitation Data for Metals

Compound	Concentrations in Groundwater, mg/l			
		Lime-Treated Water		
	Raw Water	pH 9.1	pH 9.9	pH 11.3
Arsenic	0.12	0.03	0.03	0.03
Barium	0.24	0.17	0.15	0.19
Cadmium	0.003	<0.001	<0.001	<0.001
Chromium (total)	0.09	0.006	0.006	0.006
Lead	0.03	0.006	0.006	0.006
Mercury	<0.001	<0.001	<0.001	<0.001
Selenium	<0.001	<0.001	<0.001	<0.001
Silver	<0.001	<0.001	<0.001	<0.001
Copper	0.10	<0.001	<0.001	<0.001
Iron	352	0.07	0.07	1.05
Manganese	90	0	0	0
Nickel	1.95	0.05	0.30	0.45
Zinc	0.69	0.36	0.09	0.61

Table 3.11 Comparison of Precipitation Treatment Processes (Brantner and Cichon, 1981)

Parameter	Influent (ppm) Mean	Range	Clarifier Effluent (ppm) Mean	Range	Filtered Effluent (ppm) Mean	Range
Hydroxide Precipitation Data Summary						
Suspended Solids	42	20 – 63	22	9 – 33	9	4 – 14
pH	7.5	7.0 – 8.1	9.9	9.7 – 10.2	9.8	9.5 – 10.4
Total Cadmium	1.66	0.13 – 4.30	0.05	0.03 – 0.10	0.04	0.02 – 0.06
Total Chromium	1.11	0.07 – 2.90	1.04	0.07 – 2.80	0.97	0.06 – 2.90
Total Copper	0.29	0.12 – 1.50	0.03	0.02 – 0.03	0.03	0.02 – 03
Total Lead	1.7	0.8 – 2.6	0.2	0.1 – 0.3	0.2	0.1 – 0.3
Total Zinc	31	6 – 91	0.40	0.23 – 0.75	0.28	0.10 – 0.66
Carbonate Precipitation Data Summary						
Suspended Solids	43	16 – 75	27	14 – 52	6	2 – 10
pH	7.1	6.7 – 7.7	8.3	8.1 – 8.4	8.1	7.8 – 8.5
Total Cadmium	1.37	0.26 – 2.90	0.14	0.02 – 0.27	0.04	0.02 – 0.06
Total Chromium	0.67	0.23 – 1.80	0.62	0.17 – 1.8	0.60	0.14 – 2.00
Total Copper	0.18	0.06 – 0.27	0.04	0.03 – 0.06	<0.03	<0.02 – 0.04
Total Lead	1.4	0.7 – 2.1	0.2	0.2 – 0.4	<0.1	<0.1 – 0.2
Total Zinc	26	1 – 67	3.2	0.37 – 5.0	1.18	0.19 – 5.00
Sulfide Precipitation Data Summary						
Suspended Solids	90	30 – 210	37	10 – 63	5	2 – 8
pH	6.8	6.4 – 7.7	8.2	8.0 – 8.4	8.1	7.8 – 8.5
Total Cadmium	3.3	0.65 – 5.4	0.18	0.09 – 0.29	0.06	0.02 – 0.12
Total Chromium	0.52	0.02 – 1.9	0.12	0.08 – 0.20	<0.05	<0.05 – 0.05
Total Copper	0.35	0.19 – 0.96	<0.04	<0.04 – 0.05	<0.03	<0.02 – 0.03
Total Lead	4.5	2.3 – 8.1	0.4	0.2 – 0.5	<0.1	<0.1 – 0.2
Total Zinc	93	5.8 – 220	3.1	2.0 – 6.2	0.68	0.11 – 1.8

Table 3.12 Ion Exchange System Cost Estimate Summary

Item	Cost
Capital cost, $	
Aerator	25,000
Wet well	9,500
Transfer pumps	12,700
Ion exchangers	137,500
Caustic soda system	5,600
Brine pit	16,300
Building addition	52,300
Outside piping and sitework	25,900
Electrical and instrumentation	34,200
General and special conditions	22,300
Construction contingencies	51,200
Total construction cost	392,500
Engineering and estimating contingencies	58,900
Financial cost	13,500
Bond interest	404,600
Total capital cost	869,500
O & M costs, $/yr	
Labor	14,000
Chemicals	23,200
Power	2,400
Waste disposal	1,200
Total annual O & M cost	40,800

OTHER TREATMENT TECHNIQUES FOR INORGANICS

Chemical precipitation has traditionally been a popular technique for the removal of heavy metals and other inorganics from wastewater streams. However, a wide variety of other techniques also exist. Listed in Table 3.16 are some available technologies for the removal of specific inorganic contaminants. The technologies in Table 3.16, which also include chemical precipitation, were compiled from Patterson (1978) and represent viable techniques. Patterson (1978) treats each inorganic in more detail by addressing such topics as: sources of the given inorganic, its major industrial uses, proven treatment technologies, proposed treatment technologies, removal efficiencies of various technologies, and costs of various treatment technologies.

Table 3.13   Lime Softening System Cost Estimate Summary

Item	Cost
Capital cost, $	
Aerator	25,000
Solids contact reactor	113,600
Gravity filter	222,900
Recarbonation system	51,900
Transfer pumps	12,700
Alum system	10,100
Lime system	40,000
Lime sludge dewatering	93,800
Building addition	209,800
Outside piping and sitework	78,000
Electrical and instrumentation	102,900
General and special conditions	67,200
Construction contingencies	154,200
Total construction cost	1,182,100
Engineering and estimating contingencies	177,300
Financial cost	40,800
Bond interest	1,218,400
Total capital cost	2,618,600
O & M cost, $/yr	
Labor	28,000
Chemicals	27,500
Power	4,600
Waste disposal	19,100
Total annual O & M cost	79,200

TREATMENT TRAINS

Due to the complex composition of most ground waters, no one unit operation is capable of removing all of the contaminants present. It may be necessary to combine several unit operations into one treatment process to effectively remove the contaminants required. In order to simplify, and make visible, the selection of the applicable treatment trains, Table 3.17 is presented showing a number of unit operations and the waste types for which they are effective. Results presented by Kincannon and Stover (1983) illustrate the levels of treatment achieved by a treatment train, with these results given in Table 3.18.

Table 3.14 Chemical Precipitation-Direct Filtration System Cost Estimate Summary

Item	Cost
Capital cost, $	
Aerator	25,000
Rapid mix tank	7,900
Flocculation basin	36,100
Gravity filter	286,600
Recarbonation system	62,200
Transfer pumps	12,700
Potassium hydroxide system	25,600
Gypsum system	22,400
Polymer system	5,000
Building addition	239,000
Outside piping and sitework	70,500
Electrical and instrumentation	93,000
General and special conditions	60,800
Construction contingencies	139,300
Total construction cost	1,068,100
Engineering and estimating contingencies	160,200
Financial cost	36,800
Bond interest	1,100,900
Total capital cost	2,366,000
O & M cost, $/yr	
Labor	28,000
Chemicals	116,600
Power	4,900
Waste disposal	6,400
Total annual O & M cost	155,900

Table 3.15 Monetary Comparison of Alternative Treatment Systems

Item	Alternative Treatment Systems		
	Ion Exchange System	Lime Softening System	Chemical Precipitation – Direct Filtration System
Capital cost, $	869,500	2,618,600	2,366,000
Capital annual equivalent cost, $/yr	116,400	350,000	316,800
Annual O & M cost, $/yr	40,800	79,200	155,900
Total annual equivalent cost, $/yr	157,200	429,800	472,700

Table 3.16   Treatment Alternatives for Inorganics

Inorganic	Treatment Method
Arsenic	Charcoal Filtration
	Lime Softening
	Precipitation with lime + iron
	Precipitation with alum
	Precipitation with ferric sulfate
	Precipitation with ferric chloride
	Precipitation with ferric hydroxide
	Precipitation with sulfide
	Ferric Sulfide Filter Bed
	Iron or Lime Coagulation + settling + dual media filtration + carbon adsorption
Barium	Iron or Lime Coagulation + settling + dual media filtration + carbon adsorption
	Precipitation as sulfate
	Precipitation as carbonate
	Precipitation as hydroxide
	Ion Exchange
Boron	Evaporation
	Reverse Osmosis
	Ion Exchange
Cadmium	Precipitation as hydroxide
	Precipitation as hydroxide + filtration
	Precipitation as sulfide
	Coprecipitation with ferrous hydroxide
	Reverse Osmosis
	Freeze Concentration
Chloride	Ion Exchange
	Electrodialysis
	Reverse Osmosis
	Other (holding basins, evaporative ponds, deep well injection)
Chromium (hexavalent)	Ion Exchange
	Freeze Concentration
	Activated Carbon
	Cementation
	Reduction (Reduce Cr^{+6} to Cr^{+3} and precipitation of hydroxide)
	Reduction with sulfur dioxide
	Reduction with bisulfite
	Reduction with bisulfite + hydrazine
	Reduction with metabisulfite
	Reduction with ferrous sulfate
Chromium (trivalent)	Precipitation (see above)
	Ion Exchange

Table 3.16, continued

Inorganic	Treatment Method
Copper	Precipitation with lime
	Ion Exchange
	Evaporative Recovery
	Electrolytic Recovery
	Cementation
	Reverse Osmosis
Cyanide	Alkaline Chlorination
	Electrolysis
	Ozonation
	Evaporation
Fluoride	Precipitation by lime addition
	Precipitation by magnesium addition
	Precipitation by alum addition
	Adsorption on hydroxlapatite beds
	Adsorption on alumina contact beds
Iron	Oxidation-Precipitation by aeration, sand filtration
	Oxidation-Precipitation by aeration, lime, sand filtration
	Oxidation-Precipitation by aeration, coke bed filtration, sedimentation, sand filtration
	Oxidation-Precipitation by lime, aeration, diatomite filtration
	Oxidation-Precipitation by chlorination, alum-lime-sodium silicate precipitation, sand filtration
	(deep well disposal)
Lead	Ion Exchange
	Precipitation by lime + sedimentation
	Precipitation by caustic + sedimentation
	Precipitation by ammonium hydroxide
	Precipitation by dolomite + sedimentation
	Precipitation by sodium carbonate + filtration
	Precipitation by sodium phosphate + filtration
	Precipitation by ferric sulfate + sedimentation
	Precipitation by ferrous sulfate + sedimentation
Manganese	Aeration
	Ion exchange
	Catalysis
	Oxidation-Precipitation by chlorine dioxide addition
	Oxidation-Precipitation by manganese dioxide addition
	Oxidation-Precipitation by potassium permanganate addition
Mercury	Precipitation by sodium sulfide addition
	Precipitation by sodium hydrosulfide addition
	Precipitation by magnesium sulfide addition
	Precipitation by sulfide addition
	Ion Exchange
	Coagulation with alum

Table 3.16, continued

Inorganic	Treatment Method
	Coagulation with iron
	Activated Carbon
	Reduction to Metallic Form by zinc
	Reduction to Metallic Form by stannous chloride
	Reduction to Metallic Form by sodium borohydride
Nickel	Precipitation by lime
	Precipitation by sulfide
	Precipitation by alum
	Ion Exchange
	Reverse Osmosis
	Evaporative Recovery
Selenium	Coagulation with lime
	Coagulation with ferric sulfate
	Coagulation with alum
	Activate Carbon + Cation Exchange + Anion Exchange
Silver	Precipitation with ferric chloride
	Ion Exchange
	Reductive Exchange with zinc or iron
	Electrolytic Recovery
Total Dissolved Solids	Reverse Osmosis
	Electrodialysis
	Distillation
	Ion Exchange
Zinc	Precipitation by lime addition
	Precipitation by caustic addition
	Ion Exchange
	Evaporation

Table 3.17   Summary of Suitability of Treatment Processes

	Volatile Organics	Non-Volatile Organics	Inorganics
Air Stripping	Suitable for most cases	Not Suitable	Not Suitable
Steam Stripping	Effective Concentrated Technique	Not Suitable	Not Suitable
Carbon Adsorption	Inadequate Removal	Effective Removal Technique	Not Suitable
Biological	Effective Removal Technique	Effective Removal Technique	Not Suitable Metals Toxic
pH Adjustment Precipitation	Not Applicable	Not Applicable	Effective Removal Technology
Electrodialysis	Not Applicable	Not Applicable	Inefficient Operation/ Inadequate Removal
Ion Exchange	Not Applicable	Not Applicable	Inappropriate Technology— Difficult Operation

Table 3.18   Ground Water Treatment Results From a Selected Treatment Train

Parameter	Raw Water	Metals Treatment	Steam Stripping	Parameter's Value After Treatment — Activated Carbon Adsorption		Biological Treatment
				6000 mg/l	24,000 mg/l	
pH-units	6.0	10.0 to 6.5		6.5	6.6	7.3
Specific conductance-μg/l	2800			2450	2450	
TOC-mg/l	4000	3685	1700	1575	1130	34
COD-mg/l	11750		3600	2400	1950	125
BOD₅-mg/l	7025					5.5
Phenols-mg/l	26			0.85	0.12	0.12
SS-mg/l	360					
VSS-mg/l	142					
TDS-mg/l	3050					
TVS-mg/l	1485					
Arsenic-mg/l	0.12	0.03		0.02	0.02	
Barium-mg/l	0.25	0.15		0.09	0.03	
Cadmium-mg/l	0.003	<0.001		<0.001	<0.001	
Chromium-mg/l	0.09	0.006		0.006	0.006	
Lead-mg/l	0.03	0.006		0.006	0.006	
Mercury-mg/l	<0.001	<0.001		<0.001	<0.001	
Selenium-mg/l	<0.001	<0.001		<0.001	<0.001	
Silver-mg/l	0.001	<0.001		<0.001	<0.001	
Copper-mg/l	0.10	<0.001		<0.001	<0.001	
Iron-mg/l	352	0.07		0.07	0.07	
Manganese-mg/l	90	0		0	0	
Nickel-mg/l	1.95	1.58		0.89	0.28	
Zinc-mg/l	0.69	0.09		0.09	0.09	

BOD_5

Table 3.18, continued

Parameter	Raw Water	Metals Treatment	Steam Stripping	Activated Carbon Adsorption		Biological Treatment
				6000 mg/l	24,000 mg/l	
Tetrahydrofuran- µg/l	22000		a			a
1,1,1-Trichloroethane- µg/l	150000					a
Benzene- µg/l	68750		a			1897[b]
Trichloroethylene- µg/l	338000		a			
Methyl isobutyl ketone- µg/l	76400		a			1098
Xylenes- µg/l	a		a			a
Toluene-µg/l	92000		a			25
Ethylbenzene- µg/l	23500		17			a
1,4-Dichlorobenzene- µg/l	35	35	a	a	a	a
1,2-Dichlorobenzene-µg/l	5	5		a	a	a
Naphthalene- µg/l	≤1	≤1	≤1	≤1	a	a
2-Chlorophenol-µg/l	540	540	40	≤1	≤1	a
2-Nitrophenol-µg/l	15	a	6	a	a	a
Phenol- µg/l	370	a	20	a	a	a
2,4-Dimethylphenol-µg/l	20	80	a	a	a	a
o-Cresol-µg/l	80	220	25	≤1	a	a
m-Cresol-µg/l	220	1230	a	a	a	a
Benzoic acid-µg/l	1230	40	a	8	2	a
Pentachlorophenol-µg/l	40	40	40			a

a No peaks on chromatograms.
b Benzene and Trichloroethylene peaks combined into one peak.

SELECTED REFERENCES

Brantner, Karl A. and Edward J. Cichon, "Heavy Metals Removal: Comparison of Alternative Precipitation Processes", Industrial Waste – Proceedings of the Thirteenth Mid-Atlantic Conference, 1981.

Dyksen, John E., et al., "The Use of Aeration to Remove Volatile Organics from Ground Water", Presented at 1982 Annual Conference of the American Water Works Association held May 16–20, 1982, Miami Beach, Florida.

Kaufmann, Henry G., "Granular Carbon Treatment of Contaminated Supplies", *Proceedings of the Second National Symposium on Aquifer Restoration and Ground Water Monitoring*, May 26–28, 1982, The Fawcett Center, Columbus, Ohio.

Kincannon, D. F., and E. L. Stover, "Biological Treatability of Specific Organic Compounds Found in Chemical Industry Wastewaters", Presented at the 36th Purdue Industrial Waste Conference, West Lafayette, Indiana, 1981a.

Kincannon, D. F. and E. L. Stover, "Fate of Organic Compounds During Biological Treatment", Presented at the 1981 National Conference on Environmental Engineering, ASCE Environmental Engineering Division, Atlanta, Georgia, 1981b.

Kincannon, D. F. and E. L. Stover, "Stripping Characteristics of Priority Pollutants During Biological Treatment", Presented at the 74th Annual AIChE Meeting, New Orleans, Louisiana (November 1981), 1981c.

Kincannon, D. F. and E. L. Stover, Final Report EPA Cooperative Agreement CR 806843-01-02, "Determination of Activated Sludge Biokinetic Constants for Chemical and Plastic Industrial Wastewaters".

Kincannon, D. F., and E. L. Stover. "Contaminated Groundwater Treatability – A Case Study", *Journ. American Water Works Association*, Vol. 75, June 1983.

Kincannon, D. F., E. L. Stover, and Y. P. Chung, "Biological Treatment of Organic Compounds Found in Industrial Aqueous Effluents", Presented at the ACS National Meeting, Atlanta, Georgia (March 1981).

Kincannon, D. F., et al., "Removal Mechanisms for Biodegradable and Non-Biodegradable Toxic Priority Pollutants in Industrial Wastewaters", Presented at the 54th Annual Water Pollution Control Federation Conference, Detroit, Michigan (October 1981).

Kincannon, D. F., et al., "Variability Analysis During Biological Treatability of Complex Industrial Wastewaters for Design", Presented at the 37th Purdue Industrial Waste Conference, West Lafayette, Indiana, 1982a.

Kincannon, D. F., et al., "Predicting Treatability of Multiple Organic Priority Pollutant Wastewaters from Single Pollutant Treatability Studies", Presented at the 37th Purdue Industrial Waste Conference, West Lafayette, Indiana, 1982b.

Krause, Terry L. and Enos L. Stover, "Evaluating Water Treatment Techniques for Barium Removal", *Journal American Water Works Association*, September 1982.

Lamarre, Bruce L., Frederick J. McGarry, and Enos L. Stover, "Design, Operation, and Results of a Pilot Plant for Removal of Contaminants from Ground Water", *Third National Symposium and Exposition on Aquifer Restoration and Ground Water Monitoring*, The Fawcett Center, Columbus, Ohio, May 25–26, 1983.

McDougall, W. Joseph, Richard A. Fusco, and Robert P. O'Brien, "Containment and Treatment of the Love Canal Landfill Leachate", *Journal WPCF*, Vol. 52, No. 12, Dec. 1980.

McKinnon, Ronald J. and John E. Dyksen, "Aeration Plus Carbon Adsorption Remove Organics from Rockaway Township (New Jersey) Ground Water Supply", Presented at the American Society of Civil Engineers 1982 Annual Convention held October 25–27, 1982, New Orleans, Louisiana.

O'Brien, Robert P. and J. L. Fisher, "There is an Answer to Ground-water Contamination", *Water Engineering and Management*, Vol. 130, No. 5, May 1983.

Patterson, J. W., *Wastewater Treatment Technology*, 3rd ed., Ann Arbor Science, 1978.

Pekin, Tarik and Alan Moore, "Air Stripping of Trace Volatile Organics from Watewater", *Proceedings of the 37th Industrial Waste Conference*, Purdue University, May 11, 12 and 13, 1982, West Lafayette, Indiana.

Stover, Enos L., "Removal of Volatile Organics from Contaminated Ground Water", *Proceedings of the Second National Symposium on Aquifer Restoration and Ground Water Monitoring*, May 26–28, 1982, The Fawcett Center, Columbus, Ohio.

Tabak, H. H., et al., "Biodegrad-ability Studies with Organic Priority Pollutant Compounds", *JWPCF*, 53(10), 1503, 1981.

Thomas, H.C., "Chromatography: A Problem in Kinetics", Annals of the New York Academy of Science, 49:161, 1948.

CHAPTER **4**

In Situ Technologies

C.H. Ward and M.D. Lee, Rice University

This chapter addresses aquifer restoration measures involving treatment in-place. Included is brief information on in situ chemical treatment along with more extensive information on the use of both natural and enhanced microbiological processes. In situ treatment can be used singly or in combination with other aquifer restoration strategies.

IN SITU CHEMICAL TREATMENT

In situ chemical treatment of contaminated ground water can be considered only in cases where contaminants are known, and the levels and extent of contaminated water in the aquifer are defined. In situ treatment generally involves the installation of a bank of injection wells at the head of, or within the plume of, contaminated ground water. A treatment agent is then pumped into the aquifer. The treatment agent must be specific for different classes of contamination. For example, heavy metals may be made insoluble and rendered immobile with alkalies or sulfides, cyanides can be destroyed with oxidizing agents, cations may be precipitated with various anions or by aeration, and hexavalent chromium could be made insoluble with reducing agents.

Description of Process

The process of in situ treatment can be accomplished by many means. Two examples will be described herein. The first is a process to remove iron and manganese called the Vyredox Method (Hallberg and Martinelli, 1976). This method was developed in Finland and is now used in several European countries. The aim of the Vyredox Method is to

127

achieve a high degree of oxidation in the strata around a well. In other words, the Eh and pH are kept so high that the iron and manganese are precipitated and retained in the strata. The Vyredox concept is illustrated in Figure 4.1. The water that enters the well is thus free of iron and manganese. Iron is precipitated first in the zone furthest from the withdrawal (supply) well. The number of living bacteria increases as the withdrawal well is approached, and so does the number of dead bacteria. The organic matter contained in the dead cells becomes a carbon source for the bacteria that preferentially oxidize manganese. This process takes place nearer the well, where the Eh is higher. Thus, the Vyredox Method provides favorable conditions for removing iron first, and then manganese.

Figure 4.2 illustrates a Vyredox plant for treating contaminated ground water. It consists of one or more wells for supplying water. Each well has a pipe fitted with a screen. A number of injection wells (aeration wells) are situated in a ring around each supply well. The number depends on the hydrogeological and geochemical conditions. Conduits connect each well to a nearby building. The water forced into the injection wells must first be degassed and enriched with oxygen. This is done in a special aerator, the oxygenator.

A procedure of containment of a contaminant by polymerizing the contaminant in the ground has been reported by Williams (1982). The procedure has been used to contain approximately 4200 gal of acrylate monomer which had leaked from a corroded underground pipeline. The

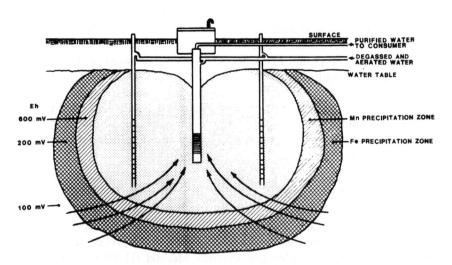

Figure 4.1. Vyredox concept.

acrylate monomer is a colorless mobile liquid less heavy than and only moderately soluble in water. Vapors from the monomer are odorous at 5 ppm, and in high concentrations can result in skin irritation, as well as explosive potential. Under favorable conditions, the monomer will polymerize to a soft rubbery texture.

Because of imposed limitations, both by the hydrogeologic conditions and by the practical characteristics of the site, there were few viable options for recovering or containing the monomer. Therefore, a system of four exfiltration galleries was designed and installed. Two-inch-diameter perforated PVC casings, in 40-ft lengths, were buried in narrow trenches 2 ft below ground surface across the shallow contaminated zone. A riser pipe and manifold header connected each gallery to the solution tanks. Four thousand gallons of catalyst and activator solution per treatment were used. Two separate treatments four days apart were used. It is estimated overall, that 85–90% of the liquid monomer contaminant was effectively converted to a solidified polymer.

LAYOUT OF VYREDOX PLANT

①-⑧ SUPPLY WELLS AND AERATION WELLS

② AERATOR, THE OXYGENATOR

③ DEGASSING TANK

④ PUMP FOR AERATED WATER

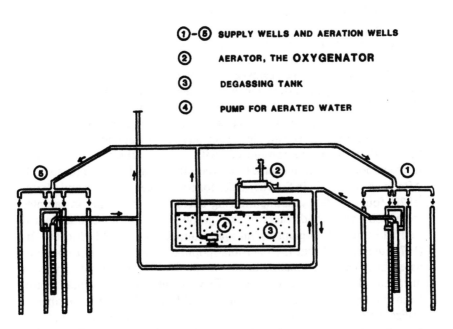

Figure 4.2 Vyredox plant with two supply wells complete with aeration wells and oxygenator building.

Advantages and Disadvantages

In situ chemical treatment is viable only under particular hydrogeological and geochemical conditions. Other aquifer restoration measures, such as withdrawal and treatment, may be more appropriate for consideration in meeting a given need.

IN SITU BIOLOGICAL STABILIZATION

Biological treatment can be used to clean up contaminated aquifers. Microbes can degrade most organic compounds; however, many manmade compounds are relatively refractive (Kobayoshi and Rittman, 1982). Compounds which possess amine, methoxy, and sulfonate groups, ether linkages, halogens, branched carbon chains, and substitutions at the meta position of benzene rings are generally persistent. Environmental factors such as dissolved oxygen level, pH, temperature, oxidation-reduction potential, availability of nutrients, salinity, and the concentration of the compounds often control the biodegradation of the compounds. The number and type of organisms present also play an important role.

Microbial activity is likely to exist in most subsurface regions where ground water is important (McNabb and Dunlap, 1975). Bacterial levels typically around 10^6 organisms/g dry soil have been found for several shallow water table aquifers which have been investigated (Wilson et al., 1983b). The potential for biodegradation of a variety of compounds in the subsurface has been extensively studied. Litchfield and Clark (1973) analyzed ground water samples from aquifers throughout the United States that were contaminated with hydrocarbons, and found hydrocarbon-utilizing bacteria in all the samples at levels up to 10^6/ml. After a gasoline spill in Southern California, McKee, Laverty and Hertel (1972) found 50,000 gasoline-utilizing bacteria/ml or higher in samples from wells which had traces of free gasoline while a noncontaminated well had only 200 organisms/ml. Jamison, Raymond and Hudson, Jr. (1975) determined that the microbial population in an aquifer contaminated by a gasoline pipeline break was limited by inadequate levels of nitrogen, phosphorus, and oxygen. Experiments with microbial isolates from the ground water showed that the different isolates were able to utilize certain components of the gasoline and suggested that co-oxidation played an important role in the biodegradation of the other compounds (Jamison, Raymond and Hudson, Jr., 1976). Complete conversion of hydrocarbons leads to carbon dioxide, water, and new cell biomass; various intermediates are also formed (Vanloocke et al., 1975). In their column studies, Kappler and Wuhrman (1978a, b) noted a lag period of 1 to 5 days (depending on temperature) before the ground water microbial flora was able to measurably degrade dissolved hydrocarbons. Addition of nitrogen increased degradation. Stimulation of the microbial population by the addition of nitrogen, phosphorus, and dissolved oxygen has been shown to be effective in restoring hydrocarbon-contaminated aquifers (American Petroleum Institute, 1980). Yang and Bye (1979) suggested that the levels of trace elements are usually sufficient to support microbial growth.

Many organic compounds which have contaminated ground water are subject to microbial action. Wilson et al. (1981, 1983a and b) studied the degradation of several halogenated volatile organics commonly found as ground water contaminants and determined that some of the compounds were degraded, although often at very slow rates. Differences were noted between the activities of the microbial populations from two sites in degrading the synthetic organics. The degradation of naphthalene has been studied in several systems. Naphthalene has a number of uses and is considered to be hazardous (Windholz, 1976). It should be a representative model for other compounds. Slavnia (1965) and Naumova (1960) found naphthalene-utilizing organisms in the ground waters of the U.S.S.R. near oil and gas beds. Ogawa, Junk and Svec (1981) and Lee and Ward (1984) reported that naphthalene was degraded rapidly in aerobic ground waters contaminated by polynuclear aromatic hydrocarbons. However, Erlich et al. (1982) concluded that there was no evidence of anaerobic degradation of naphthalene in ground water samples from an area contaminated by wood creosoting products, although it was disappearing at a faster rate in the aquifer than if only dilution were occurring. Naphthalene was slightly sorbed onto sediments. It was biodegraded in an aquifer recharged with reclaimed water from wastewater treatment after an initial lag (Roberts et al., 1980). The importance of acclimation to the contaminant was demonstrated by these results.

Many ground water systems are deficient in oxygen and consequently anaerobic degradation is important. Erlich et al. (1982) and Rees and King (1980) reported evidence of anaerobic degradation of phenolics under aquifer conditions. Bouwer, Rittman and McCarty (1981) and Bouwer and McCarty (1983a,b) found degradation of several halogenated aliphatic organic compounds under methanogenic and denitrifying anaerobic conditions that were not degraded under aerobic conditions. Wood et al. (1980) reported that anaerobic metabolism of some of these compounds led to the formation of similar contaminants such as the production of vinyl chloride, vinylidene chloride, cis- and trans-1,2-dichloroethene from trichloroethylene and/or tetrachloroethylene. None of the aromatic compounds tested by Bouwer and McCarty (1983b) were significantly utilized under anaerobic conditions. DiTommaso and Elkan (1973) and Leenheer, Malcolm, and White (1976) disclosed that methane was produced by the degradation of a high-organic-content wastewater after deep well injection. This provides evidence for anaerobic degradation in the deep subsurface.

The potential for microbial degradation of contaminants has been utilized to restore polluted environments. Atlas (1977) outlined the following approaches that can be used to enhance the degradation of petroleum: the microbial population can be altered by seeding; the environment can be modified to enhance microbial activity; or the petroleum can be modified to make it more susceptible to biodegradation. Seed microorganisms would probably have to be a mixture of microorganisms that collectively can degrade all the components of the contaminant. Different mixtures of seed organisms will have to be used for different environments and contaminant problems. The environment can be modified to enhance microbial activity by the addition of dissolved oxygen, nitrogen, phosphorus, sulfur, iron, magnesium, calcium, sodium, and water which are not present in sufficient quantities. Modification of the contaminant could involve increasing the surface area available for biodegradation by dispersing the contaminant or by emulsification. However, the emulsifying agent or dispersant may prove toxic to the microbial population (Atlas, 1977).

Biological restoration of contaminated aquifers has been accomplished using several techniques. Suntech pioneered the use of "bioreclamation" to restore petroleum-contaminated aquifers (Raymond, 1974, 1978; Raymond, Hudson and Jamison, 1976; Jamison, Raymond and Hudson, 1975). This process enhances the natural population by providing dissolved oxygen and nutrients to accelerate biodegradation. This process has been modified for use at hazardous waste sites (J.R.B. Associates, 1982; Jhaveria and Mazzacca, 1982). Another technique which has been employed is the addition of acclimated microbes and nutrients to stimulate degradation (Quince and Gardner, 1982). Yet another option is withdrawal and biological treatment by conventional wastewater treatment processes, including activated sludge, lagoons (facultative, anaerobic, or aerobic), rotating biological discs, and trickling filters (J.R.B. Associates, 1982). Combinations of these biological techniques with physical-chemical treatment have also been used effectively (Shuckrow and Pajak, 1981).

Enhancement of the Indigenous Population

Stimulation of the native microbial population by the addition of nutrients has long been suggested as a means of increasing the degradation of hydrocarbons in aquatic and soil environments. In marine environments, hydrocarbon degraders have been found in higher concentrations in regions chronically polluted with petroleum (Atlas, 1977). Nitrogen and phosporus levels are often inadequate and must be supplemented. Use of an oleophilic fertilizer that concentrates the nutrients at the water surface has been shown to increase the degradation of oil slicks (Atlas and Bartha, 1973b). Levels of iron may also be deficient (Dibble and Bartha, 1976). The temperature of the sea water is a controlling factor in microbial degradation (Mulkin-Phillips and Stewart, 1974b), especially in colder areas like the Arctic (Atlas and Schofield, 1975). Seasonal and climatic variability, the types of hydrocarbons involved, and the nature of the site are also important (Colwell and Walker, 1977). Other factors to be considered include the dispersion of the hydrocarbon turbulence, the concentration of the hydrocarbon, and microbial predation (Zobell, 1973). The effectiveness of nutrient supplementation has also been demonstrated in freshwater systems; nutrient addition increased the degradation of gasoline in a lake from 90% (control) to 96% (nutrient supplemented) after five weeks (Horowitz and Atlas, 1977).

Addition of nutrients to hydrocarbon-contaminated surface soil has proven useful in increasing microbial degradation. The total numbers of microbes increase greatly after a petroleum spill. Odu (1972) noted an increase from 10^6 organisms/g to 10^8 organisms/g after an oil well blowout. Application of fertilizer stimulated greater microbial growth and utilization of some components of oil (Jobson et al., 1974). Other components of oil are not attacked by the microbes and remain in the soil. Saturated fractions are highly degraded while asphaltenes and aromatics are often resistant to microbial attack (Jobson, Cook and Westlake, 1972). Lehtomaki and Niemela (1975) found that the addition of yeast cells also served as a nutrient source. The soil must be tilled to distribute the hydrocarbons through the soil and provide oxygen (Raymond, Hudson and Jamison, 1976). Wentsel et al. (1981) attempted to enhance the biodegradation of monochlorobenzene and Ethion by the addition of nutrients (Difco nutrient broth) and aeration of the soil. No evidence for significant microbial degradation of monochlorobenzene or Ethion was observed, thus showing that stimulation of the native microbial population will be ineffective against compounds which the microbes cannot readily attack, such as the water-insoluble Ethion or highly volatile monochlorobenzene.

Enhancement of the natural microflora has been used in a number of cases to degrade hydrocarbons in surface soils. Dibble and Bartha (1979c) documented the cleanup of a large kerosene spill (1.9 million liters) in New Jersey. Much of the kerosene was recovered by physical means and by removing 200 m³ of contaminated soil. Following stimulation of microbial degradation by liming, fertilization, and tillage, phytotoxicity was reduced. Microbial degradation was slowed by the lower winter temperatures. Guidin and Syratt (1975) suggested that a black covering be placed over the soil to increase the soil temperature during the winter to overcome this problem. Landfarming has been tested as a method of disposing of waste oil and oil sludges. Landfarming relies on the soil organisms to degrade the waste applied to the soil. Brown et al. (1981) investigated the factors which influenced biodegradation rates of oil sludges at a land farm site. Only under excessively wet or dry conditions did the moisture content become a dominant factor. The nitrogen level was critical since no additional increase in the biodegradation rate occurred when potassium and phosphorus were also added. The optimal carbon-to-nitrogen ratio ranged from 150:1 to 10:1 depending upon the nature of the sludge. Dibble and Bartha (1979a) conducted similar experiments on landfarming waste oil and determined that carbon-to-nitrogen and carbon-to-phosphorus ratios of 60:1 and 800:1 were optimal under the conditions they used. Addition of micronutrients and organic supplements in the form of yeast extract and trace elements or domestic sewage did not prove beneficial. Other experiments showed that urea formaldehyde was the most satisfactory nitrogen source tested since it effectively stimulated biodegradation and did not leach nitrogen which could contaminate the ground water (Dibble and Bartha, 1979b). Weldon (1979) found no adverse impact on ground water quality from landfarming refinery oil sludges, although heavy metals were concentrated in the soil and could be leached away. Another problem was runoff water from the site could contain high amounts of oil and fertilizer (Kincannon, 1972).

These examples illustrate the effectiveness of nutrient supplementation in aquatic and soil systems. Many of the same procedures can be used to restore contaminated aquifers. Among the first to suggest such actions were Williams and Wilder (1971) and McKee, Laverty and Hertel (1972) who investigated a large gasoline spill (estimated to be 250,000 gal) in Glendale, California, that had contaminated an irrigation well. An investigation into the microbial degradation of the gasoline showed that several bacterial species of the genera *Pseudomonas* and *Arthrobacter* could utilize the gasoline when supplied with trace nutrients and adequate dissolved oxygen (McKee, Laverty and Hertel, 1972). Bacterial degradation of trapped gasoline was more rapid in the zone of aeration above the water table than in the water-saturated zone. The levels of gasoline-utilizing bacteria in contaminated wells were 50,000/ml or higher and gradually fell to the background level of 200/ml as the gasoline disappeared. The authors thought that biodegradation could eventually restore the aquifer to service. At roughly the same time, Davis et al. (1972) recommended that the addition of nutrients to stimulate degradation of hydrocarbons in ground water might be a feasible solution to ground water pollution problems.

Raymond, Jamison and their co-workers at Suntech were probably the first to put these suggestions into practice. Raymond (1974) received a patent entitled "Reclamation of Hydrocarbon Contaminated Ground Waters" on a process to eliminate hydrocarbon contaminants in aquifers by providing nutrients and oxygen to the hydrocarbon-utilizing microorganisms present in the ground water. The nutrients and oxygen are to be introduced through wells and circulated through the contaminated zone by pumping one or more producing wells. Supplementation of the nutrient and oxygen levels and recirculation of the water would allow the normal microbial flora to decompose the hydrocarbons more rapidly than under natural conditions. The nutrient solution would contain sources of nitrogen and phosphorus and other inorganic salts, if necessary, at concentrations of 0.005–0.02% by weight for each of the nutrients. Oxygen would be supplied by sparging air into the ground water. The process was expected to be largely complete within six months. Return to the normal levels of bacteria would occur after nutrient addition was stopped, and since no organisms were added, the normal flora would be maintained.

Suntech's process has been largely used to clean up gasoline-contaminated aquifers. The following steps are involved. Physical methods are employed to recover as much of the gasoline as possible (Suntech, 1977). If practical, it may be wise to continue to pump contaminated wells to contain the gasoline (Raymond, Jamison and Hudson, 1976). Then investigations of the hydrogeology and the extent of contamination should be made (Suntech, 1977). A laboratory study must then be conducted to determine if the native microbial population can degrade the components of the spill and what kinds and amounts of inorganic salts are required to stimulate degradation (Raymond, 1978). The laboratory study determines what combination of nutrients gives the maximal cell growth on gasoline in 96 hr at the ambient temperature of the ground water. Considerable variation in the nutrient requirements has been found between aquifers. One system required only the addition of nitrogen and phosphorus sources (Raymond, Jamison and Hudson, 1976), while the growth of microbes in another aquifer was stimulated best by the addition of ammonium sulfate, mono- and disodium phosphate, magnesium sulfate, sodium carbonate, calcium chloride, manganese sulfate, and ferrous sulfate (Raymond et al., 1978). The form of the nutrient which must be added also varies; ammonium sulfate gave much greater growth than ammonium nitrate in one aquifer system. Chemical analyses of the ground water provide little information as to the nutrient requirements for the system. After the microbial investigation has established the optimal growth conditions, the system for injecting the nutrients and oxygen and producing water to circulate them in the formation must be designed and built (Raymond, 1978). This work should be under the direction of a competent ground water geologist since controlling the ground water flow is critical to the success of the operation. Placement of the injection and production wells such that ground water flow goes through the contaminated zone is required. Recycling the contaminated water from the producing wells is suggested, since it eliminates the problem of waste disposal and allows the recirculation of unused nutrients. The screens for the wells should be large enough to permit fluctuations of the ground water table due to weather conditions or operation of the system.

Once actual operations are underway, nutrient addition can be by batch or continuous feed (Raymond, 1978). Batch addition gives satisfactory results and is more economical. For one system, the nutrients were prepared as a 30% concentrate in a tank truck and injected into the aquifer (Raymond, Jamison and Hudson, 1976). This may have resulted in osmotic shock to the microorganisms which come into contact with the concentrate before dilution. Large amounts of nutrients may be required; at one site, 16.65 tons of chemicals were added (Minugh et al., 1983), while a total of 87 tons of food grade quality chemicals were purchased to clean up another site (Raymond, Jamison and Hudson, 1976). Oxygen can be supplied to the aquifer by sparging air into wells with Carborundum diffusers powered by paint sprayer-type compressors (Raymond, Jamison and Hudson, 1976) or by diffusers made from a short piece of DuPont Viaflo® tubing for smaller wells (Raymond et al., 1978). The large diffusers can provide up to 10 ft^3 of air per minute (scfm) while the smaller tubing diffusers can provide only 1 scfm. Another approach was the use of diffusers spaced along air lines buried in the injection trench (Minugh et al., 1983). The size of the compressor and the number of diffusers are determined by the extent of contamination and the period allowed for treatment (Raymond, 1978). The supply of dissolved oxygen may be the limiting factor in the biostimulation process, especially in low-permeability aquifers (Raymond et al., 1978). As the levels of contamination decrease, the biochemical oxygen demand lessens and the DO level in the water rises. The system must be monitored to insure that the levels of nutrients are at their optimal concentrations and are being evenly distributed, and that the discharge water meets state or federal requirements (Raymond, 1978). Adequate supplies of nitrogen and phosphate can be readily maintained once breakthrough has occurred (Raymond et al., 1978).

The bacterial population's response to nutrient addition should also be monitored. The highest microbial population is likely to be in the area of greatest contamination (up to 10^7 organisms/ml have been reported); a map of the bacterial counts following a gasoline spill in Ambler, Pennsylvania, resembled the contours of gasoline contamination (Raymond, Jamison and Hudson, 1975). During the course of the biostimulation program at this site, 32 cultures thought to be the predominant gasoline-utilizing bacteria were isolated from the ground water. These cultures included 10 cultures assigned to *Nocardia*, two *Micrococcus*, four cultures of *Actinetobacter*, eight *Pseudomonas*, *Flavobacterium devorars*, and seven unidentified cultures. *Nocardia* cultures were probably responsible for the major paraffinic hydrocarbon degradation, and the *Pseudomonas* cultures were responsible for much of the aromatic degradation (Jamison, Raymond and Hudson, 1976). Cooxidation seemed to have played a major role in the degradation since none of the isolates could grow on branched paraffins, olefins, or cyclic alkanes as sole carbon sources. Bacteria capable of degrading these compounds may also not have been isolated. Similar results from the microbiological studies were found during the biostimulation program at Millville, New Jersey (Raymond et al., 1978). The preliminary tests before the bioreclamation operation began showed the presence of from 10^2 to 10^5 gasoline-utilizing organisms/ml in the ground water. The microbial population responded to the addition of nutrients and oxygen with a ten- to thousandfold increase in the numbers of gasoline-utilizing and total bacteria in the vicinity of the spill, with levels of hydrocarbon utilizers in excess of 10^6/ml in several wells. The microbial response was an order of magnitude greater in the sand than the ground water. Forty-one cultures were isolated from the soil and ground water at this site, with 17 considered to be *Pseudomonas*, four *Flavobacterium*, eleven *Nocardia*, and nine were not assigned to any genus. Several of the *Pseudomonas* species were fluorescent, in contrast to the Ambler bacteria which did not include fluorescent pseudomonads. Many of the cultures were composed of very small cells.

After biostimulation at a LaGrange, Oregon site, bacterial levels increased up to six million times the initial levels.

Suntech's bioreclamation process has met with reasonable success when applied to gasoline spills in the subsurface. After the biostimulation program ended at Ambler, Pennsylvania, the gasoline-utilizing bacterial levels declined from the high levels present during the period of nutrient addition. This suggested a depletion of nutrients and gasoline (Raymond, Jamison and Hudson, 1975). The concentration of gasoline in the produced water was not reduced during the period of nutrient addition, but no gasoline was found in the produced water 10 months after nutrient addition ceased. Estimates based on the amount of nitrogen and phosphate retained in the system suggested that 744 to 944 barrels of gasoline were degraded. Residual gasoline was found at the last sampling period at the Millville, New Jersey, site, but no free hydrocarbon was observed in any of the wells after the biostimulation program (Raymond et al., 1978). The gasoline concentrations in cores taken from the aquifer did not seem to change substantially. Initially, the level of phenol in the ground water was a problem, but it decreased to acceptable values after more aerobic conditions were achieved in the aquifer. In the produced water which had been circulated through the site, gasoline levels remained low and fairly constant. The operation was successful in cleaning up the aquifer to the point where no free gasoline was present and the site met state approval. The operation was terminated and all well casings and injection equipment removed. The nutrient supplementation program succeeded in removing all the free product in the wells at the LaGrange, Oregon site (Minugh et al., 1983); however, gasoline odors and a cloudy sheen were detected in some of the pits dug after the cleanup operation ended. Gasoline concentrations of 100–500 ppm were found in the areas where the pits were dug with the average level of DOC in the ground water at 20 ppm. Samples taken later showed continued improvement. Vapor problems which had threatened two restaurants were mitigated.

The advantages of the biostimulation process include:

(1) It is useful for the removal of hydrocarbons and certain organic compounds, especially water-soluble pollutants and low levels of other compounds which would be difficult to remove by any other means.

(2) It is environmentally sound since no waste products are generated and no ecological changes result because it utilizes the indigenous microbial flora.

(3) It is fast, safe, and generally economical.

(4) Treatment moves with the plume.

(5) It is good for short-term treatment of contaminated ground water (Yang and Bye, 1979; J.R.B. Associates, 1982).

Its disadvantages are:

(1) It does not work with heavy metals and some organics.
(2) Bacteria can plug the soil and reduce circulation.
(3) Introduction of nutrients could adversely affect nearby surface water.
(4) Residues may cause taste and odor problems.
(5) It could be expensive since equipment maintenance may be high and long-term injection of oxygen and nutrients may be necessary to sustain high rates of degradation.
(6) Under certain conditions, such as high concentration of pollutants, it may be slower than physical recovery methods.
(7) Long-term effects are unknown.

The bioreclamation process has been demonstrated to be useful in the cleanup of hydrocarbon-contaminated aquifers of varying properties. Concern was once expressed that the process might not be effective in aquifers with low permeabilities. A laboratory study showed that gasoline-utilizing bacteria could penetrate sand columns with effective permeabilities ranging from 200 darcys to 3.5 darcys (sand packs of coarse 20 mesh sand to very fine 80+ mesh sand) and consolidated sandstone cores with effective permeabilities of 19 and 75 millidarcys (Raymond, Jamison and Hudson, 1975). The process has been used to restore aquifers of dolomite (Raymond, Jamison and Hudson, 1976), a highly permeable sand (Raymond et al., 1978), and alluvial fan deposits composed of sand, gravel, cobbles, and some clay and silt (Minugh et al., 1983).

Alternative sources of oxygen have been suggested as a means to increase the degradative activity in contaminated aquifers (Texas Research Institute, 1982a). For example, hydrogen peroxide could be injected into the contaminated soil above the water table in conjunction with nutrients where it would decompose naturally or by enzymatic action to increase the dissolved oxygen content. Hydrogen peroxide is relatively inexpensive and does not present a persistent hazard in ground waters (Texas Research Institute, 1982b). It is cytotoxic, but many bacterial cells have enzymatic defenses (hydroperoxidases) against it. Hydrogen peroxide was shown to be toxic to fresh bacterial cultures at levels greater than 100 ppm, although mature cultures suffered less and could function at levels as high as 10,000 ppm (Texas Research Institute, 1982a). Another concern was that the hydrogen peroxide would decompose before it reached the ground water; however, phosphate buffered solutions at a pH of 7.0 and moderate flow rates were effective for maintaining hydrogen peroxide levels (Texas Research Institute, 1982c). Subsequent experimentation with sand columns inoculated with gasoline and gasoline-degrading bacteria showed that 1.0, 0.5, and 0.25% hydrogen peroxide solutions were toxic to the bacteria (Texas Research Institute, 1983). Nagel et al. (1982) documented the use of ozone to treat petroleum contamination in Karlsruhe, Germany, that threatened a drinking water supply. The polluted ground water was withdrawn, treated with ozone, and infiltrated back into the system, via three infiltration wells. About 1 g ozone/g DOC was added to the ground water, which increased the biodegradability of the petroleum contaminants and added dissolved oxygen; the dissolved oxygen levels increased in the ground water and reached equilibrium at about 80% of the initial concentration injected. The oxygen consumption peaked at about 40 kg/day during the initial infiltration period. The DOC values in the wells fell from a range of 2.5 to 5.5 g/m^3 to a steady state of about 1.5 g/m^3. Levels of cyanide, a contaminant identified after the treatment began, also decreased, although biodegradation was not shown to be the cause. Total bacterial counts in the ground water increased tenfold, but potentially harmful bacteria did not increase. The drinking water from this aquifer contained no trace of contaminants after 1^1/2 years of ozone treatment.

Jhaveria and Mazzacca (1982) of Groundwater Decontamination Systems (GDS) adapted the biostimulation process to the cleanup of an aquifer contaminated with compounds other than petroleum hydrocarbons; the pollutants were solvents, including methylene chloride, acetone, n-butyl alcohol, and dimethylaniline. The contamination problem resulted from a spill from a pharmaceutical company, Biocraft Laboratories, in Waldwick, New Jersey, where nearly 136,500 kg of the solvents were estimated to have been lost. After consultation with Geraghty and Miller, Inc., the initial consulting firm in Princeton-Aqua Science, and Suntech, Inc., it was decided to contain the contaminated ground water on site and treat it by stimulating the natural microbial population. The contaminated ground water was pumped from dewatering trenches and wells to two activating tanks where nutrients and oxygen were supplied. The water was then pumped into two settling tanks and later injected into the subsurface enriched with nutrients, oxygen, and microorganisms. A bacterial culture acclimated to the contaminants was established before the biostimulation process began.

The biodegradation process was studied before the system was designed (Jhaveria and Mazzacca, 1982). The wells had a population of 10^3–10^4 colonies/ml prior to the biostimulation programs, and the addition of nitrogen and phosphate increased the growth of the organisms as high as four times the control. A pilot study where nutrients and oxygen were supplied to the aquifer showed that the microbial population increased from 10^3–10^6/ml after seven days and remained constant at that level. A batch process study using a fermenter demonstrated that after an acclimation period of eight days, a rapid decrease in the COD of the ground water was noted; essentially all the compounds were degraded. Additional air gave a quicker response. Two continuous process studies were initiated to test the design of the plant. They demonstrated that the addition of oxygen and nutrients to the acclimated bacteria led to rapid removal rates of 60-90% of the methylene chloride, acetone, dimethylaniline, and butyl alcohol.

These successful tests led to the design and construction of the system (Jhaveria and Mazzacca, 1982). The aboveground portion consisted of two activating tanks and two settling tanks. The activated tanks were maintained at 20°C and supplied with air and the following nutrients: ammonium chloride, monopotassium phosphate, dipotassium phosphate, magnesium sulfate, sodium carbonate, calcium chloride, manganese sulfate, and iron sulfate. The levels of the nutrients were adjusted to provide the necessary nitrogen and phosphorus to stimulate in situ degradation. Good results were achieved by the system; removal of organics exceeded 95.9% in the aboveground systems, and the levels of contamination in one of the producing wells fell from 91 ppm of methylene chloride and 54 ppm of acetone, to less than 1 ppm after a year. Similar degradation efficiencies were noted for the monitoring wells. Sludge production was minimal since some of the sludge was recycled from the settling tanks to the activating tanks and some was allowed to pass to the recharge trenches to inoculate the soil with acclimated microorganisms. An independent study by Professor W. W. Umbreit of Rutgers University confirmed that methylene chloride was oxidized by the GDS culture.

Enhancement of the microbial population has also been reportedly used to reduce levels of iron and manganese in the ground water (Hallberg and Martinelli, 1976). The process known as the Vyredox method was developed in Finland and has been used in Sweden and other locales where high levels of iron and manganese are found in the ground water. Iron bacteria and manganese bacteria oxidize the soluble forms of iron and manganese to insoluble forms; the bacteria use the electrons adsorbed from the oxidation process as sources of energy. The Vyredox method works by adding dissolved oxygen to the ground water to stimulate the iron and manganese to first remove the iron and later the manganese. As the iron bacteria population builds up and begins to die, it supplies the organic carbon necessary for the manganese bacteria. The efficiency of the process increases with the number of aerations. Since no data were presented that show an increase in the number of iron and manganese bacteria or in their activity, it is difficult to attribute the reduced level of iron and manganese in the ground water to microbial activity instead of chemical oxidation from the introduction of dissolved oxygen.

Microbial activity in contaminated aquifers can also be enhanced by altering the contaminant chemically or physically to make it more degradable. One way to do that is to spread the contaminant over a larger area by adding dispersants. Several dispersants were tested on marine samples to determine their effects on the degradation of oil (Atlas and Bartha, 1973a). They were found to increase the rate of mineralization provided the sea water was amended with nitrate and phosphate, but were not able to increase the extent of degradation. Mulkin-Phillips and Stewart (1974a) tested four dispersants and found that only one stimulated biodegradation. The four dispersants supported microbial growth and were not toxic to the microbes, but did cause population changes. Two oil herders tested by Atlas and Bartha (1973a) were determined to increase the mineralization rate, but not the extent of degradation. Application of surfactant has also been advocated to clean-up contaminated aquifers, but they may be toxic and nonbiodegradable (Texas Research Institute, 1979).

Addition of Acclimated Microorganisms

Another approach for restoring contaminated aquifers is the addition of microbes to degrade the pollutants. Usually microbes that have been acclimated to degrade the contaminants are used as "seed"; the microorganisms may have been selected by enrichment culturing or genetic manipulation. A mixed microbial population is often necessary to degrade all the contaminants (Zajic and Daugulis, 1975). Microorganisms can become acclimated to the degradation of compounds by repeated exposure to that substance. Felsot, Maddox and Bruce (1981) reported that bacteria exposed to carbofuran yearly became so active in degrading it, that it was no longer effective against the western corn rootworm. Spain and Van Veld (1983) went so far as to suggest that the microbial community near a pollutant spill might be primed with small amounts of the pollutant to ensure a rapid microbial response. Seed organisms can be isolated by enrichment culturing; the type of microorganisms which are isolated are dependent on the source of the inoculum, the conditions used for the enrichment, and the substrate (Atlas, 1977). Sequential enrichment is a modification of enrichment culturing that allows microorganisms that can more fully degrade the substrate to be isolated. The substrate is inoculated with a microbial population and the organisms which can degrade it are isolated. The undergraded substrate is then used to isolate another set of organisms which are capable of degrading the residual components. This process is continued until none of the substrate remains or no new isolates are made. A potential problem with such enrichments is that the organisms may interfere with each other.

A new approach has recently been developed to induce acclimation by genetically manipulating the microbes. Zitrides (1978) and McDowell, Bourgeois and Zitrides (1980) used radiation to increase the genetic variability of an adapted microbial population in hopes of producing strains that could better degrade the contaminant. Selected strains of bacteria chosen for their known ability to degrade similar compounds are exposed to successively increasing concentrations of substrate. The fastest growing strains are irradiated. The genetic alterations should increase the growth rate and fix the desired biochemical capability. The adapted mutant will probably be at a disadvantage in the competition with the native microbial population, and may only be able to proliferate on the substrate upon which it was isolated.

Genetic engineering can be used to increase the degradation capacity of microbes by improving the stability of the organisms, enhancing their activity, providing them with multiple degradative activities, and ensuring that they are safe both to the environment and human health (Pierce, 1982). The ability to degrade some hydrocarbons can be encoded on bits of extrachromosomal DNA known as plasmids. Plasmids can be transferred to an organism to increase the number of compounds which can be attacked. The plasmids can also be fused together to provide multiple degradative traits or to produce a novel or previously unexpressed degradative pathway. Stable strains can be engineered that will not pass on their plasmids or will be capable of growth only under restricted conditions; this should limit their potential for escape into the environment. Environmental conditions such as temperature, pH, substrate concentration, oxygen tension, and competition with the native microbial population may prevent the genetically engineered organism from reaching its degradative potential.

One successful application of genetic engineering is the development of a bacterial strain capable of degrading 2,4,5-trichlorophenoxyacetic acid (2,4,5-T). This strain was developed using the technique of plasmid-assisted molecular breeding. This herbicide is normally degraded very slowly and only by cooxidative metabolism (Kilbane et al., 1982). Kellogg, Chatterjee and Chakrabarty (1981) isolated an organism that could degrade 2,4,5-T as its sole carbon source from a chemostat inoculated with microorganisms from a variety of hazardous waste dumping sites and microorganisms harboring a variety of plasmids controlling the degradation of several hydrocarbons. The chemostat was fed with low concentrations of 2,4,5-T and higher concentrations of the plasmid substrates for 8 to 10 months. A 2,4,5-T degrading population was isolated which could degrade up to 70 percent of the 2,4,5-T supplied at 1.5 mg/ml. The strain was identified as *Pseudomonas cepacia* AC1100 (Chatterjee, Kilbane and Chakrabarty, 1982). The ability of *P. cepacia* AC1100 to degrade 2,4,5-T is soil was tested and it was shown to be able to degrade more than 95% of 1 mg/g of 2,4,5-T within a week. Soil samples contaminated with from 1000–20,000 μg/g were inoculated with 5×10^7 AC1100 cells/g. Complete degradation occurred in samples with 5000 μg/g, and more than 90% of the 2,4,5-T was degraded after six weekly applications of AC1100 for the 10,000-μg/g samples (Kilbane, Chatterjee and Chakrabarty, 1983). Other experiments were conducted to determine the survival of AC1100 in the soil. When added to soil without 2,4,5-T the levels of AC1100 dropped to negligible values after 12 weeks, but following addition of 2,4,5-T the population increased until the compound was exhausted. No appreciable effects on the number or types of indigenous bacteria were noted when 10^7 AC1100 cells were added. This indicates that no adverse ecological effects will occur when AC1100 is used as a seed to remove 2,4,5-T from soil. Until results from field tests are available, it will be difficult to determine how well the organism will work under actual conditions.

There are problems with the use of genetically engineered microorganisms to degrade contaminants. Johnston and Robinson (1982) could find no conclusive evidence that commercially available genetically engineered organisms were effective in establishing themselves or significantly enhancing biodegradation of pollutants in aeration basins or natural environments having an active native microbial population. Most of the bacteria typically used in microbial genetic work are eutrophs of the families *Enterbacteriaceae* and *Pseudomonodoceae* which may not be able to attack substrates in the parts-per-billion range that are often found in environmental samples. Environmental stress such as unsuitable water quantities or the presence of toxicants may also affect an introduced microorganism. However, genetic engineering holds great promise, especially in treatment facilities where conditions can be controlled. These facilities will probably be the first area where gene manipulation will be used.

Seeding microorganisms has been used in a number of different environments to degrade organics. Gutnick and Rosenberg (1977) argued that microbial seeding was ineffective in reducing oil contamination in the sea, but that microbial seeding and nutrient supplementation would work in contained environments like oil tankers. Miget (1972) found that the effectiveness of microbial seeding in simulated marine environments varied with the type and quantity of crude oil more than with the inoculum density or nutrient salts concentration. Several petroleum-degrading seed cultures have been developed for marine settings (Anonymous, 1970; Pacific Northwest Laboratories, 1970), but the effectiveness of some of the agents is in doubt. Atlas and Bartha (1973a) tested two bacterial inocula and found them to be totally ineffective; the rates of crude oil mineralization in the inoculated and control systems were identical and when the cultures were added to sterile sea water, the rates of degradation were slower than in natural sea water. A mutant bacterial culture used to degrade oil at a spill-contaminated beach was somewhat effective (McDowell, Bourgeois and Zitrides, 1980).

Schwendinger (1968) was one of the first to attempt to seed microorganisms onto soil to effect the removal of oil. An inoculum of *Cellumonas* sp. and nutrients was able to degrade the hydrocarbons more effectively than just the addition of fertilizer alone. Other researchers have met with varying success when seeding microorganisms onto petroleum-contaminated soils. Lehtomaki and Niemela (1975) found that the addition of 10^8 cells/g soil of two hydrocarbon-degrading isolates did not significantly influence oil concentrations. The application of 10^6 oil-utilizing bacteria/cm^2 resulted in a slight additional degradation of the C20 to C25 group of n-saturated compounds (Jobson et al., 1974). Hunt et al. (1973) advocated the use of seeding to stimulate the degradation of oil in arctic climates where short summers and low soil temperatures may limit the activity and rate of growth of the indigenous population. The addition of 10^4 microorganisms/gram dry soil was tested as a means of increasing this activity. The inoculum was isolated from a soil taken around oil seeps and enriched with Prudhoe Bay crude oil. The addition of nitrogen and phosphorus at 300 and 100 ppm, inoculation and adjustment of the pH to 7 increased microbial activity over the controls by at least a factor of 4 after 40 days. On the other hand, Westlake, Jobson and Cook (1978) determined that the addition of oil-degrading bacteria to soil of the boreal region of the arctic resulted in no increase in the changes of the recovered oil, but this may have been due to insufficient application.

Microbial seeding of soil environments has been applied to the degradation of organics other than petroleum. Inoculation of cultures acclimated to parathion could remove almost all the parathion within 11 days, while in the control, only 14% was removed (Daughton and Hsieh, 1977). The acclimated culture could rapidly degrade the parathion at levels up to 5000 ppm in non-flooded soils, and was able to retain this capacity for 8 to 14 days after inoculation onto soils with parathion.

Edgehill and Finn (1983) isolated a strain of *Arthrobacter* (ATCC3379) that could utilize pentachlorophenol as its sole carbon source. In laboratory tests, inoculation of 10^6 organisms/g dry soil reduced the half life of the pesticide from 14 days to less than one. Thorough mixing, larger inoculum sizes, and higher temperatures increased the effectiveness of the treatment. A mixed microbial population from primary sewage and another culture containing primarily *Pseudomonas* were tested to determine their effects on the degradation of formaldehyde and aniline in soil (Wentsel et al., 1981). This treatment was moderately successful in removing formaldehyde at levels less than 2000 ppm. Aniline was not degraded by the primary sewage inoculum, but was removed when a mixed population acclimated to it was added along with nutrients and yeast extract. Application of microorganisms from sewage effluent was a low-cost, fairly effective method for the removal of water-soluble biodegradable organics. The use of adapted mutant cultures to degrade chlordane and dinitrobenzene was only partially effective with some loss of chlordane was noted. The rates of dinitrobenzene loss were accelerated slightly in the chambers inoculated with adapted microbes, but microbial activity was not shown to be the cause.

The addition of adapted mutant microbes was not completely successful in these cases, but has great potential. At least six companies are involved in the production of microbial strains to be used to treat abandoned hazardous waste sites and chemical spills (Anonymous, 1981 and Anonymous, 1982). They have had problems marketing their bacterial products partly due to the number of irresponsible salesmen who once sold enzymes as a potential cure-all. Polybac Corporation's (Allentown, PA) products have been used to reduce formaldehyde levels in the ground from 1000 ppm to under 50 ppm within 50 days, to clean up oil spilled onto surface soils and a beach (McDowell, Bourgeois and Zitrides, 1980), to treat wastewater containing hazardous organics (Wilkinson, Kelso and Hopkins, 1978), to restore soils and ground water contaminated by acrylonitrile (Polybac Corporation, 1983), and remove orthochlorophenol which polluted soil and a pond (Anonymous, 1981). O.H. Materials (Findley, OH) has used mutant bacteria to treat incidents involving spills to soils and ground water of acrylonitrile; phenol and its chlorinated derivatives (Walton and Dobbs, 1980); ethylene glycol, propyl acetate and other compounds; dichlorobenzene, trichlorobenzene, and methylene chloride (Quince and Gardner, 1982) and other compounds. Other companies identified (Anonymous 1981, 1982) which produce mutant bacteria to detoxify soils contaminated with hazardous organics include Sybron Biochemicals (Salem, VA), Flow Laboratories (Englewood, NJ) and General Environmental Sciences (Cleveland, OH).

Literature written by the scientists at these companies indicate that their products have been effective in restoring contaminated soils. An oil pollution problem at a fuel transfer depot was treated with the addition of 500 lb of nutrients, 80 lb of emulsifier, and rehydrated mutant bacterial degraders along with tilling to provide aeration and thorough mixing (McDowell, Bourgeois and Zitrides, 1980). An orthochlorophenol spill in Missouri contaminated soil and a pond (Polybac Corporation, 1983). A spray/injection leachate system was built using the pond as a treatment reactor. The pond was seeded with Polybac's products and the concentration of orthochlorophenol fell from 15,000 ppm to less than 1 ppm within nine months. Polybac's products were able to reduce acrylonitrile concentrations from 1000 to 1 ppm within three months following a spill in Ohio. Surface foaming and a spray/leachate recovery system with a portable bioreactor were employed in this cleanup.

Several factors must be considered before a cleanup system employing acclimated bacteria can be implemented to restore contaminated aquifers. The biodegradability of the contaminants must be assessed; the source, quantity, and nature of the spilled material and the environmental conditions of the site must be considered (McDowell, Bourgeois and Zitrides, 1980). A laboratory investigation of the kinetics of biodegradation for the acclimated bacteria, the potential for inhibition under various conditions, the oxygen and nutrient requirements, and effects of temperature should be done. The solubility of the contaminant may need to be increased by adding emulsifiers to allow microbial activity on the contaminants. The geology of the site and the extent of contamination must be investigated. The stratigraphy beneath the site and a description of the soils and bedrock should be detailed (Quince and Gardner, 1982). The hydrogeologic data that are needed are formation porosity, hydraulic gradient, depth to water, permeability, ground water velocity and direction, and recharge/discharge information. The quantity and character of the contaminants and their location in the aquifer determines what containment technique, recovery method, and treatment system should be used. The systems which have used acclimated bacteria to restore contaminated aquifers typically have relied on biological wastewater treatment techniques such as activated sludge, aeration lagoons, trickling filters, aerobic digestion, composting, and waste stabilization. Many of the systems recharge the effluent from biological treatment to the aquifer to create a closed loop of recovery, treatment, and recharge. This flushes the contaminants out of the soil rapidly and establishes hydrodynamic control separating the contaminated zone from the rest of the aquifer. Another benefit is that the acclimated bacteria can be added to the aquifer and can act in situ to degrade the contaminant. The recharge water can be adjusted to provide optimal conditions for the growth of the acclimated bacteria and the indigenous populations which may also act on the contaminants.

Several examples illustrate the effectiveness of acclimated bacteria for cleanup of contaminated aquifers. An acrylonitrile spill was handled by removing the ground water, treating it with mutant bacteria in a small reactor, and recharging it (Polybac Corporation, 1983). The levels of acrylonitrile fell from 1000 ppm to 1 ppm in three months. For another acrylonitrile spill, the result of a tank car leak (Walton and Dobbs, 1980), the primary treatment was air stripping, but after the levels of acrylonitrile dropped enough to permit microbial growth, mutant bacteria were seeded in the spill site. Following a period of adaptation and growth, the bacteria were able to reduce the levels of acrylonitrile to less than 2000 ppb. After treatment by clarification and air stripping had reduced levels of the spilled organics (including propyl acetate and ethylene glycol) at another site to concentrations less than 200 ppm, bacteria, nutrients, and air were injected into the subsurface (Quince and Gardner, 1982). This method reduced the levels of the contaminants to below that required by the regulatory agencies, with a reduction in the TOC from 40,000 to less than 1 ppm. A similar treatment system was used to restore an aquifer contaminated by organics lost from storage tanks. Air stripping and inoculation with hydrocarbon-degrading bacteria were able to remove 95% of the contamination.

At a site where 130,000 gal of organic chemicals were spilled, treatment was by clarification, adsorption onto granular activated carbon (GAC), air stripping, and then recharge (Ohneck and Gardner, 1982). Once contamination levels fell below 1000 ppm, a biodegradation program employing facultative hydrocarbon-degrading bacteria, nutrients, and oxygen was initiated. The goal for this portion of the treatment was to establish colonies to bacteria in the vadose zone to degrade the contaminants trapped in the soil. The indigenous bacteria probably were as important or more so for the degradation of the contaminants since laboratory experiments showed that when they were supplied with nutrients they were able to metabolize the contaminants at the same or greater rates than the hydrocarbon-degrading bacteria inocula. Levels of contamination decreased in one set of soil cores from about 25,000 mg/l to about 2000 mg/l in two months. Monitoring wells showed no chemical contamination above the allowable criteria of 1 mg/l at the end of the treatment program. It was concluded that complete cleanup had been achieved and had restored the quality of the aquifer to a level equal to that existing prior to the spill. Shuckrow and Pajak (1981) were less successful in their bench-scale studies at a site in Muskegon, Michigan, contaminated by several priority pollutants and at least 70 other organics. The priority pollutants included solvents, degreasers, and vinyl chloride. Attempts to acclimate an activated sludge culture to the new ground water were only minimally successful. A commercial microbial culture was not effective either. However, coupling an activated sludge process with treatment by granular activated carbon proved beneficial since the activated sludge unit removed the organics that passed through the GAC column. Up to 95% of the TOC was removed so long as the GAC continued to function well. Microbial growth on the activated carbon may also play a role in the removal of contaminants (Werner, 1982).

Conclusions

Of the three techniques discussed in this chapter, enhancement of the native population and withdrawal and treatment seem to currently be the most effective methods for biological restoration of aquifers contaminated by organic compounds. Biostimulation by the addition of oxygen, nitrogen, phosphorus, and other inorganic nutrients has been chiefly used to reclaim aquifers contaminated by gasoline, and has been effective in reducing the quantity of gasoline although not completely eliminating it. Enhancing the indigenous microbial population was also fairly effective in treating organic solvents which contaminated ground water. Although alternative sources of oxygen such as ozone or hydrogen peroxide increase the dissolved oxygen available for microbial activity, they have the disadvantages of being toxic and being largely unproven. Withdrawal and biological treatment of contaminated ground water has been shown to be an effective method for restoration of aquifers. Activated sludge treatment, often in conjunction with physical treatment processes such as activated carbon adsorption or air stripping, has been demonstrated to build up an acclimated microbial population that can degrade the contaminants. While seeding of an acclimated or mutant microbial population holds a great deal of potential since it reduces the period needed for microbial acclimation to what are often recalcitrant substances, results from previous attempts have not proven it to be responsible for the removal of the contaminants. Further work needs to be done to demonstrate that seeding microbes is a viable technique for the restoration of contaminated aquifers.

SELECTED REFERENCES

American Petroleum Institute, "Underground Spill Cleanup Manual", American Petroleum Institute Publication No. 1628, 1980, Washington, D.C.

Anonymous, "Microorganisms Consume Oil in Test Spills", *Chemical and Engineering News*, Vol. 48, 1970, pp. 48–49.

Anonymous, "It's Hard to Find Jobs for Industrious Bugs", *Chemical Week*, Vol. 128, 1981, pp. 40–41.

Anonymous, "Bugs Tame Hazardous Spills and Dumpsites", *Chemical Week*, Vol. 130, 1982, p. 49.

Atlas, R. M., "Stimulated Petroleum Biodegradation", *C.R.C. Critical Reviews in Microbiology*, Vol. 5, 1977, pp. 371–386.

Atlas, R. M. and Bartha, R., "Effects of Some Commercial Oil Herders, Dispersants, and Bacterial Inocula on Biodegradation of Oil in Sea Water", In D. G. Ahern, W. L. Cook and S. P. Meyers, eds., *Proceedings*, The Microbial Degradation of Oil Pollutants, Atlanta, Georgia, December 4–6, 1972, Publication No. LSU-SG-73-01, 1973a, Louisiana State University, Center for Wetland Resources, Baton Rouge, Louisiana, pp. 32–34.

Atlas, R. M. and Schofield, E. A. "Petroleum Biodegradation in the Arctic", In A. W. Bourquin, D. G. Ahearn and S. P. Meyers, eds., *Proceedings*, Impact of the Use of Microorganisms on the Aquatic Environment, EPA 660/3-75-001, 1975, U.S. Environmental Protection Agency, Corvallis, Oregon, pp. 183–198.

Bouwer, E. J. and McCarty, P. L., "Transformation of 1- and 2- Carbon Halogenated Aliphatic Organic Compounds Under Methanogenic Conditions", *Applied and Environmental Microbiology*, Vol. 45, 1983a, pp. 1286–1294.

Bouwer, E. J. and McCarty, P. L., "Transformations of Halogenated Organics Compounds Under Denitrification Conditions", *Applied and Environmental Microbiology*, Vol. 45, 1983b, pp. 1295–1299.

Bouwer, E. J., Rittman, B. E. and McCarty, P. L., "Anaerobic Degradation of Halogenated 1- and 2- Carbon Organic Compounds", *Environmental Science and Technology*, Vol. 15, 1981, pp. 496–599.

Brown, K. W., et al., "Factors Influencing the Biodegradation of API Separator Sludges Applied to Soil", In D. W. Shultz, ed., *Proceedings*, Land Disposal: Hazardous Waste, Philadelphia, Pennsylvania, March 16–18, 1981, EPA-600/9-81-002b, U.S. Environmental Protection Agency, Cincinnati, Ohio, pp. 19:88–199.

Chatterjee, D. K., Kilbane, J. J. and Chakrabarty,A. M., "Biodegradation of 2,4,5-Trichloroacetic Acid in Soil by a Pure Culture of *Pseudomonas cepacia*", *Applied and Environmental Microbiology*, Vol. 44, 1982, pp. 514–516.

Colwell, R. R. and Walker, J. D., "Ecological Aspects of Microbial Degradation of Petroleum in the Marine Environment", *C.R.C. Critical Reviews in Microbiology*, Vol. 5, 1977, pp. 423–445.

Daughton, C. G. and Hsieh, D. P. H., "Accelerated Parathion Degradation in Soil by Inoculation with Parathion-Utilizing Bacteria", *Bulletin of Environmental Contamination and Toxicology*, Vol. 18, 1977, pp. 48–56.

Davis, J. B., et al., "The Migration of Petroleum Products in Soil Ground Water: Principles of Countermeasures", American Petroleum Institute Publication No. 4149, 1972, Washington, D.C.

Dibble, J. T. and Bartha, R., "Effect of Iron on the Biodegradation of Petroleum in Seawater", *Applied and Environmental Microbiology*, Vol. 31, 1976, pp. 544–550.

Dibble, J. T. and Bartha, R., "Effect of Environmental Parameters on the Biodegradation of Oil Sludge", *Applied and Environmental Microbiology*, Vol. 37, 1979a, pp. 729–739.

Dibble, J. T. and Bartha, R., "Leaching Aspects of Oil Sludge Biodegradation in Soil", *Soil Science*, Vol. 127, 1979b, pp. 365–370.

Dibble, J. T. and Bartha, R., "Rehabilitation of Oil Inundated Agricultural Land: A Case History", *Soil Science*, Vol. 128, 1979c, pp. 56–60.

DiTommaso, A. and Elkan, G. H., "Role of Bacteria in Decomposition of Injected Liquid Waste at Wilmington, North Carolina", In J. Braunstein, ed., *Proceedings*, Underground Water Management and Artificial Recharge, Vol. 1, New Orleans, Louisiana, Sept. 1973, pp. 585–595.

Edgehill, R. U. and Finn, R. K., "Microbial Treatment of Soil to Remove Pentachlorophenol", *Applied and Environmental Microbiology*, Vol. 45, 1983, pp. 1122–1125.

Erlich, G. G., et al., "Degradation of Phenolic Contaminants in Ground Water by Anaerobic Bacteria, St. Louis Park, Minnesota", *Ground Water*, Vol. 20, 1982, pp. 702–710.

Felsot, A., Maddox, J. V. and Bruce, W., "Enhanced Microbial Degradation of Carbofuran in Soils and Histories of Furadan Use", *Bulletin of Environmental Contamination and Toxicology*, Vol. 26, 1981, pp. 781–788.

Guidin, C. and Syratt, W. J., "Biological Aspects of Land Rehabilitation Following Hydrocarbon Contamination", *Environmental Pollution*, Vol. 8, 1975, pp. 107–112.

Gutnick, D. L. and Rosenberg, E., "Oil Tankers and Pollution: A Microbiological Approach", *Annual Review of Microbiology*, Vol. 31, 1979, pp. 379–396.

Hallberg, R. O. and Martinelli, R., "Vyredox — *In-Situ* Purification of Ground Water", *Ground Water*, Vol. 14, 1976, pp. 88–93.

Horowitz, A. and Atlas, R. M., "Response of Microorganisms to an Accidental Gasoline Spillage in an Arctic Freshwater Ecosystem", *Applied and Environmental Microbiology*, Vol. 33, 1977, pp. 1252–1258.

Hunt, P. G., et al., "Terrestrial Oil Spills and Alaska: Environmental Effects and Recovery", *Proceedings*, Joint Conference on Prevention and Control of Oil Slicks, 1973, Washington, D.C., American Petroleum Institute, pp. 773–740.

Jamison, V. W., Raymond, R. L. and Hudson, Jr., J. O., "Biodegradation of High-Octane Gasoline in Groundwater", *Developments in Industrial Microbiology*, Vol. 16, 1975, pp. 305–311.

Jamison, V. W., Raymond, R. L. and Hudson, Jr., J. O., "Biodegradation of High Octane Gasoline", in J. M. Sharpley and A. M. Kaplan, ed., *Proceed-*

ings, 1976, Third International Biodegradation Symposium, Applied Science Publishers, pp. 187–196.

Jhaveria, V. and Mazzacca, A. J., "Bioreclamation of Ground and Groundwater by the GDS Process", Groundwater Decontamination Systems, Inc., 1982, Waldwick, New Jersey.

Jobson, A., Cook, F. D. and Westlake, D. W. S., "Microbial Utilization of Crude Oil", *Applied Microbiology*, Vol. 23, 1972, pp. 1082–1089.

Jobson, A., et al., "Effects of Amendments on the Microbial Utilization of Oil Applied to Soil", *Applied Microbiology*, Vol. 26, 1974, pp. 166–171.

Johnston, J. B. and Robinson, S. G., "Opportunities for Development of New Detoxification Process through Genetic Engineering", In J. H. Exner, ed., *Detoxification of Hazardous Waste*, Ann Arbor Science Publishers, 1982, Ann Arbor, Michigan, pp. 301–314.

J.R.B. Associates, "Handbook for Remedial Action at Waste Disposal Sites", Final Report, EPA-625/6-82-006, 1982, U.S. Environmental Protection Agency, Cincinnati, Ohio.

Kappler, T. and Wuhrman, K., "Microbial Degradation of the Water-Soluble Fraction of Gas-Oil, Part One", *Water Research*, Vol. 12, 1978a, pp. 327–333.

Kappler, T. and Wuhrman, K., "Microbial Degradation of the Water-Soluble Fraction of Gas-Oil, Part Two", *Water Research*, Vol. 12, 1978b, pp. 335–342.

Kellogg, S. J., Chatterjee, D. K. and Chakrabarty, A. M., "Plasmid-Assisted Molecular Breeding: New Technique for Enhanced Biodegradation of Persistent Toxic Chemical", *Science*, Vol. 214, 1981, pp. 1133–1135.

Kilbane, J. J., et al., "Biodegradation of 2,4,5-Trichlorophenoxy-Acetic Acid by a Pure Culture of *Pseudomonas cepacia*", *Applied and Environmental Microbiology*, Vol. 44, 1982, pp. 72–78.

Kilbane, J. J., Chatterjee, D. K. and Chakrabarty, A. M., "Detoxification of 2,4,5-Trichlorophenoxy-Acetic Acid from Contaminated Soil by *Pseudomonas cepacia*", *Applied and Environmental Microbiology*, Vol. 45, 1983, pp. 1697–1700.

Kincannon, C. B., "Oily Waste Disposal by Soil Cultivation Process", EPA-R2-72-110, 1972, U.S. Environmental Protection Agency, Washington, D.C.

Kobayoshi, H. and Rittman, B. E., "Microbial Removal of Hazardous Organic Compounds", *Environmental Science and Technology*, Vol. 16, 1982, pp. 170–183.

Lee, M. D. and Ward, C. H., "Microbial Degradation of Selected Aromatics at a Hazardous Waste Site", *Developments in Industrial Microbiology*, Vol. 25, 1984, (in press).

Leenheer, J. A., Malcolm, R. L. and White, W. R., "Investigation of the Reactivity and Fate of Certain Organic Components of an Industrial Waste after Deep-Well Injection", *Environmental Science and Technology*, Vol. 10, 1976, pp. 445–451.

Lehtomaki, M. and Niemela, S., "Improving Microbial Degradation of Oil in Soil", *Ambio*, Vol. 4, 1975, pp. 126–129.

Litchfield, J. H. and Clark, L. C., "Bacterial Activities in Ground Waters Containing Petroleum Products", American Petroleum Institute, 1973, Publication No. 4211.

McDowell, C. S., Bourgeois, Jr., H. J. and Zitrides, T. G., "Biological Methods for the *In-Situ* Cleanup of Oil Spill Residues", Presented at: Coastal and Off-shore Oil Pollution Conference, The French/American Experience, Sept. 10-12, 1980, New Orleans, Louisiana.

McKee, J. E., Laverty, F. B. and Hertel, R. M., "Gasoline in Ground Water", *Journal of Water Pollution Control Federation*, Vol. 44, 1972, pp. 293-302.

McNabb, J. E. and Dunlap, W. J., "Subsurface Biological Activity in Relation to Ground Water Pollution", *Ground Water*, Vol. 13, 1975, pp. 33-44.

Miget, R. J., "Bacterial Seeding to Enhance Biodegradation of Oil Slicks", In D. G. Ahearn, W. L. Cook and S. P. Meyers, eds., *Proceedings*, Microbial Degradation of Oil Pollutants, Atlanta, Georgia, Dec. 4-6, 1972, Publication No. LSU-SG-73-01, Louisiana State University, Center for Wetland Resources, Baton Rouge, Louisiana, p. 32.

Minugh, E. M., et al., "A Case History: Cleanup of a Subsurface Leak of Refined Product", In *Proceedings*, 1983 Oil Spill Conference: Prevention, Behavior, Control and Cleanup, San Antonio, Texas, Feb. 28-March. 3, 1983, pp. 397-403.

Mulkin-Phillips, G. J. and Stewart, J. E., "Effect of Four Dispersants on Biodegradation and Growth of Bacteria on Crude Oil", *Applied Microbiology*, Vol. 28, 1974a, pp. 547-552.

Mulkin-Phillips, G. J. and Stewart, J. E., "Effect of Environmental Parameter on Bacterial Degradation of Bunker C Oil, Crude Oils, and Hydrocarbons", *Applied Microbiology*, Vol. 18, 1974b, pp. 915-922.

Nagel, G., et al., "Sanitation of Ground Water by Infiltration of Ozone Treated Water", *GWF-Wasser/Abwasser*, Vol. 123, 1982, pp. 399-407.

Naumova, R. P., "A Comparative Study of Naphthalene-Oxidizing Organisms in Underground Waters", *Mikrobiologiyia*, Vol. 29, 1960, pp. 415-418.

Odu, C. T. I., "Microbiology of Soils Contaminated with Petroleum Hydrocarbons, Part One — Extent of Contamination and Some Soil and Microbial Properties after Contamination", *Journal of the Institute of Petroleum*, Vol. 58, 1972, pp. 201-208.

Ohneck, R. J. and Gardner, G. L., "Restoration of an Aquifer Contaminated by an Accidental Spill of Organic Chemicals", *Ground Water Monitoring Review*, Fall 1982.

Ogawa, I., Junk, G. A. and Svec, H. J., "Degradation of Aromatic Compounds in Groundwater and Methods of Sample Presentation", *Talanta*, Vol. 28, 1981, pp. 725-729.

Pacific Northwest Laboratories, "Oil Spill Treatment Agents — A Compendium", American Petroleum Institute, Richland, Washington, pp. 170-178.

Pierce, G. C., "Development of Genetically Engineered Microorganisms to Degrade Hazardous Organic Compounds", In T. L. Sweeny, H. G. Bhatt, R. N. Sykes and O. J. Sproul, eds., *Hazardous Waste Management for the 1980's*, 1982, Ann Arbor Science, Ann Arbor, Michigan, pp. 431-439.

Polybac Corporation, "Product Information Packet", 1983, Allentown, Pennsylvania.

Quince, J. R. and Gardner, G. L., "Recovery and Treatment of Contaminated Ground Water, Part I", *Ground Water Monitoring Review*, Fall 1982.

Raymond, R. L., "Reclamation of Hydrocarbon Contaminated Ground Waters", U.S. Patent Office 3,846,290, Patented Nov. 5, 1974.

Raymond, R. L., "Environmental Bioreclamation", For 1978 Mid-Continent Conference on Exhibition on Control of Chemicals and Oil Spills, Sept. 1978.

Raymond, R. L., Hudson, J. O. and Jamison, V. W., "Oil Degradation in Soil", *Applied and Environmental Microbiology*, Vol. 31, 1976, pp. 522-535.

Raymond, R. L., Jamison, V. W. and Hudson, J. O., "Final Report on Beneficial Stimulation of Bacterial Activity in Ground Waters Containing Petroleum Products", Committee on Environmental Affairs, American Petroleum Institute, 1975, Washington, D.C.

Raymond, R. L., Jamison, V. W. and Hudson, J. O., "Beneficial Stimulation of Bacterial Activity in Ground Waters Containing Petroleum Products", *AIChE Symposium Series*, Vol. 73, 1976, pp. 390-404.

Raymond, R. L., et al., "Final Report—Field Application of Subsurface Biodegradation of Gasoline in a Sand Formation", American Petroleum Institute, 1978, Project No. 307-77.

Rees, J. F. and King, J. W., "The Dynamics of Anaerobic Phenol Biodegradation in Lower Greensand", *Journal of Chemical Technology and Biotechnology*, Vol. 31, 1980, pp. 306-310.

Roberts, P. V., et al., "Organic Contaminant Behavior During Groundwater Recharge", *Journal of Water Pollution Control Federation*, Vol. 52, 1980, pp. 161-172.

Schwendinger, R. B., "Reclamation of Soil Contaminated with Oil", *Journal of Institute of Petroleum*, Vol. 54, 1968, pp. 182-197.

Shuckrow, A. J. and Pajak, A. P., "Bench Scale Assessment of Concentration Technologies for Hazardous Aqueous Waste Treatment", In D. W. Shultz, ed., *Proceedings*, Land Disposal Hazardous Waste, Philadelphia, Pennsylvania, Mar. 16-18, 1981, EPA-600/9-81-002b, U.S. Environmental Protection Agency, Cincinnati, Ohio, pp. 341-351.

Slavnia, G. P., "Naphthalene-Oxidizing Bacteria in Ground Water of Oil Beds", *Mikrobiologiya*, Vol. 34, 1965, pp. 103-106.

Spain, J. C. and Van Veld, P. A., "Adaptation of Natural Microbial Communities to Degradation of Xenobiotic Compounds: Effects of Concentration, Exposure, Time, Inoculum, and Chemical Structure", *Applied and Environmental Microbiology*, Vol. 45, 1983, pp. 428-435.

Suntech, Inc., "Environmental Bioreclamation Brochure", 1977.

Texas Research Institute, Inc., "Underground Movement of Gasoline on Groundwater and Enhanced Recovery by Surfactants", Final Report, American Petroleum Institute, 1979, Washington, D.C.

Texas Research Institute, Inc., "Enhancing the Microbial Degradation of Underground Gasoline by Increasing Available Oxygen", Final Report, American

Petroleum Institute, 1982a, Washington, D.C.

Texas Research Institute, Inc., "Feasibility Studies on the Use of Hydrogen Peroxide to Enhance Microbial Degradation of Gasoline", American Petroleum Institute, 1982b, Washington, D.C.

Texas Research Institute, Inc., "Feasibility Studies on the Use of Hydrogen Peroxide to Enhance Microbial Degradation of Gasoline", Progress Report July 1982, American Petroleum Institute, 1982c, Washington, D.C.

Texas Research Institute, Inc., "Progress Report: Biostimulation Study", Feb. 1983.

Vanloocke, R., et al., "Soil and Groundwater Contamination by Oil Spills: Problems and Remedies", *International Journal of Environmental Studies*, Vol. 8, 1975, pp. 99–111.

Walton, G. C. and Dobbs, D., "Biodegradation of Hazardous Materials in Spill Situations", *Proceedings*, 1980 National Conference on Control of Hazardous Material Spills, Louisville, Kentucky, pp. 23–29.

Weldon, R. A., "Biodisposal Forming of Refinery Oil Wastes", *Proceedings*, Oil Spill Conference, Mar. 19–22, 1979, Reprint.

Wentsel, R. S., et al., "Restoring Hazardous Spill-Damaged Areas – Technique Identification/Assessment", Final Report, EPA-600/2-7-81-1-208, 1981, U.S. Environmental Protection Agency, Cincinnati, Ohio.

Werner, P., "Mikrobiologische Untersuchungen zur Chemish-Biologischen Aufbereitung eines Humensaurehaltigen Grund-Wassers", *Vom Wasser*, Vol. 57, 1982, pp. 157.

Westlake, D. W. S., Jobson, A. M. and Cook, F. D., "*In Situ* Degradation of Oil in a Soil of the Boreal Region of the Northwest Territories", *Canadian Journal of Microbiology*, Vol. 24, 1978, pp. 254–260.

Wilkinson, R. R., Kelso, G. L. and Hopkins, F. C., "State of the Art Report – Pesticide Disposal Research", EPA-600/2-78-183, 1978, U.S. Environmental Protection Agency, Cincinnati, Ohio.

Williams, D. E. and Wilder, D. G., "Gasoline Pollution of a Ground-Water Reservoir – A Case History", *Ground Water*, Vol. 9, 1971, pp. 50–56.

Williams, E. B., "Contamination Containment by In-Situ Polymerization", *Aquifer Restoration and Ground Water Rehabilitation*, Proceedings, Second National Symposium on Aquifer Restoration and Ground Water Monitoring, May 26–28, 1982.

Wilson, J. T., et al., "Transport and Fate of Selected Organic Pollutants in a Sandy Soil", *Journal of Environmental Quality*, Vol. 10, 1981, pp. 501–506.

Wilson, J. T., et al., "Enumeration and Characterization of Bacteria Indigenous to a Shallow Water Table Aquifer", *Ground Water*, Vol. 21, 1983a, pp. 134–142.

Wilson, J. T., et al., "Biotransformations of Selected Organic Pollutants in Ground Water", *Developments in Industrial Microbiology*, Vol. 24, 1983b.

Windholz, M., ed., "The Merck Index", Merck and Co., Inc., 1976, Rahway, New Jersey.

Wood, P. R., et al., "Introductory Study of the Biodegradation of the Chlorinated Methane, Ethane, and Ethene Compounds", 1980.

Yang, J. T., and Bye, W. E., "Protection of Ground Water Resources from the Effects of Accidental Spills of Hydrocarbons and Other Hazardous Substances, (Guidance Document)", Technical Report, EPA-570/9-79-017, 1979, U.S. Environmental Protection Agency, Washington, D.C.

Zajik, J. E. and Daugulis, A. J., "Selective Enrichment Processes in Resolving Hydrocarbon Pollution Problems", In A. W. Bourquin, D. G. Ahearn and S. P. Meyers, eds., *Proceedings*, Impact of the Use of Microorganisms on the Aquatic Environment, EPA-660/3-75-001, 1975, U.S. Environmental Protection Agency, Corvallis, Oregon, pp. 169–182.

Zitrides, T. G., "Mutant Bacteria for the Disposal of Hazardous Organic Wastewaters", Presented at Pesticide Disposal Research and Development Symposium, Reston, Virginia, Sept. 6–7, 1978.

Zobell, C.E., "Microbial Degradation of Oil: Present Status, Problems, and Pespectives", In D. G. Ahearn, F. D. Cook, and S. P. Meyers, eds., *Proceedings*, The Microbial Degradation of Oil Pollutants, Atlanta, Georgia, Dec. 4–6, 1973, Publication No. LSU-SG-73-01, Louisiana State University, Center for Wetlands Resources, Baton Rouge, Louisiana, pp. 3–16.

DECISION-MAKING IN AQUIFER RESTORATION PROJECTS

CHAPTER **5**

Protocol for Aquifer Restoration Decision-Making

The general approach for developing aquifer restoration strategies and selecting the most appropriate one for meeting a given need is, for the most part, intuitively obvious. A logical first step is a preliminary assessment of the nature of the problem. Based on the preliminary assessment, a number of alternative strategies (remedial measures) are developed. From the list of possible alternatives, an optimum would be selected through systematically considering a series of decision factors, environmental impact and cost-effectiveness analysis. Implementation and construction of the chosen alternative would be next, followed by monitoring of the effectiveness of the measure.

The objective of this chapter is to present a structured protocol that can be followed to develop aquifer restoration strategies. Emphasis is placed on the actual procedure for developing a list of technical alternatives based on appropriate consideration of numerous decision factors. The procedure is not intended to be a set of explicit instructions, but rather a general approach which, when modified, could be applied to a wide variety of ground water pollution problems. The chapter begins with a review of pertinent literature. This is followed by information on preliminary activities, and development and evaluation of alternatives.

BACKGROUND INFORMATION

Most of the work in developing structured approaches or protocols for addressing ground water pollution problems has been conducted in response to the Comprehensive Environmental Response, Compensation and Liability Act, PL 96-510 (known as CERCLA or "Superfund").

Specifically, Superfund sites require the development of a "remedial action master plan" (RAMP). The purpose of a RAMP is to identify the type, scope, sequence and schedule of remedial projects which may be appropriate in meeting an identified need (Kaschak and Nadeau, 1982). Figures 5.1 and 5.2 are attempts to represent the RAMP process in the form of flowcharts. Additionally, Table 5.1 lists the phases of a site contamination and liability audit. Two features are common to all three of these outlines. First, they all have the same general intuitive pattern. Second, they all have provisions for immediate remedial actions to enable the conduction of studies to develop longer-term solutions.

The RAMP is designed as an approach for developing an optimal solution for meeting a given need. Incorporated within the RAMP is the analysis of alternative remedial measures in order to decide on an optimum strategy. The analysis involves consideration of three aspects: (1) environmental impacts, 2) costs, and 3) risks. Risk assessment is an area of study that is now receiving increased attention; Chapter 7 contains some summary information on risk assessment as related to aquifer restoration. The main problem faced by all risk assessment techniques is that a large portion of the needed information, such as risk pathways or acceptable concentrations, is unknown.

Table 5.1 **Site Contamination and Liability Audit Phased Structure (Housman, Brandwein, and Unites, 1981)**

Screening Phases	Phase 1	Initial Property Inventory
	Phase 2	Classification and Identification of Potential Problem Properties
	Phase 3	Preliminary Field Screening
	Phase 4	Prioritization of Problem Properties
Emergency Action Phase	Phase 5	Immediate Emergency Stop Action Response
Detailed Site Investigation and Remedial Phases	Phase 6	Detailed Site Field Investigation
	Phase 7	Definition of Remedial Strategies, Risk and Financial Liability Assessment and Remedial Cost Effectiveness
	Phase 8	Selection of Preferred Remedial Strategy
	Phase 9	Implementation of Remedial Action
	Phase 10	Certification of Performance and Addressing Future Potential Liability Issues

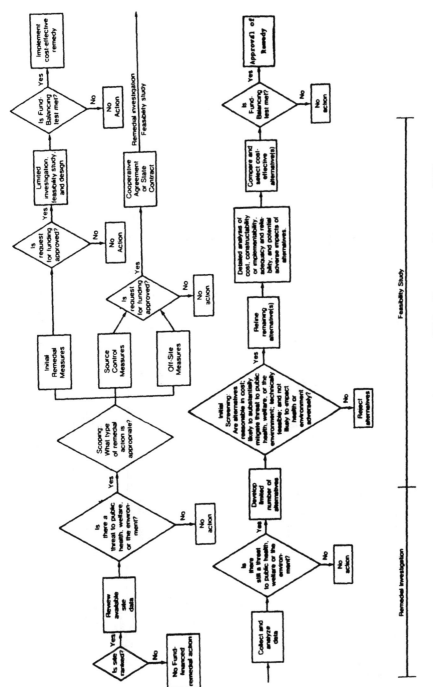

Figure 5.1. Remedial action process (Bixler, Hanson, and Langner, 1982).

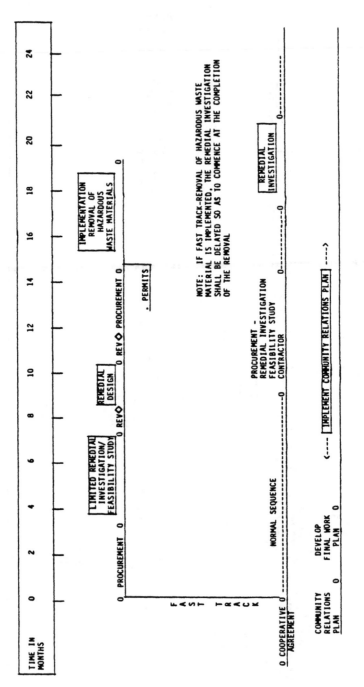

Figure 5.2. Master site schedule (Kaschak and Nadeau, 1982).

Because ground water cleanup activities are, in general, expensive, considerable interest exists in analyzing costs. St. Clair, McCloskey and Sherman (1982) discuss the advantages and disadvantages of risk assessment, cost/benefit analysis, cost-effectiveness analysis, decision-tree analysis, trade-off matrices, and sensitivity analysis for alternatives evaluation. Using certain elements of these techniques, they developed a framework for evaluating the cost-effectiveness of remedial actions. Evans, Benson and Rizzo (1982) describe an integrated, three-phased approach for cost-effective preliminary assessments at hazardous waste sites. Dawson and Brown (1981) have developed an integrated site restoration process, as outlined in Figure 5.3, which includes cost-effectiveness consideration.

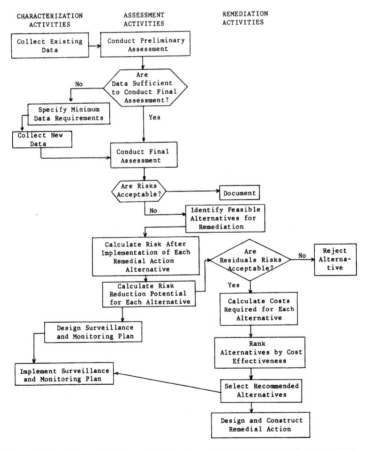

Figure 5.3. Integrated site restoration process (Dawson and Brown, 1981).

CONCEPTUAL FRAMEWORK FOR AQUIFER
RESTORATION DECISION-MAKING

Depicted in Figure 5.4 is a flowchart for aquifer restoration decision-making. Each of the steps in Figure 5.4 is discussed in detail in the following sections. Although the general order of the steps in Figure 5.4 is important (and to some degree dictated), it is emphasized that none of the steps is completely independent of the others. Additionally, the procedure is meant to be iterative, thus allowing for refinement in both judgment and design.

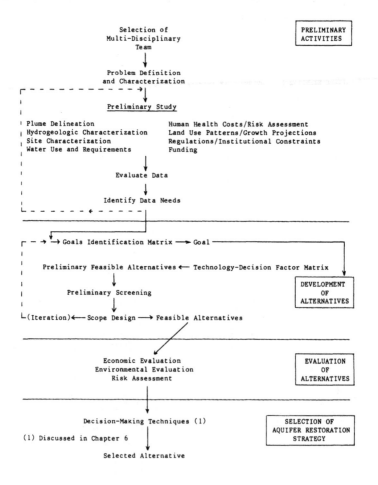

Figure 5.4. Flowchart for aquifer restoration decision-making.

Preliminary Activities

Preliminary activities associated with aquifer restoration decision-making include the assemblage of a multidisciplinary team and problem definition and characterization through the conduction of a preliminary study and evaluation of data and data needs.

Multidisciplinary Team

Ground water pollution is not strictly a hydrogeological problem. The solution to a ground water pollution problem will require involvement from a number of entities and disciplines. Managerial personnel will be involved in the overall planning and development of the project. Technical personnel, both on-site and off-site, will aid in the design of the remedial measure. Remedial-related personnel from the construction industry will have involvement. Also, institutional personnel from different levels of government will almost invariably have involvement. Within each group, a number of different disciplines might be required. For example, the technical personnel required could include hydrogeologists, environmental engineers, soil scientists, microbiologists, chemists, and toxicologists, to name a few. Formulation and implementation of a solution for meeting a particular need will require a multidisciplinary approach involving a multidisciplinary team.

Problem Definition and Characterization

The obvious first step in actually dealing with a ground water pollution concern is to define and characterize the problem. The problem will need to be defined in terms of its temporal and areal release patterns, and its urgency for formulating and implementing a solution. The two temporal categories are "anticipated problems" and "existing problems." Anticipated problems most usually will result from planned facilities that have the potential to threaten ground water supplies. The other type of problem is the existing problem. This is the situation where a facility or activity with ground water threatening potential is already in effect. Existing problems can be further subdivided into those that have already degraded ground water, and those that have not yet disturbed ground water but are expected to do so. The approach needed in identifying potential solutions to either of the two types of problems is similar, but because of their "temporal" differences, the proposed solutions may differ significantly.

Anticipated Problems

Having identified an anticipated problem, the next step would be to identify the areal release characteristics, and the potential duration and urgency of the problem. The release characteristics of the problem will be directly related to the source. The source could be a point source (disposal well), an area source (fertilizers in agriculture), a line source (highway deicing salts), or a regional source (increasing number of septic tanks in a region).

The duration of the problem can be classified as either acute or chronic. Will the anticipated problem be short-term such as lowering of water levels due to construction dewatering, or will it be a long-term problem that will require long-term solutions such as a series of injection wells to prevent salt water intrusion?

The urgency of the anticipated problem will be a function of the potential contaminant(s), local hydrogeological characteristics, and the criticality of the threatened aquifer. Anticipated problems usually will have no urgent need for a solution. Ground water pollution potential can be minimized by incorporating certain controls in the design stage.

Existing Problems

With existing pollution problems the information needed for characterization is straightforward. Initial attention should be given to source identification. The contaminant release characteristics and duration of the problem can be characterized as outlined above. The urgency or need for an immediate solution can sometimes be the critical factor with an existing problem. If the polluted aquifer is a source of drinking water, or the problem was discovered due to an adverse public health reaction, there may be a need for an immediate (if only temporary) solution.

A general description of the problem is necessary for defining the scope and extent of further studies and ultimate remedial actions. For example, if a hazardous substance is detected in a water supply well, the immediate solution may be to provide an alternative supply of water. This temporary measure may allow time for study and development of a permanent solution.

Preliminary Study

After identifying an existing or proposed ground water threatening activity or facility, the next effort should be toward further information gathering and more detailed problem characterization. The detail and duration of the preliminary study will be determined by the urgency of the problem and funds available.

The information needs listed below are intended to be comprehensive even though they are described in general terms. Not all information needs will pertain to every problem. Probably the most useful function of these lists will be to aid in developing a list of needed information for specific problems. The following information needs lists are not in any particular order, although some of the needs are dependent on others. An important issue that needs to be considered prior to initiation of any detailed study would be that of funding. Once funding possibilities have been determined, the scope of the analysis can be more appropriately delineated. The other areas of information could probably be approached in groups such as problem-specific information (plume delineation and hydrogeologic characteristics), site-specific information (site characterization, water use and requirements, and land use patterns and growth projections), and others (human health costs and risk assessment, and regulations and institutional constraints). Should it be decided to go into more than just a cursory review of the problem, sources of information may become a question. Some of the agencies identified in considering regulations and institutional constraints most probably will have information on other issues. Listed in Table 5.2 are some other possible sources of information as related to the different groups.

Plume Delineation

Plume delineation is the step in which the amount, nature, and extent of the plume and the source of the ground water pollution are characterized. The information obtained in this step will not only help determine feasible aquifer restoration strategies but may also aid in assessing the

Table 5.2 Potential Sources of Information

Group I (problem-specific)	Federal or state geological surveys, university libraries, geology and engineering departments, state health departments, property owner, county records, well drillers.
Group II (site-specific)	Weather bureaus, state water resources boards, census bureaus, soil and water conservation districts, employment commissions, corporation commissions, 208 studies, Department of Agriculture, Forest Service.
Group III (other)	Medical libraries, state or federal environmental protection agencies, state attorney general's office.

possibilities of pretreatment or economic recovery. Information categories of interest are delineated as follows:

(1) Physical/Chemical Characterization of Pollutant(s)—A complete physical/chemical characterization of the pollutant(s) is essential. The physical/chemical characterization will determine which surface treatment technology will be required should the pollutant(s) be removed. Some pollutants can be contained by subsurface impermeable barriers, such as slurry walls, while others (those with high ionic strengths) have been shown to actually reduce the impermeability of such structures. A physical/chemical characterization of the pollutant(s) can also give information as to the possibilities for economic recovery, i.e., will the pollutant(s) recovered be salvageable or will it require treatment and disposal? In those cases where a pollutant(s) is anticipated but not yet detected in ground water, a physical/chemical characterization should be done on the known or suspected source of pollution. A physical/chemical characterization of a waste can give an idea as to what pollutants to expect.

(2) Information on Transport and Fate of Pollutant(s)—Having characterized the pollutant(s) it is then necessary to obtain any and all information on the transport and fate of the pollutant(s) in the subsurface environment. Information on the attenuating capacity of the given soils for the pollutant(s) is important. The behavior of the pollutant(s) under the different pH environments of the subsurface is important in that some pollutants can precipitate or solubilize given the correct conditions. The final outcome of this step should be information on the ability of the particular pollutant(s) to actually migrate through the soil structure and reach the ground water. Feasible mitigation measures will be a function of this ability.

(3) Toxicity and Health Risks of Pollutant(s)—Information on the toxicity and health risks of the pollutant(s) relative to not only humans but livestock and vegetation should be obtained. The ultimate use of the polluted ground water will be a function of its toxicity and health risks. For example, an aquifer that contains high concentrations of heavy metals will have to be abandoned from further use, while one that has elevated nitrate concentrations might be usable for irrigation with minimal treatment. Specifically, information should be obtained on toxic levels of the pollutant(s) perhaps in the form of a standard. It is also desirable to obtain information on the actual side effects of toxic levels of the pollutant(s), i.e., does it cause temporary acute discomfort; is it a potential carcinogen; are the effects long term or life threatening?

(4) Areal Extent, Depth, Amount of Pollutant(s)—If possible, an estimate of the areal extent, depth, and amount of pollutant(s) in the subsurface environment needs to be obtained. The objective is to obtain an estimate of the magnitude of the problem. The "size" of the problem to be rectified, in itself, may limit the number of feasible solutions. Ideally, data from existing monitoring wells, well logs, etc., should be used to estimate the magnitude of the problem. However, this information is often not available. In most cases an "educated guess" of the magnitude of the problem will have to be made by considering broad data such as soil types, typography, climate, and duration of the problem.

An important set of information pertinent to plume delineation is the identification and characterization of the source of pollution. The source will not always be readily identifiable and might require extensive field surveys to be determined. Once the source has been identified, information specific to it should be obtained. These information categories are as follows:

(1) Physical/Chemical Characterization of Source—As mentioned previously, a complete physical/chemical characterization of a waste might be needed in order to predict potential pollutants. However, this information can also be used for reducing the problem at the source. A physical/chemical characterization can aid in assessing the possibilities for waste pretreatment, in-place stabilization of the waste, recycling, or some other form of economic recovery.

(2) Variability of the Wastes—The variability of the wastes must be considered in order that a pollutant-specific mitigation strategy is not implemented for a highly variable waste. A highly variable waste source would include a landfill that accepted both hazardous and non-hazardous constituents. Other highly variable waste sources, which are quite common today, are those situations in which records of exactly what was disposed do not exist. When the materials which have been disposed cannot be identified, it is difficult to design anything but a general, all-inclusive type of cleanup measure.

(3) Time Factors—Information on the time of existence of the waste source is also needed. Data on the period a waste source has been in existence can give insight to the magnitude of the problem. Specifically, it is desirable to know how long a given waste has been in-place, or how long a certain activity has been operating.

(4) Previous Waste Disposal Practices—It is important to know historical information on the practice of waste disposal at the source. Was the waste indiscriminately dumped and covered, or were there certain precautions taken? The feasible solutions to the problem of leachates from an abandoned open dump may differ from those of a sanitary landfill with a leaking liner. Specifically, information is needed on the exact disposal practices, the length of these operations, any special ground water protection strategies employed, and any or all daily records.

Hydrogeologic Characteristics

Assemblage of information on the hydrogeologic characteristics of a site is essential for a successful aquifer restoration program. In essence, this step involves characterization of the subsurface where the problem exists. Subsurface characterization serves two main purposes. First, a description of the hydrogeologic characteristics of the site will allow for a better understanding of the magnitude of the problem. A thorough hydrogeologic investigation will not totally delineate the limits of the problem, but it will aid in estimating the transport and fate of the pollutant in the subsurface. Second, a thorough hydrogeologic investigation will aid in identifying and designing potential aquifer restoration strategies. Information from this step will also be valuable in the plume delineation step. Some areas of needed information and reasons for their interest are as follows:

(1) Geologic Setting and Generalized Soil Profiles—Determination of the types of soils is important for determining both the capacity of the pollutant(s) to move through the subsurface and the feasible restoration measures. Certain soils will possess higher tendencies to attenuate the pollutant(s) (through adsorption, precipitation, filtration, etc.) than others. Likewise, certain soils will be amenable to certain restoration strategies while others are not.

(2) Soil Physical/Chemical Characteristics—Once a general soil type has been identified, it is then necessary to characterize this type both physically and chemically. Physical characterization of the soil type will provide information on the ability of the soil to filter the pollutant(s). The physical characterization will also give an idea as to the "workability" of the soil for different cleanup measures. Chemical characterization will provide information on the ability of the soil to chemically remove a given pollutant through adsorption, precipitation, etc. In some cases, the chemical composition of the soil will be needed to determine the feasibility of a particular cleanup strategy, especially in situ technologies.

(3) Depth to Ground Water and Bedrock—The depth to ground water in conjunction with the soil physical/chemical characterization will give insight as to how long it will take a pollutant(s) to actually reach the aquifer, if in fact it will. If sufficient depth to the ground water exists in a highly attenuating soil, minimal ground water pollution can be expected. The depth to bedrock is needed to assess the feasibility of some pollutant containment strategies such as slurry walls, grout cutoffs, or sheet piles.

(4) Ground Water Flow Patterns and Volumes—The flow patterns and volume of ground water threatened will play a vital role in determining the feasible solutions to the problem. Obviously the direction of flow will dictate the actual physical placement of any proposed cleanup measures. Similarly the volume of water affected will dictate the scope of potential cleanup measures.

(5) Recharge Areas and Rates—Identifying recharge areas and rates will play an important role in aquifer protection plans. The use of institutional measures (such as zoning) to minimize ground water threatening activities in aquifer recharge areas is one such plan. Information on recharge areas and rates for the cause of an existing problem will be important for a number of reasons. First, it should be determined whether or not the source(s) is in a recharge area. Solutions to problems located in recharge areas will most likely be more elaborate than those not located in recharge areas, and should include source removal if possible. Second, recharge rates will give insight into the rate of pollutant(s) movement and pollutant(s) dilution.

(6) Aquifer Characteristics—Identification of aquifer characteristics will be essential for any analysis of ground water flow and pollutant transport. This information becomes extremely important if ground water modeling studies are to be initiated. Listed in Table 5.3 are some of the characteristics that need to be determined, either through the use of extant data or the conduction of field studies.

(7) Existing Monitoring Well Locations and Procedures—Identification of existing monitoring well locations and the parameters monitored can save both study time and costs. In addition to providing immediate data, existing monitoring wells can, in some cases, become permanent parts of a monitoring network or converted to removal wells.

(8) Background Water Quality Data—Background water quality data are important in determining the severity of the problem and the appropriate remedial actions.

Table 5.3   Aquifer Characteristics

confined-unconfined	hydraulic conductivity
isotropic-anisotropic	dispersion coefficients
homogeneous-nonhomogeneous	specific yield (storage coefficient)

Site Characterization

A description of the general characteristics of the site of the problem is also an important step. Because surface attributes of the site will directly and indirectly affect the subsurface environment, these attributes need to be identified. Information is needed on the climatic factors of precipitation, temperature, and evapotranspiration. Locational factors for which information is needed include topography, accessibility, site size, proximity to surface water, and proximity to population centers. Additional comments on each of these factors are as follows:

(1) Precipitation—Precipitation in most cases will dictate both the amount and rate at which pollutants from a site "leach" to the ground water. This is especially true for pollutants leaching from solid waste disposal areas. Precipitation also affects the recharge rate of an aquifer.

(2) Temperature—Surface temperature will become an important factor in determining the feasibility of certain surface treatment strategies. For example, the transfer of organics from ground water in an air stripping operation will be affected by temperature.

(3) Evapotranspiration—The amount of water lost to the atmosphere through transpiration and evaporation can be important when considering the ultimate use of a site. For example, surface capping of a landfill relies in part on evapotranspiration from the vegetative cover to reduce the amount of water infiltrating to the solid waste.

(4) Topography—The general topography of the site will affect the infiltration rate and feasible solutions. Areas of steep gradients will have little infiltration and will be subject to high erosion rates. This could be important when designing surface disposal sites.

(5) Accessibility—Related to topography is the accessibility of the site. Areas of rugged terrain or limited access will present not only construction problems but could also hamper any subsequent operation and maintenance activities.

(6) Site Size—The size of the site refers to the actual surface areal extent of the problem. This factor will affect the feasible solutions in that many of the technologies are size specific. For example, soil removal is probably not economically feasible for a multiacre contaminated site.

(7) Proximity to Surface Water—Locations of surface water relative to the site are important for a number of reasons. First, the solution to the ground water problem should not create a surface water problem. In addition, schemes that involve surface treatment may require surface discharge. A location for this

discharge needs to be determined. Second, water may play an important role in the construction (slurry walls) and/or operation and maintenance (revegetation) of a site. A source of water may be vital to the success of a given restoration scheme.

(8) Proximity to Population Centers — The location of the nearest population center relative to the site will help determine both the magnitude of the problem and the potential solutions. If the site is located near a large population center supplied by a well field, the problem may be more urgent than one located in a relatively isolated area.

Water Use and Requirements

Determination of current ground water usage and future requirements in the study area will aid in determining the criticality of the threatened or polluted ground water resource. Categories of information needed and the reasons for their importance are as follows:

(1) Current Use — Information on the amount of water used for domestic, agricultural and industrial purposes should be obtained. Furthermore, this information should be divided into those uses supplied by surface water and ground water. Specifically, information on the amount of water withdrawn from the aquifer in question and what it is used for should be obtained. This information could become critical should abandonment of the aquifer be considered.

(2) Future Use — An estimate of predicted future water requirements for domestic, agricultural and industrial purposes should also be obtained. Those areas anticipating large-scale growth accompanied by high water demands may need to clean up existing ground water problems and institute aquifer protection measures (institutional measures).

(3) Current/Future Water Quality Standards — Beneficial Use Standards — Procurement of currently available or proposed future beneficial use standards for local ground water could play an important role in determining a feasible aquifer restoration strategy. More specifically, beneficial use standards may dictate the degree of treatment polluted ground water must receive before subsequent use.

(4) Costs — An estimate of the current costs of supplying water is needed, especially the costs of obtaining ground water from the aquifer under study. The costs associated with changing sources of supply should also be estimated. If changing to a different source of water supply becomes prohibitively expensive, it may be necessary to consider some more elaborate treatment schemes for restoring the aquifer.

Human Health Costs and Risk Assessment

An assessment of risks and human health costs associated with the polluted ground water is desirable. Having identified the toxicity and health hazards of the given pollutant during the plume delineation step, an assessment needs to be made of the actual potential for human health impacts. This step is essential in evaluating the alternative of taking no action on aquifer restoration. Examples of needed information are as follows:

(1) Potential Health Problems—From the plume delineation step there should be information on the health problems associated with the given pollutant(s). Specifically, information is needed on whether the pollutant is a toxicant, carcinogen, or threat to the general health of the population.

(2) Risk Assessment—Assessing the risks of illness is desirable. Obviously, if a certain water is in violation of a given health or water supply standard the risk is high. The question of interest here is what are the risks of using a potentially polluted yet acceptable (in accordance with the standards or lack thereof) water supply? Ideally, it would be desirable to have quantitative information on the number and frequency of health incidents to be expected. A more feasible solution would be to assign a subjective probability of occurrence for different levels of contamination.

Land Use Patterns and Growth Projections

The objective of this information gathering step is to insure that potential restoration strategies do not become exercises in futility. Specifically, it would be unwise to spend millions of dollars cleaning up a particular problem that is just one of many currently or potentially polluting a given aquifer. Useful information in this regard includes current and future land use patterns and activity growth in the study area. Examples of needed information are as follows:

a. Current Land Use Patterns—A determination of the existing land use patterns will give considerable insight into the type and degree of treatment a particular problem should be given. The percentage of land devoted to urban, rural and industrial uses in and around the site should be determined. A removal and treatment system in a highly industrialized area could lead to one group removing, treating and paying for pollution it did not cause.

b. Growth Projections—A projection of growth in activities that possess the potential for degrading ground water is essential. If a certain ground water formation is already degraded in quality, and the projections for future growth include increased activity in the area, then containment rather than cleanup may be desirable. Listed in Table 5.4 are some of the more immediate ground water threatening activities which should be explored.

Regulations and Institutional Constraints

A key step in any aquifer restoration strategy is identification of institutional constraints and enforcement programs. A variety of regulatory agencies for ground water quality control may exist within a given state,

Table 5.4   Examples of Ground Water Threatening Activities

Industrial landfills and impoundments

Municipal landfills

Septic tanks

Mining, minerals extraction, petroleum production

Agricultural chemicals

and determining who is in charge may be somewhat of a problem. Identification of all constraints and enforcement agencies is necessary so that a proposed solution satisfactory to one agency does not violate the requirements of another agency. Examples of necessary information in this regard are as follows:

(1) Regulatory Agencies—As early as possible the regulatory agencies in charge of ground water quality should be identified. Areas of interest pertaining to ground water would include monitoring of ground water, waste disposal and construction permitting, and enforcement of standards.

(2) Standards and Guidelines—All pertinent water quality and effluent standards should be obtained. In addition, some states might have standards or guidelines for waste disposal facilities, minimum construction standards, siting of specific facilities, or required environmental studies.

(3) Legislation—There has been a good deal of recent environmental legislation which could apply to a wide variety of ground water pollution problems. Thorough knowledge of pertinent legislation will not only provide goals and guidelines for mitigation measures, but may also provide insight as to possible sources of funding. For example, consideration should be given to the requirements of the Resource Conservation and Recovery Act (RCRA), the Surface Mining Control and Reclamation Act (SMCRA), the "Superfund" program (Comprehensive Environmental Response, Compensation and Liability Act—CERCLA) for hazardous waste disposal facilities, and the Safe Drinking Water Act for "sole source" aquifers. In addition, review of current state legislation would be warranted.

Funding

Probably the biggest obstacle to be overcome in any aquifer restoration program will be funding. In simple terms, cleaning up ground water can be a very expensive venture. Not only are the technologies expensive to implement and operate, but they may be difficult to design. If questions arise over who will pay for the cleanup (as they often do), the pollution problem may end up in litigation, thus adding to the expense and delaying the implementation of aquifer restoration strategies. These delays exacerbate the problem and escalate the costs. A preliminary review of the availability of funds should most probably be done immediately after identification of the problem. Based on the availability of funds, the scope of the study can be determined. If sufficient funds are not available, detailed analysis and design will probably have to be foregone for a small-scale aquifer management or policy analysis.

One of the key funding issues is related to responsibility for cleanup costs. The pattern to date has been for the suspected polluters to balk at accepting responsibility for a given case of ground water pollution. The case then goes to litigation and the suspected polluter, if found guilty, will often file bankruptcy rather than accept the immense financial burden associated with ground water pollution cleanup. This leaves the state or federal government with the satisfaction of having identified the polluter, but lacking in terms of a responsible party for cleaning up. This type of arrangement is not the most desirable for both sides and should be avoided if possible. The fact that legal recourse can be taken should be recognized, but an effort towards joint cooperation is more desirable.

Data Evaluation

A critical aspect of the information gathering step will be evaluating the quality of the information gathered. Kaschak and Nadeau (1982) list three issues to be concerned with once available information has been gathered as: (1) how good are the data today; (2) can an engineering solution be properly developed and designed; and (3) will the data be defensible in court? Consideration of the age of the information, sampling and analysis protocols, and the chain of custody of the information can aid in assessing whether or not the information is accurate and/or useful.

Data Needs

Having completed the gathering and evaluation of available information, the next step is to identify those areas where the search for information was unsuccessful. Kaschak and Nadeau (1982) identify this as the most difficult step, i.e., to decide what additional information is necessary to be able to identify and evaluate alternative restoration strategies without "studying the site to death." Information voids should be identified and categorized. One categorization scheme is as follows:

(1) Criticality—The relative importance needs to be considered for each of the individual data gaps. It would probably not be justified to put as much effort into obtaining "regional population growth patterns" as that of obtaining "toxicity" information for the pollutants of concern. Some type of priority ranking should be assigned to the missing information. Chapter 6 describes several techniques which could aid in this ranking.

(2) Availability—For each of the information gaps consideration should be given to whether or not the information is obtainable. For example, records of disposal at a landfill are desirable but are not always kept. If these records are not available, there is no way of obtaining them. Conversely, background water quality data may not be presently available but are certainly obtainable through the drilling of monitoring wells. In those cases where the information is not obtainable, it could be excluded from the decision-making process or a best estimate could be developed and utilized.

(3) Time—Certain information may be obtainable, but only after extended periods of time. The background water quality data mentioned previously are certainly obtainable, but monitoring should occur over a period of time. Conversely, certain aquifer characteristics can be obtained in just a few days. If the problem being studied is of an emergency nature, then data that require long periods of time to gather should be considered nonobtainable.

(4) Costs—For all the information voids, consideration must be given to the cost of searching for and/or acquiring that information. Obviously, an estimate of aquifer characteristics from soils and geological maps will be less expensive than extensive monitoring networks.

DEVELOPMENT OF ALTERNATIVES

The development of alternatives stage of the aquifer restoration decision-making process shown in Figure 5.5 can best be described as an iterative process. The goals for addressing the established need should be delineated, and potential technological approaches for meeting the goals/need should be identified. These potential approaches (strategies or alternatives) should be screened and scope designs developed for potentially feasible alternatives. This proces may require several iterations as additional information is developed.

Definition of Goals

The term "restoration" seems to imply that any ground water remediation activity will attempt to return the aquifer to its original condition. Although desirable, this is not always feasible. Therefore, a subjective decision to be made in any aquifer restoration program is related to the

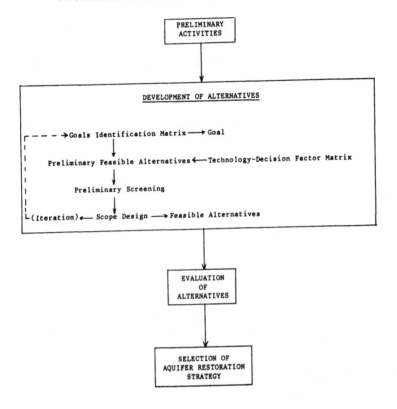

Figure 5.5. Flowchart for development of alternatives.

"goal" of the program. Specific strategies to achieve the goal(s) will be dictated, for the most part, by physical, chemical and monetary constraints.

This decision-making process includes four different goals: prevention, abatement, cleanup, and restoration. Prevention, as the name implies, means that pollution is not allowed to occur. The context of "prevention" herein will be taken to mean "not allowing pollutants to reach ground water." Abatement means "putting an end to." Hence, abatement of ground water pollution will include the "cessation of pollutants moving into the ground water, and the curtailment of the movement of pollutants having already reached the ground water." Cleanup refers to "elimination of pollutants through removal and treatment or in situ immobilization or treatment." Restoration will include those measures that attempt to return the aquifer to its original state. This most often will involve a cleanup strategy plus recharge of treated or fresh water. It should be noted that these goals are not totally independent of each other. More specifically, a truly effective "cleanup" strategy may also include "prevention" and "abatement" steps. Table 5.5 contains a listing of various strategies (detailed in Chapters 2, 3, and 4) available for obtaining various goals.

The definition of the goals of a ground water management strategy does not represent a totally isolated evaluation. In some instances, the goals may be limited by problem-specific conditions. Thorough review of the results of the preliminary activities should be undertaken prior to establishment of goals.

Goals Identification Matrix

Table 5.6 is a goals identification matrix. The possible goals are listed along with some decision factors and conditions under which the goals might be appropriate. Review of the limiting factors for each of the possible goals will enable a more rapid identification of feasible and obtainable goals for an identified need. Not all possible influencing fac-

Table 5.5   Potential Aquifer Restoration Strategies for Chronic Pollution Problems

Goal	Strategy Type	Action
Prevention	Institutional Measures	1. Aquifer or effluent standards, effluent charges or credits, land zoning.
	Source Control	1. Source reduction or removal
		2. Optimum site selection
		3. Man-made control options
		a. Impermeable membranes
		b. Impermeable materials
		c. Surface capping
		d. Collector drains
		e. Interceptor trenches
Abatement	Waste Management	1. Modification of pumping
		2. Removal Wells
		3. Pressure ridges
		4. Subsurface barriers
Cleanup	Waste Treatment and Disposal	1. Aboveground treatment
		2. In situ methods
Restoration	Cleanup Plus Recharge	

Table 5.6 Goals Identification Matrix

Goals	Temporal Situation	Funding Costs	Water Use Requirements	Hydrogeological Characteristics
Restoration	Restoration is a possible goal when considering existing problems.	Restoration schemes are the most expensive management strategies. However, restoration will also have the added benefits of returning the aquifer to a beneficial use.	Restoration should be the primary goal when dealing with ground water that is or will be the main source of water supply for a given area.	Restoration is most feasible for shallow, unconfined aquifers of originally high-quality water.
Cleanup	Cleanup is a goal when considering existing problems.	Cleanup strategies are expensive. Sometimes the aquifer can be returned to a beneficial use and there is also the possibility for economic recovery from certain pollutants.	Cleanup should be considered when the ground water serves a beneficial use other than human consumption such as agricultural or industrial purposes.	Cleanup measures are most feasible for shallow ground water sources, but do have some applicability to deeper, confined sources by removal and treatment.
Abatement	Abatement can be considered when dealing with existing problems and can also be included as a safeguard for future facilities or anticipated problems.	Abatement strategies range from moderate to very expensive with little chance for economic recovery.	Abatement measures are usually employed to save the unpolluted portions of an aquifer while conceding the polluted portion as lost.	Abatement measures are most applicable to shallow sources with cessation being the most applicable abatement measure for deep resources.
Prevention	Prevention measures are only applicable to future facilities or anticipated problems.	Prevention strategies range from low to moderately expensive with little chance for economic recovery.	Prevention measures should be applied to protect any currently used or potentially useable ground water resource.	No restrictions.

Table 5.6   continued

Goals	Areal Extent	Human Health Risks	Land Use Patterns
Restoration	As the size of the problem increases the attractiveness of the restoration schemes decreases due to increased costs. Problems of regional nature most probably will be amenable to abatement measures.	If human health risks are high, restoration and cleanup should be given top priority if technically feasible. If not technically feasible, abandonment and development of new source should be considered.	Restoration and cleanup measures should not be given high priority in areas that are or will continue to be subjected to ground water quality stresses such as industrial waste disposal areas.
Cleanup	Cleanup measures are not economically attractive for small, localized problems and do not have economies of scale for larger problems. However, if the problem becomes extremely large, such as regional, these economies of scale disappear.	See above.	See above.
Abatement	The abatement measures to be employed are size specific and cover the whole range from very small, localized problems to large, regional problems.	For low to moderate health risks abatement measures can be employed to prevent further degradation of an aquifer.	Abatement measures are the most attractive alternatives for ground water pollution control in areas subjected to continued threats if anything at all is to be done.
Prevention	Physical prevention measures are size specific. Some institutional measures are applicable on a regional level.	No restrictions.	No restrictions.

tors have been listed in Table 5.6. Each identified need will have its own influencing factors which should be considered in the selection of the specific goal(s).

Technology-Decision Factor Matrix

Having identified a preliminary goal for an identified need, the technology-decision factor matrix in Table 5.7 can be utilized. The technology (technologies) that can be considered for achieving various goals are identified in the first two columns. After identifying the feasible technology (technologies), the decision factors and conditions under which the technology might be applied can be reviewed from the remaining columns in Table 5.7. The end result should be a series of technologies or technology combinations that are potentially applicable for meeting an identified need. Table 5.7 is not totally comprehensive since there are many additional decision factors associated with each of the listed technologies. However, Table 5.7 does have general utility in that major decision factors are outlined. Additionally, Table 5.7 should trigger information exchange among members of the multidisciplinary team. The team could develop an expanded version of the technology-decision factor matrix specific to the identified need under examination.

Preliminary Screening

Having developing a preliminary list of feasible technologies (alternatives), a preliminary screening process can be used to narrow the choices. The preliminary screening process will necessarily rely heavily on professional judgment; however, listed below are some factors that should be considered:

(1) Technical Feasibility—The technical feasibility of all potential alternatives should be considered. Some technologies have been widely used and their success documented; others are still in the early stages of development, or at least their application to ground water pollution control has been limited. Another consideration would be the technical capabilities of the personnel that are to design and operate the proposed technology. Some technologies are complex in design and require extensive monitoring for successful operation and maintenance. If trained personnel will not be available to handle these responsibilities, perhaps a more passive system should be considered.

(2) Public Acceptance—One important aspect of any aquifer restoration scheme will be its acceptance by the public. If the proposed solution will involve surface "eyesores" or highly technical operational requirements, it might be difficult to gain the acceptance of the general public.

Table 5.7 Technology-Decision Factor Matrix

Technology	Management Strategy and Complementary Technologies	Temporal Situation	Pollutant Type	Areal Extent
Institutional Measures	Implementation of institutional measures will most probably be for the prevention of ground water pollution but could also be used as abatement measures.	Applicability to existing problems is probably minimal except for orders to cease a given activity. Widest application is for preservation of current high-quality ground water.		
Source Control Strategies	Source control measures are most applicable as preventive techniques, however, they are often incorporated as part of overall cleanup measures at uncontrolled sites.	Source control strategies are applicable to existing problems and should be included in the design of all future disposal facilities.	No restrictions.	No restriction.
Well Systems	Well systems are applicable to all management strategies including restoration. Well systems in all cases require complementary technologies, usually surface treatment and discharge.	Well systems are applicable to both existing and anticipated problems and have been used to control emergency situations in certain cases.	Well systems are applicable to any pollutant that flows with or on top of the ground water.	Well systems are probably not economically feasible for small, localized pollution sources. Conversely, for large sources of pollution well systems may become uneconomical in that they by necessity draw in both polluted and nonpolluted water.
Interceptor Systems	Interceptor systems can be employed for cleanup, abatement or prevention of ground water pollution. However, in all cases a complementary technology is required, usually in the form of surface treatment and discharge.	When dealing with existing problems, interceptor systems will take the form of interceptor trenches, while anticipated problems are usually prevented by employing collector drain systems. There is some applicability of interceptor trenches to emergency spills involving floating hydrocarbons.	Interceptor systems are applicable to any pollutant that flows with or on top of the ground water.	Interceptor systems are feasible for small, localized pollution sources.

Surface Water Control	Surface water control technologies are most often used as preventive measures but can also be used as part of abatement schemes.	Surface water control can be applied to existing problems as part of an overall abatement scheme and should be considered in the design of all future land disposal facilities.	Surface water control is applicable to most any pollutant, however, the possibility of gas production should be considered when designing "surface capping" measures for organic wastes.	No restrictions.
Impermeable Liners	Liners are usually placed on a site prior to the initiation of a given activity, hence they are most applicable as prevention steps. Liners themselves are not totally effective and are usually used in conjunction with surface water control, source control, or underdrain systems.	The applicability of liners to existing problems is limited in that is usually involves removal of the waste or polluted soil.	Some asphaltic and treated soil liners become more permeable upon exposure to leachates. Some synthetic liners show poor resistance to hydrocarbons.	Because installation of liners involves excavation and specialized construction, economies of scale occur for larger problems. However, if the size of the problem is more than 5 to 10 acres, liners may become uneconomical due to material costs.
Sheet-piling	Sheet-piling follows the attributes of grouting and slurry systems in that its most practical applications will be for abatement and prevention of ground water pollution. However, the number of cases where sheet-piling will be the best technology will be minimal.	Sheet-piling is similar to grouting. Probably the most applicable situation will be as emergency measures for containment of spills.	Sheet-piles are generally corrosion resistant.	As the size of a project increases, sheet piling will become uneconomical because of high material and shipping costs.
Grouting	Grouting is similar to slurry systems in that it will often require complementary technologies to accomplish restoration or cleanup. Grouting can be individually applicable as an abatement measure.	Grouting is also a versatile technology in that it has wide application to both existing and anticipated pollution problems. In addition grouting is being promoted as a procedure for containing emergency spills.	Grouts containing bentonite clay are subject to limitations similar to those for slurry systems. Additionally, chemical grouts are not successful in highly acidic or alkaline environments.	Costs for grout cutoff systems are high, hence they will be applicable to small localized cases of pollution.

Table 5.7, continued

Technology	Management Strategy and Complementary Technologies	Temporal Situation	Pollutant Type	Areal Extent
Slurry Walls	Slurry walls can only be parts of overall restoration or cleanup schemes. Slurry walls are individually applicable as abatement measures, however, some sort of ground water removal system is usually included. Slurry walls are also individually applicable as prevention measures, but once again are usually used in conjunction with other measures.	Slurry walls are very versatile in that they can be applied to existing problems as abatement measures or they can be incorporated in the design of future facilities for prevention of ground water pollution. Slurry walls are probably not applicable to emergency situations such as spills.	Slurry systems are sensitive to specific pollutants, especially those of high ionic strength.	Slurry systems are more applicable to larger problems. Slurry systems become more economical as the "length to depth" ratio increases.
Surface Treatment	Surface treatment technologies can be used in systems designed to prevent, abate, cleanup or restore polluted aquifers. Treatment technologies will always require some sort of pollutant removal system.	Surface treatment can be used for acute and chronic cases of ground water pollution. Treatment can be used at existing sites and at anticipated problems in a preventive mode.	No restrictions.	Surface treatment has both upper and lower size limitations. Surface treatment can be too elaborate for small sources yet uneconomical for extremely large areas.
In Situ Treatment	In situ treatment technologies can be used in systems designed to abate, cleanup or restore polluted aquifers. In-situ technologies will always require some sort of nutrient/chemical injection system.	In-situ technologies can be applied to both acute and chronic cases of existing pollution. The application of in-situ technologies for treatment of anticipated problems is not possible.	In-situ treatment is extremely pollutant specific. Biological methods are applicable mainly to organics and chemical methods are applicable mainly to inorganic compounds.	Same as for surface treatment.

Table 5.7, continued

Technology	Soils	Aquifer Characteristics	Site Characteristics	Regulations
Institutional Measures				
Source Control Strategies	No restrictions.	No restrictions.	Since most source control strategies will involve surface activities, the site should be readily accessible.	
Well Systems	Well systems require soils that readily transmit ground water and any associated pollutants.	Well systems are amenable to most aquifer types; however, they may not be ideal for extremely shallow aquifers (<5m).	Sites must be accessible to drilling equipment. Well systems usually will require monitoring and a receiving source for their discharge. If the removed water requires treatment the site should be accessible to or in the proximity of surface treatment facilities.	
Interceptor Systems	Interceptor systems require fairly loose, pervious soils that readily transmit ground water and any associated pollutants.	Areas with high water tables or high impervious soils or rock layers are desirable.	Because the collected water will need to be removed, treated, and discharged, the site should be readily accessible.	
Surface Water Control	Surface water control measures are applicable to any soil. Surface capping becomes economical when there exist native clays readily available.	No restrictions.	Regrading to maximize runoff is an important aspect. Hence areas with extreme topographic relief should probably be considered less desirable.	
Impermeable Liners	Soils do not limit the application of liners, except when native clays are to be used as the liner. In this case the soils should be of sufficient impermeability and sufficient thickness.	Very high water tables may limit the use of liners in that they may cause the system to "float".	The site should be accessible to a variety of construction vehicles, including some specialized liner installation machines.	Some states now have regulations governing the in-place characteristics which a liner must possess.

Table 5.7, continued

Technology	Soils	Aquifer Characteristics	Site Characteristics	Regulations
Sheet-piling	Pile driving requires a relatively uniform, loose, boulder-free soil for ease of construction.	Sheet-piling will be limited to shallow (<100 ft.) aquifers.	Similar to those for slurry systems except the operation is not water intensive.	
Grouting	Grouting is limited to granular types of soils that have a pore size large enough to accept grout fluids under pressure yet small enough to prevent significant pollutant migration before implementation. Grouting in a highly layered soil profile may result in incomplete formation of grout envelope.	Presence of high water tables and rapidly flowing ground water limits grout ability due to dilution of grouts and rapid transport of contaminants.	Similar to those for slurry walls.	Some grouting procedures are proprietary and some chemicals may be controlled under the Underground Injection Program.
Slurry Walls	Slurry systems are applicable to most all soil types. However, in loose soils such as sand or gravel there can be significant losses of slurry which may make the system uneconomical.	Slurry systems are generally applicable to shallow (<150 ft.) aquifers.	Sites with general accessibility, moderate topographic relief and a ready source of water are most desirable. Construction of slurry systems is an equipment intensive operation which requires large amounts of water.	Some construction procedures are patented and will require a license.
Surface Treatment	No restrictions.	No restrictions.	On-site surface treatment will require extremely accessible sites that can adequately handle construction traffic and equipment. Once built, surface treatment facilities will only occupy a finite space, but will have to remain accessible for monitoring activities.	
In Situ Treatment	Most probably applicable only to permeable soils which will readily accept injected materials.	Shallow aquifers are probably more desirable.	Same as for well systems.	

(3) Physical Constraints—Some of the potential alternatives might be eliminated by obvious physical constraints. For example, placement of impermeable barriers might not be feasible in a densely populated area.

Scope Design

Probably the best means of eliminating possible but nonfeasible alternatives would be through the use of a scope design. The scope design is a small-scale analysis of the proposed alternatives for preliminary estimates of size, costs, life expectancy, efficiency, etc. The scope design will rely on estimates or rules of thumb figures for costs and other factors to generate an approximation for each of the potential alternatives. The scope design will be less expensive, less time consuming, and inherently less accurate than a complete engineering and hydrogeological analysis. The objective of the scope design should be to eliminate the obviously too expensive or too land-intensive alternatives.

Iteration

With preliminary size and cost data from the scope design, an iteration through the procedure outlined above would be appropriate. Perhaps the data generated might indicate that all the possible alternatives exceed the available funding. In this case, it might be necessary to redefine the goals of the ground water management strategy, and attempt to develop new possible technologies for achieving these modified goals. The objective of iteration is to further narrow the list of potential alternatives to those that are economically and technically feasible.

EVALUATION OF ALTERNATIVES AND SELECTION OF AQUIFER RESTORATION STRATEGY

Selection of a single alternative from the list of feasible alternatives should come from balanced consideration of technical, economic and environmental factors. A philosophy often quoted and generally accepted is that the "most inexpensive remedial action that reduces risks to an acceptable level can be considered the most cost-effective" (St. Clair, McCloskey and Sherman, 1982). This statement raises a number of issues, such as:

(1) What is an acceptable level?

(2) Acceptable to whom?

(3) Is the most cost-effective the optimum?

(4) Are there environmental implications in addition to economic and risk implications?

(5) Are cost and risk equally important?

(6) Are confidence levels the same for cost and risk calculations?

The structure of this section will be to discuss methodologies for and issues associated with evaluating alternatives based on their economic, environmental and risk implications. Chapter 6 presents some decision-making techniques for integrating these evaluations and developing an optimum strategy.

Economic Considerations

General statements concerning the costs of aquifer restoration are difficult to make. One widely accepted statement is that the cost of cleaning up a polluted aquifer is extremely expensive. This is true for many but not all cases. Expense can be considered relative and not all ground water remediation activities are designed with cleanup as the ultimate goal. Difficulties arise in any attempt to quantify or collate information on the costs of aquifer restoration. First, minimal detailed cost data is being or has been reported. This is due in part to the confidentiality and litigation associated with many ground water pollution problems. Additionally, the data that have been reported show an extremely wide range, are highly problem specific, and sometimes are estimates at best.

The fact that ground water clean-up costs exhibit a wide range can be exemplified by citing some examples. Kaufman (1982) has shown that removing TCE and other organic compounds from the water supply of a small New Jersey city can be accomplished for under $140,000/yr using granular activated carbon. Neely et al. (1981), on the other hand, have estimated the cost of cleaning up ground water at a site polluted by industrial solvents and acid sludge to be over $7 million. Hittman Associates (1981) have preliminarily assessed the cleanup of ground water polluted from abandoned lead and zinc mines in northeastern Oklahoma to be in excess of $20 million.

The cost of a ground water remediation activity is second in importance only to public health concerns. The technologies for cleaning polluted ground water exist, but may not be economical. In a recent study, Neely et al. (1981) found that cost was the prime determinant of the type of remedial technology employed at uncontrolled hazardous waste sites. Since elaborate cleanup schemes are expensive, companies may have opted for less expensive but environmentally more threatening prevention and abatement measures.

Any ground water cleanup activity will require funding for feasibility and engineering studies, construction, operation and maintenance, and other costs. Any one of these areas can become a major cost factor. For example, Roux Associates recently began a $1 million study just to develop a list of recommendations for remedial actions at Massachusetts toxic waste sites. In contrast, Kaufmann (1982) has shown operating costs to be the major cost component of a granular activated carbon cleanup system.

The generic assignment of costs to various aquifer restoration technologies (strategies) is difficult. Individual technologies can vary in cost depending on the type of pollutant, nature of the aquifer, the ultimate goal of the technology, and many other factors. A good deal of work has been done in assessing the costs of various components of individual technologies. For example, extensive information has become available on the cost of installation and operation of monitoring wells. Monitoring wells should be an integral part of any aquifer restoration technology. Minning (1982) provides a cost comparison for five different drilling methods. Rehtlane and Patton (1982) examine the economic tradeoffs of multiple port versus standpipe piezometers. Ward (1982) examines the costs of in situ leach wells for uranium extraction. As noted in Chapter 2, Campbell and Lehr (1977) give an excellent discussion of the costs of completing wells under a variety of geological conditions.

There have been several recent studies to assign unit costs, or at least comparative cost figures, for some of the physical measures for minimizing, abating or controlling ground water pollution. Rishel et al. (1983) have estimated costs for remedial response action at two types of uncontrolled and abandoned hazardous waste disposal sites. Table 5.8 is a listing of costs for a landfill site, and Table 5.9 summarizes costs for a surface impoundment site.

Tolman et al. (1978) developed a summary of estimated costs and characteristics of remedial methods for a hypothetical, single size (10-acre) landfill; the summary is shown in Table 5.10. Additionally, some qualitative comments can be made concerning some of the technologies listed in Table 5.10. First, surface water control technologies are usually only used as preventive and/or abatement measures when dealing with ground water pollution. Hence, these operations will require some complementary technology to clean up an aquifer. Of the four ground water control technologies listed in Table 5.10, a bentonite slurry trench must

Table 5.8   Average U.S. Low and High Costs of Unit Operations for Medium-Sized Landfill Sites (Rishel et al., 1983)

Unit Operations	Unit	Average Cost, $/Unit[a]				Total Units Used[b]
		Initial Capital		Life Cycle Costs		
		Low	High	Low	High	
Contour grading and surface water diversion	Site area, ha	15,300	17,900	16,300	19,900	5.4-ha site area
Bituminous concrete surface sealing	Site area, ha	67,300	92,700	67,300	92,700	5.4-ha site area
Revegetation	Site area, ha	3,450	16,500	14,300	18,100	5.4-ha site area
Bentonite slurry trench	Wall face area, m²	54.5	96.1	61.2	103	10,800-m² wall face area
Grout curtain	Wall face area, m²	600	1,209	937	1,880	10,800-m² wall face area
Sheet piling cutoff wall	Wall face area, m²	73	108	73	108	10,800-m² wall face area
Grout bottom sealing	Site area, ha	5,282,000	10,209,000	5,296,000	10,224,000	5.4-ha site area
Drains	Pipe length, m	72.7	106	357	416	260-m pipe length
Well point system	Intercept face area, m²	62.5	105	107	153	2,000-m² intercept face area
Deep well system	Intercept face area, m²	11.6	18.3	28.6	37.2	4,800-m² intercept face area
Injection	Intercept face area, m²	77	90	1,760	1,785	550-m² intercept face area
Leachate recirculation by subgrade irrigation	Site area, ha	5,270	8,360	19,700	24,000	5.4-ha site area

Technology	Basis parameter					Basis
Chemical fixation	Site area, ha	69,100	130,000	82,500	145,000	5.4-ha site area
Chemical injection	Landfill volume, m³	1.67	3.28	2.16	3.81	150,000-m³ landfill volume
Excavation and reburial	Landfill volume, m³	116	120	116	120	596,000-m³ landfill volume
Ponding	Site area, ha	647	1,028	647	1,028	5.4-ha site area
Trench construction	Trench length, m	12.2	14.34	15.11	20.32	930-m trench length
Perimeter gravel trench vents	Trench length, m	99.2	144	100	146	935-m trench length
Treatment of contaminated ground water	Contaminated water, L/d	1.52	2.57	2.52	4.38	440,740 L/d contaminated water
Gas migration control—passive	Site perimeter, m	161	241	168	256	935-m site perimeter
Gas migration control—active	Site perimeter, m	113	173	167	279	935-m site perimeter

[a] Mid-1980 dollars, 10-yr life cycle, O & M costs are discounted at 11.4% to present value, capital costs are not amortized.
[b] For 5.4-ha site.

Table 5.9 Average U.S. Low and High Costs of Unit Operations for Medium-Sized Surface Impoundment Sites (Rishel et al., 1983)

Unit Operations	Unit	Average Cost, $/Unit[a]				Total Units Used[b]
		Initial Capital		Life Cycle Costs		
		Low	High	Low	High	
Pond closure and contour grading of surface	Site area, ha	26,900	35,100	35,900	53,500	0.47-ha site area
Bituminous concrete surface	Site area, ha	48,500	70,700	48,500	70,700	0.47-ha site area
Revegetation	Site area, ha	2,540	3,820	3,970	5,450	0.47-ha site area
Slurry trench cutoff wall	Wall face area, m²	60.1	106	60.1	106	4,165-m² wall face area
Grout curtain	Wall face area, m²	326	631	343	649	4,104-m² wall face area
Sheet piling cutoff wall	Wall face area, m²	76.8	115	94.6	135	4,100-m² wall face area
Grout bottom seal	Site area, ha	868,000	1,621,000	1,024,000	1,792,000	0.47-ha site area
Toe and underdrains	Pipe length, m	316	609	1,550	1,960	60-m pipe length
Well point system	Intercept face area, m²	62.3	117	321	398	300-m² intercept face area
Deep well system	Intercept face area, m²	33.2	60.3	114.4	149	950-m² intercept face area
Well injection system	Intercept face area, m²	31.3	55.5	109	141	950-m² intercept face area
Leachate treatment	Contaminated water, L/d	1.16	1.96	4.49	8.14	51,870 L/d contaminated water
Berm reconstruction	Replaced berm, m³	2.98	3.80	4.00	5.85	410-m³ berm
Excavation and disposal at secure landfill	Impoundment volume, m³	260	268	260	268	5,000 impoundment volume

[a] Mid-1980 dollars, 10-yr life cycle, O & M costs are discounted at 11.4% to present value, capital costs are not amortized.
[b] For 0.47-ha impoundment.

Table 5.10   Summary of Estimated Costs and Characteristics of Remedial Methods (Tolman et al., 1978)

Method	Average Estimated Costs[a] ($ in Thousands)	Characteristics/Remarks
		Surface Water Control
Contour Grading	184	Increases runoff, reduces infiltration.
Surface Water Diversion	20	Diverts surface water from fill.
Surface Sealing Clay, 14 – 46 cm (6 – 18 in.)	234	If locally available, native clay is economical means of retarding infiltration.
Bituminous Concrete, 4 – 13 cm (1.5 – 5 in.)	315	Rapid coverage; can eliminate infiltration.
Fly Ash, 30 – 60 cm (12 – 24 in.)	235	Material may leach metals; may be available free.
PVC, 30 mil	482	Very impermeable; expensive seal; careful subgrade preparation is necessary.
Drainage Field (if required)	65	Carries infiltrated water off seal; increases effectiveness of seal.
Revegetation on Slopes <12%	10	Stabilizes cover material; seasonally increases transpiration; provides aesthetic benefit.
on Slopes >12%	19	
		Groundwater Control
Bentonite Slurry Trench	670	Simple construction methods; retards groundwater flow.
Grout Curtain	1,400	Very effective in permeable soils.
Sheet Piling	800	Widely used for shoring.
Bottom Sealing	4,000	Leachate collection may be needed, acts as a liner; difficult drilling through refuse.
		Plume Management[b]
Drains	23	Effective in lowering water table a few meters in unconsolidated materials; can be used to collect shallow leachate.
Well Point Dewatering	185	Suction lift limits depth to 7 – 9 m (20 – 30 ft); inexpensive installation; uses only one pump; can be used to collect shallow leachate.

Table 5.10 **continued**

Method	Average Estimated Costs[a] ($ in Thousands)	Characteristics/Remarks
Deep Well Dewatering	183	Used in lowering deep water tables; one pump needed per well; high maintenance costs.
Injection/Extraction Barrier	199	Creates a hydraulic barrier to stop leachate movement; operation and maintenance costs are high.
Spray Irrigation	366	Spreads leachate over the landfill for recycling; potential odor problem.
At-grade Irrigation	32	Grated pipe with ridge and furrow irrigation; potential odor problem; recycles leachate.
Subgrade Irrigation	28	Large-scale drainage field; recycles leachate.
		Chemical Immobilization
Chemical Fixation of Cover	145	Uses chemically fixed sludge to provide a top seal; provides means of disposal for sludge; helps stabilize landfill.
Chemical Injection	86	Immobilizes a single pollutant; in most cases not feasible.
		Excavation and Reburial
Excavation and Reburial	4,570	Very expensive; difficult construction.

[a]Costs for hypothetical 4-ha (10-acre) landfill. High and low estimates were averaged to determine these costs.
[b]Costs include present worth of 20 yr, operation, maintenance, and, where applicable, power for 4-ha (10-acre) landfill.

be considered the most cost-effective. Construction of grout curtains is slow, labor intensive and expensive, and as such this process is probably only economically feasible on small, localized cases of pollution. The plume management strategies listed in Table 5.10 all will incur long-term operating and maintenance costs. These costs are associated with manpower to oversee operations, power supplies for pumps, and possibly accelerated maintenance due to the nature of pollutants being pumped

(Glover, 1982). Also, the passive management systems (drains, interceptor trenches) generally have lower operating costs than the active (pumped) systems.

In summary, most of the work to date in the area of costs of aquifer restoration technologies has been in developing data for the components of the technologies. Minimal information has been reported in terms of costs of overall aquifer restoration strategies. Most studies will require the analysis of individual cost components as outlined in the following section.

Economic Evaluation

Economic evaluation of feasible alternatives can follow the traditional procedure of itemizing all the costs for a given project, amortizing these costs for the life of the project, developing measures of the benefits of the project, and comparing these data for each of the alternatives. A number of unique concerns arise when considering an economic evaluation of ground water contamination problems.

The first major concern is associated with the comparison of alternatives on a common basis. Two alternatives may be projected to achieve the same result through radically different methods. Relevant issues under this heading are as follows:

(1) The first issue is the level of treatment or efficiency of the system. One alternative might reduce the contaminant concentration to parts per billion, another to parts per million. Is the increased reduction in concentration needed and/or economically justifiable?

(2) Some alternatives may be operation- and maintenance-intensive while others, providing the same solution, might be less intensive. Comparison of these types of alternatives is difficult due to constantly changing interest rates.

(3) Alternatives providing similar solutions might have radically different time frames or life expectancies. All projects should be compared on a common project period. Bixler, Hanson and Langner (1982) have identified this period to be the lesser of: the period of potential exposure to the contaminated materials in the absence of remedial action; or 20 years.

The second major concern is the cost of specific items associated with alternatives. It is important that all cost items be identified. As an example of how extensive these lists can be, Table 5.11 summarizes the cost components for a well system (Lundy and Mahan, 1982). Bixler, Hanson and Langner (1982) note that cost items can be monetary and nonmonetary, direct or indirect. In addition to the obvious direct monetary costs such as materials, equipment, etc., nonmonetary costs must be included

Table 5.11   Cost Components of Well Systems (Lundy and Mahan, 1982)

Cost Component [a]	Cost Element	Principal Factor Affecting Cost
Plume Delineation (K)	Soil borings Monitor wells Data analysis Laboratory analysis Reporting	Plume area and depth Complexity of hydrogeology

System Design (K)	Consulting fees
	Computer time
		Complexity of
		  hydrogeplogy
		Size of containment
		  system

Well System (K)	Construction engineering Well construction Materials	Plume size Aquifer flux Transmissivity
Surface Infrastructure (K)	Access roads Power transmission Piping	Plume size Configuration of wells
Treatment Facility (K)	Construction engineering Treatment system	System discharge Composition/concentration of recovered water
Well System (O&M)	Pump operation (power) System maintenance	Aquifer flux/ pumping depth Fluid corrosivity
Treatment Facility (O&M)	Chemicals Labor Power	System discharge Composition/concentration of recovered water
Monitoring	Sampling Analysis Reporting	Complexity of system

[a]K denotes capital costs; O and M denote operation and maintenance costs.

such as relocation costs, loss of revenues, decreased property values and others. It is important that the list of cost items be complete and include all pertinent costs (Thorsen, 1981).

Another concern associated with the economic evaluation of alternatives is that of developing a measure of the benefits of an alternative. Some benefits will be very difficult to assign dollar values, for example, reduced health risks. Additionally, some benefits may be hard to quantify, much less assign dollar values.

The final and most important issue associated with economic analysis is the generic approach to be utilized. St. Clair, McCloskey and Sherman (1982) analyzed the advantages and disadvantages of various approaches, including risk assessment, cost-benefit analysis, cost-effectiveness analysis, trade-off matrices, and sensitivity analysis. Utilizing the positive attributes of these approaches, St. Clair, McCloskey and Sherman (1982) have developed a framework for evaluating cost-effectiveness of remedial actions at uncontrolled hazardous waste sites. Table 5.12 displays an example trade-off utilizing the developed framework.

In summary, there exists no ideal methodology for economic evaluations of aquifer restoration strategies. The best approach is to develop a modified version of an existing methodology that is applicable to the problem of concern. The methodology utilized should have the following basic attributes: (1) it compares alternatives on a common basis; (2) it possesses some measure of assessing benefits; (3) it is comprehensive and considers all relevant factors; and (4) its results are replicable and are not subject to bias.

Environmental Evaluation

Each of the proposed alternatives must also be analyzed for their potential impact on the environment. There are a variety of approaches that can be used to assess these impacts, including empirical assessment methodologies, interaction matrices, network analyses, checklists, and others (Canter, 1977). The purpose of this section is to identify the need for environmental impact assessment in aquifer restoration projects, and to identify some unique issues associated with it.

The first unique issue to be identified in association with ground water cleanup activities is that some alternatives may represent "An irretrievable commitment of natural resources," i.e., some alternatives may require that the aquifer remain permanently altered. This is especially true of technologies such as impermeable barriers (slurry walls, grouts). Secondly, because the unique aspects of ground water remediation activities usually lie underground, the aboveground consequences of the remedial action should be relatively easy to assess. More specifically, ground water remedial actions, for the most part, do not represent large, land intensive undertakings. Depending on the technology employed, there will exist a potential to affect several environmental categories. Examples include air quality, surface water quality, and noise. There should exist ample information on related activities so that the assessment of the

Table 5.12 Example Trade-Off Matrix for Cost-Effectiveness Assessment of Remedial Actions at Uncontrolled Hazardous Waste Sites (St. Clair, McCloskey, and Sherman, 1982)

Alternatives	Cost Ratings				Generic Effectiveness Measures								Site-Specific Effectiveness Measures				Σ Effectiveness Ratings	Σ Effectiveness Ratings / Σ Cost Ratings
	Construction	Operation & Maintenance	Other	Σ Cost Ratings	Level of Cleanup/ Isolation Achievable	Time to Achieve Cleanup/Isolation	Technology Status	Usability of Land After Action	Capability of Action to Minimize Community Impacts During Implementation	Capability of Action to Minimize Adverse Health & Environmental Impacts During Implementation	Risk of Failure	Impact of Failure	Ability to Minimize Impact on Adjacent Residents	Usability of Downstream Lake After Action				
Weighting Factors																		
Encompassing Slurry Wall with Cap																		
Downgradient Slurry Wall with Extraction Wells, Treat with Carbon																		
Extraction Wells, Treat with Carbon																		
Extraction Wells, Treat with Biox																		

aboveground impacts of ground water remedial actions is simple, straightforward and less time-consuming than the assessment of subsurface effects.

The last issue associated with environmental evaluation is the importance of public participation. Because of the recent attention afforded ground water pollution episodes by the media, the public has become ill-informed and frightened. It is desirable to avoid public hysteria in any project, and this is especially true for aquifer restoration projects. Because ground water is a "hidden resource," improving the public understanding may be especially difficult. Careful planning and increased attention should be applied to the role of public participation in any aquifer restoration studies. Freudenthal and Calender (1981) note that public participation programs promote conflict resolution by providing opportunities for individuals and opposing groups to explore compromise solutions. Detailed information on public participation in aquifer restoration decision-making is included in Chapter 8.

Risk Assessment

Cornaby et al. (1982) stated:

"The paramount concern is that no known general methodology is available for conducting overall environmental risk assessment, risk assessment that includes both humans and especially non-human or ecological receptors. In fact, the terminology in the literature is not always clear, suggesting that our views and knowledge of environmental risk assessment are still evolving."

This statement should serve as adequate notice that assessing the risks of aquifer restoration alternatives will not be easy. Berger (1982) states that risk assessment has been defined as, "The identification of hazards, the allocation of cause, the estimation of probability that harm will result, and the balancing of harm with benefit." Moreover, Berger (1982) states that selecting an effective remedial technique involves the balancing of the need to contain contaminants within acceptable levels against the costs associated with the cleanup measures. St. Clair, McCloskey and Sherman (1982) state that "a risk assessment involves the definition of the risks to the environment and human health of continued pollution from a site" and "the most inexpensive remedial action that reduces the risk to an acceptable level could be considered the most cost-effective." Dawson and Sanning (1982) note that the objective of a remedial action is to reduce associated risks to an acceptable level.

The most important question associated with risk assessment is — "how can the risks of aquifer restoration alternatives be assessed?" There are two generally accepted approaches to risk assessment. The first approach is to utilize criteria or standards for a pollutant and work backwards, utilizing intrinsic properties of the pollutant and aquifer, to develop a list of possible alternatives. The second approach is to analyze the effectiveness of various alternatives and compare their resultant concentrations with a given standard. No matter which approach is used, a "criteria" or "standard" or "acceptable level" is involved and these data, for the most part, are nonexistent. Dawson and Sanning (1982) outline a method for setting site restoration criteria by using air or water standards and working backwards with data on dilution potential and distribution characteristics. This methodology is dependent on existing criteria for other environmental media and knowledge of transport mechanics, which may not be available.

Several risk assessment techniques have been developed for assessing the risk of hazardous waste sites. With slight modifications, some of these would be applicable to assessing different alternatives at individual sites. Unterberg, Stone and Tafuri (1981) outline such a procedure for assessing spill sites. Caldwell, Barrett and Chang (1981) have developed a ranking system for the release of hazardous substances. Nelson and Young (1981) discuss a location and prioritization scheme for future investigation of abandoned dump sites. Unites, Possidento and Housman (1980) describe the development of a site investigation manual for risk assessment. Kufs et al. (1980) have developed a methodology for selecting sites for investigation based on their adverse environmental impacts. Berger (1981) describes a general methodology for assessing risks based on a similar approach as Kufs et al. (1980). This methodology considers four characteristics: receptors, pathways, waste characteristics, and waste management practices. Additional information on risk assessment and risk assessment techniques is in Chapter 7.

Nisbet (1982) discusses some uses and limitations of risk assessments. Nisbet (1982) and Berger (1982) both point out that without detailed analytical monitoring programs, risk assessments can usually only produce a qualitative ranking of alternatives. Nisbet (1982) also points out that risk assessments are difficult to conduct and give uncertain results, because: (1) exposure is usually variable and poorly characterized; (2) toxicity information is hard to extrapolate to humans from animals; and (3) ground water transport mechanics are often unknown, thus making the population at risk difficult to estimate.

In summary, risk assessment could be the most difficult aspect of evaluating aquifer restoration alternatives. The problems to be encountered stem from the fact that most of the needed information for any comprehensive methodology will not exist. The use of estimations in any methodology will limit its replicability.

The best approach to risk assessment will be to develop a methodology suited to the identified need. Modification of existing risk assessment techniques is one approach. The best risk assessment technique will be one that best utilizes available information and requires the least amount of estimated input.

SELECTED REFERENCES

Berger, I. S., "Determination of Risk for Uncontrolled Hazardous Waste Sites", *Management of Uncontrolled Hazardous Waste Sites*, Hazardous Materials Control Research Institute, 1982, pp. 23–26.

Bixler, B., Hanson, B. and Langner, G., "Planning Superfund Remedial Actions", *Management of Uncontrolled Hazardous Waste Sites*, Hazardous Materials Control Research Institute, 1982, pp. 141–145.

Caldwell, S., Barrett, K. W. and Chang, S. S., "Ranking System for Releases of Hazardous Substances", *Management of Uncontrolled Hazardous Waste Sites*, Hazardous Materials Control Research Institute, 1981, pp. 14–20.

Canter, L. W., *Environmental Impact Assessment*, 1st ed., McGraw-Hill, 1977.

Cornaby, B. W., et al., "Application of Environmental Risk Techniques to Uncontrolled Hazardous Waste Sites", *Management of Uncontrolled Hazardous Waste Sites*, Hazardous Materials Control Research Institute, 1982, pp. 380–384.

Dawson, G. W. and Brown, S. M., "New Assessment Methods to Aid Site Restoration", *Management of Uncontrolled Hazardous Waste Sites*, Hazardous Materials Control Research Institute, 1981, pp. 79–83.

Dawson, G. W. and Sanning, D., "Exposure-Response Analysis for Setting Site Restoration Criteria", *Management of Uncontrolled Hazardous Waste Sites*, Hazardous Materials Control Research Institute, 1982, pp. 386–389.

Evans, R. B., Benson, R. C. and Rizzo, J., "Systematic Hazardous Waste Site Assessments", *Management of Uncontrolled Hazardous Waste Sites*, Hazardous Materials Control Research Institute, 1982, pp. 17–22.

Freudenthal, H. D. and Calender, J. A., "Public Involvement in Resolving Hazardous Waste Site Problems", *Management of Uncontrolled Hazardous Waste Sites*, Hazardous Materials Control Research Institute, 1982, pp. 346–349.

Glover, E. W., "Containment of Contaminated Ground Water: An Overview", *The Proceedings of the Second National Symposium on Aquifer Restoration and Ground Water Monitoring*, 1982, pp. 17–22.

Hittman Associates Inc., "Surface and Ground Water Contamination From Abandoned Lead-Zinc Mines, Picher Mining District, Ottawa County, Oklahoma", Oct. 1981.

Housman, J. J., Brandwein, D. I. and Unites, D. F., "Site Contamination and Liability Audits in the Era of Superfund", *Management of Uncontrolled Hazardous Waste Sites*, Hazardous Materials Control Research Institute, 1981, pp. 398–404.

Kaschak, W. M. and Nadeau, D. F., "Remedial Action Master Plans", *Management of Uncontrolled Hazardous Waste Sites*, Hazardous Materials Control Research Institute, 1982, pp. 124–127.

Kufs, C., et al., "Rating the Hazard Potential of Waste Disposal Facilities", *Management of Uncontrolled Hazardous Waste Sites*, Hazardous Materials Control Research Institute, 1980, pp. 30–41.

Lundy, D. A. and Mahan, J. S., "Conceptual Designs and Cost Sensitivities of Fluid Recovery Systems for Containment of Plumes of Contaminated Groundwater", *Management of Uncontrolled Hazardous Waste Sites*, Hazardous Materials Control Research Institute, 1982, pp. 136–140.

Minning, R. C., "Monitoring Well Design and Installation", *The Proceedings of the Second National Symposium on Aquifer Restoration and Ground Water Monitoring*, 1982, pp. 194–197.

Neely, N. S., et al., "Remedial Actions at Uncontrolled Hazardous Waste Sites", EPA 430/9-81-05, Jan. 1981, U.S. Environmental Protection Agency, Cincinnati, Ohio.

Nelson, A. B. and Young, R. A., "Location and Prioritization of Abandoned Dump Sites for Future Investigations", *Management of Uncontrolled Hazardous Waste Sites*, Hazardous Materials Control Research Institute, 1981, pp. 52–62.

Nisbet, I. C. T., "Uses and Limitations of Risk Assessments in Decision-Making on Hazardous Waste Sites", *Management of Uncontrolled Hazardous Waste Sites*, Hazardous Materials Control Research Institute, 1982, pp. 406–407.

Rehtlane, E. A., and Patton, F. D., "Multiple Port Piezometers vs. Standpipe Piezometers: An Economic Comparison" *The Proceedings of the Second National Symposium on Aquifer Restoration and Ground Water Monitoring*, 1982, pp. 287–295.

Rishel, H. L., et al., "Costs of Remedial Response Actions at Uncontrolled Hazardous Waste Sites", EPA-600/52-82-035, March 1983, U.S. Environmental Protection Agency, Cincinnati, Ohio.

St. Clair, A. E., McCloskey, M. H. and Sherman, J. S., "Development of a Framework for Evaluating Cost-Effectiveness of Remedial Actions at Uncontrolled Hazardous Waste Sites", *Management of Uncontrolled Hazardous Waste Sites*, Hazardous Materials Control Research Institute, 1982, pp. 372–376.

Thorsen, J. W., "Technical and Financial Aspects of Closure and Post Closure Care", *Management of Uncontrolled Hazardous Waste Sites*, Hazardous Materials Control Research Institute, 1981, pp. 259–262.

Tolman, A. L., et al., "Guidance Manual for Minimizing Pollution From Waste Disposal Sites", EPA-600/2-78-142, Aug. 1978, U.S. Environmental Protection Agency, Cincinnati, Ohio.

Unites, D., Possidento, M. and Housman, J., "Preliminary Risk Evaluation for Suspected Hazardous Waste Disposal Sites in Connecticut", *Management of Uncontrolled Hazardous Waste Sites*, Hazardous Materials Control Research Institute, 1980, pp. 25–29.

Unterberg, W., Stone, W. L. and Tafuri, A. N., "Rationale for Determining Priority of Cleanup of Uncontrolled Hazardous Waste Sites", *Management of Uncontrolled Hazardous Waste Sites*, Hazardous Materials Control Research Institute, 1981, pp. 188–197.

Ward, J. R., "Well Design and Construction for In Situ Leach Uranium Extraction", *Proceedings of the Second National Symposium on Aquifer Restoration and Ground Water Monitoring*, 1982, National Water Well Association, Columbus, Ohio, pp. 205–213.

CHAPTER 6

Techniques for Decision-Making

Numerous decisions are required in the selection of an aquifer restoration strategy for meeting a given need. Examples of required decisions include those related to selection of containment options, ground water treatment processes and design, and site closure options. Alternatives exist for each required decision, and the selection of the most appropriate one for meeting a given need can be a difficult task. The alternative with the greatest potential for meeting a given need will most likely not be the least environmentally damaging, least costly, and least risky. Therefore, trade-off analysis typically includes importance weighting of decision factors and the evaluation of a set of alternatives relative to each factor. Although the methodologies outlined in this chapter have found their greatest application in environmental impact studies, they can be utilized for decision-making related to aquifer restoration strategies.

Importance weighting is done in every decision related to the selection of an aquifer restoration strategy. For example, decisions are made on the relative importance of existing environmental resources, the importance/significance of anticipated beneficial and detrimental environmental impacts, monetary costs, and public acceptability. In most cases, this weighting is done via informal, ad hoc approaches without using available formalized techniques to provide a documented record of the considerations and rationale. Professional judgment is involved in both ad hoc and formalized approaches.

The extent to which a set of alternative strategies and combinations thereof are evaluated relative to a series of decision factors depends upon the number of strategies being evaluated and the nature and extent of the study. Thorough studies involving multiple alternatives (strategies) will tend to give greater emphasis to the comparisons of alternatives. Comparisons of alternatives are often made on an ad hoc basis in the absence

205

of quantitative information on the alternatives, and without using available structured techniques to provide a documented record of the considerations and rationale. Again, as was the case for importance weighting, professional judgment is involved in both ad hoc and structured approaches.

The primary focus of this chapter will be on the principles and practice of importance weighting in decisions related to the selection of an aquifer restoration strategy for meeting a given need. Briefer information will be presented on techniques for evaluation of alternatives relative to a series of decision factors. The reason for this balance is that importance weighting requires more judgment and has greater influence on the final selection than does the evaluation of alternatives. This is not to say that the latter is unimportant, but simply to indicate that more attention should be given to fundamental importance weighting techniques. This chapter begins with a conceptual framework for trade-off analysis and is followed by examples of both informal and structured approaches for importance weighting. The ranking/rating/scaling of alternatives are also briefly addressed, followed by information on the development of a decision matrix.

CONCEPTUAL FRAMEWORK FOR
TRADE-OFF ANALYSIS

Trade-off analysis typically involves the comparison of a set of alternatives relative to a series of decision factors. For example, for cleanup of a polluted aquifer there may be several treatment processes which could be used, including both in situ options and treatment following pumpage to the surface. Factors related to these alternatives can include the health risk, economic efficiency, social concerns (public preference), and biophysical, cultural and socioeconomic impacts (both beneficial and detrimental impacts should be considered). Table 6.1 displays a trade-off matrix for systematically comparing specific alternatives relative to a series of decision factors. The following approaches can be used to complete the trade-off matrix:

(1) qualitative approach in which descriptive information on each alternative relative to each decision factor is presented in the matrix;

(2) quantitative approach in which quantitative information on each alternative relative to each decision factor is displayed in the matrix;

(3) ranking, rating, or scaling approach in which the qualitative or quantitative information on each alternative is summarized via the assignment of a rank, or rating, or scale value relative to each decision factor (the rank, or rating, or scale value is presented in the matrix);

(4) weighting approach in which the importance weight of each decision factor relative to each other decision factor is considered, with the resultant discussion of the information on each alternative (qualitative; or quantitative; or ranking, rating, or scaling) being presented in view of the relative importance of the decision factors; and

(5) weighting-ranking/rating/scaling approach in which the importance weight for each decision factor is multiplied by the ranking/rating/scale of each alternative, then the resulting products for each alternative are summed to develop an overall composite index or score for each alternative.

Decision-making which involves the comparison of a set of alternatives relative to a series of decision factors is not unique in terms of its application to environmentally oriented problems such as aquifer restoration. This approach represents a classic decision-making problem which is often referred to as multiattribute or multicriteria decision-making. The conceptual framework for this type of decision-making appeared in the early 1960s in the U.S. Department of Defense. Since that time it has been applied to numerous situations requiring decisions, including those which involve consideration of environmental factors and impacts.

Two decision factors listed in Table 6.1 are health risk and social concerns (public preference). Public participation programs for determining public preference are highlighted in Chapter 8. One approach addressing health risk involves risk assessment, and Chapter 7 provides information on using risk assessment in decisions related to remedial actions for polluted ground water.

IMPORTANCE WEIGHTING FOR DECISION FACTORS

If the qualitative or quantitative approach is used for completion of the matrix as shown in Table 6.1, information for this approach relative to economic efficiency and environmental impacts can be found in Chapters 2 through 4. It should be noted that this information would also be needed for the approaches involving importance weighting and/or ranking/rating/scaling. If the importance weighting approach is used the critical issue is the assignment of importance weights to the individual decision factors, or at least the arrangement of them in a rank ordering of importance. Table 6.2 summarizes some structured importance weighting or ranking techniques which could be used to achieve this step. These techniques have been used in numerous environmental decision-making projects. Brief descriptions of several reference sources and techniques from Table 6.2 will be presented later as illustrations of the tech-

Table 6.1   Trade-Off Matrix

Decision Factor	Alternatives				
	1	2	3	4	5
Health Risk					
Economic Efficiency					
Social Concerns (public preference)					
Environmental Impacts Biophysical Cultural Socioeconomic					

Table 6.2   Importance Weighting Techniques

Reference Source	Importance Weighting Technique
Dalkey (1969)	Delphi
Dean and Nishry (1965)	Paired comparison
Dee et al. (1972)	Ranked pairwise comparison
Eckenrode (1965)	Ranking Rating Paired comparison (3 types) Successive comparison
Edwards (1976)	Multiattribute utility measurement
Falk (1968)	Public preference
Finsterbusch (1977)	Consensal weights Formula weights Justified subjective weights Subjective weights Inferred subjective weights Ranking Equal weight Multiple methods
Gum, Roefs and Kimball (1976)	Metfessel general allocation test
O'Connor (1972)	Multiattribute scaling
Ross (1976)	Paired comparison and checking
School of CEES and Oklahoma Biological Survey (1974)	Ranked pairwise comparison
Toussaint (1975)	Delphi
Voelker (1977)	Nominal group process

niques. In addition to the structured techniques, less formal approaches, such as reliance on public participation programs, can also be used as the basis for importance weighting.

SYSTEMATIC COMPARISON OF STRUCTURED IMPORTANCE WEIGHTING TECHNIQUES

As shown in Table 6.2 there are a number of importance weighting techniques which can be used in decision-making related to aquifer restoration strategies. As would be expected, there are advantages and limitations to specific structured importance weighting techniques. One study to compare six techniques was conducted by Eckenrode (1965). To insure that the research results would be useful in a variety of situations where comparative judgments are required, three different judgment situations were used, one in each of three experiments:

(1) A specific air defense system development, where the judges were six persons who had personally conducted analytical studies on the system in question and were thoroughly familiar with it. They judged the relative importance of six carefully defined system criteria frequently found in military air defense system specifications: economy, early availability, lethality, reliability, mobility, and troop safety.

(2) A general (hypothetical) air defense system development, where the 12 judges used were familiar with the problems of air defense system design and use in a general way from their previous experience. The same criteria were judged as in the specific system experiment.

(3) A personnel selection problem, in which six judges who had extensive experience in personnel subsystem management judged the relative value of the following carefully defined qualifications for a command and control personnel subsystem manager: previous personnel subsystem experience, technical competence, and demonstrated capabilities for fiscal planning, work planning, and maintaining good client relations and good staff relations.

It is realized that these three experiments do not address aquifer restoration strategies; however, they can be used generically since they involve importance weighting of decision factors (criteria). The six importance weighting techniques which were compared for their reliability and efficiency in collecting the judgment data were (Eckenrode, 1965):

(1) Ranking: The judge (J) was asked to place a numerical rank next to each criterion, indicating by 1 that criterion most valuable in the situation, by 2, the next most valuable, etc.

(2) Rating: The criteria were presented next to a continuous scale marked off in units from 0 to 10 (see Figure 6.1). J was asked to draw a line from each criterion to any appropriate point on the value scale. He was permitted to select points between numbers or to assign more than one criterion to a single position on the scale.

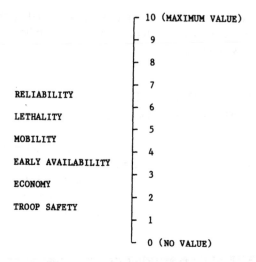

Note: J draws a line from each criterion to the appropriate
point on the scale.

Figure 6.1. Rating scale (Eckenrode, 1965).

(3) Partial Paired Comparisons I: The criteria were presented on the ordinate and abscissa of a partial matrix (see Table 6.3). J was asked to indicate in each block the number of the more valuable of the pair of criteria which were the coordinates of that block.

(4) Partial Paired Comparisons II: A list of criterion pairs was presented in the form illustrated below, and J was asked to circle the member of each pair which was more valuable for the system in question.

Troop safety — Reliability

Each criterion was paired once with every other criterion.

(5) Complete Paired Comparisons: This method was the same as partial paired comparisons II, except that the list was doubled in length by requiring that each pair appear twice, once in the order A–B, the second time (elsewhere in the list) in the order B–A, in order to eliminate any position error.

(6) Successive Comparisons: In this method the list of criteria was presented to J who proceeded as follows:

(a) He ranked criteria in order of importance (as in the ranking method above).

(b) He tentatively assigned the value (V_1) 1.0 to the most important criterion, and other values (V_i) between 0 and 1 to the other criteria in order of importance.

(c) He decided whether the criterion with value 1.0 was more important than all other criteria combined.

If so, he increased V_1 so that V_1 was greater than the sum of all subsequent V's, i.e.,

Table 6.3   Partial Paired Comparisons I (Eckenrode, 1965)

	2. TROOP SAFETY	3. MOBILITY	4. ECONOMY	5. RELIABILITY	6. LETHALITY
1. EARLY AVAILABILITY					
2. TROOP SAFETY					
3. MOBILITY					
4. ECONOMY					
5. RELIABILITY					

Note: J puts the number of the more valuable criterion of the pair being considered in each block. Example: 2 in the "early availability-troop safety" block means J judges the latter to be more important for the system in question.

$$V_1 > \Sigma\, {}_2^n\, V_i$$

If not, he adjusted V_1 (if necessary) so that V_1 was less than the sum of all subsequent V's, i.e.,

$$V_1 < \Sigma\, {}_2^n\, V_i$$

(d) He decided whether the second most important criterion with value V_2 was more important than all lower-valued criteria; he then proceeded as in c above with V_2.

(e) He continued until $n - 1$ criteria had been so evaluated.

All methods except successive comparison were used in all three experiments. The successive comparisons method was used only in the first experiment, in which the criteria were applied to a specific hardware system design. All three experiments were conducted in the same way. For each, a booklet was developed, the cover page of which contained instructions to the judge, including a definition of each criterion to be compared. Each succeeding page contained a form for recording his judgments by one of the methods. The order of the forms in the booklet was randomized and each judge was administered a booklet individually in a quiet office, under instructions to complete it rapidly but without making mistakes. The experimenter recorded the time required by the judge to complete each page of the booklet.

The consistency of importance weighting assignments among the six techniques was very high in all three experiments. This means that there is high reliability among the techniques in the importance weights produced; that is, each method produced essentially the same ordering of weights. It was also noted that each judge produced the same ordering of weights regardless of the method he used. This high within-judge reliability is to be expected when the set of items being judged is small as in the present experiments and judges behave transitively. This is the usual case in weighting multiple criteria, for sets of criteria to be weighted frequently consist of between three and a dozen items in practical situations. Taken together, this information indicates that the techniques are reliable approaches for recording judgments. Any of the six techniques could be used for importance weighting of decision factors associated with the selection of an aquifer restoration strategy. The individuals doing the weighting could be technical specialists or public officials, or both.

RANKING TECHNIQUES FOR IMPORTANCE WEIGHTING

Ranking techniques for importance weighting basically involve the rank ordering of decision factors in their relative order of importance. If there are "n" decision factors, rank ordering would involve assigning 1 to the most important factor, 2 to the second most important factor, and so forth until "n" is assigned to the least important factor. It should be noted that the rank order numbers could be reversed; that is, "n" could be assigned to the most important factor, "n-1" to the second most important factor, and so forth until 1 is assigned to the least important factor. Two illustrations of ranking techniques will be briefly described herein: (1) a public preference approach (Falk, 1968); and (2) the use of the Nominal Group Process technique (Voelker, 1977). Both of these techniques could be used for importance weighting of decision factors associated with the selection of an aquifer restoration strategy for meeting a given need.

Public Preference Approach

Falk (1968) described the use of public participation in the assignment of relative importance weights to a series of factors. A pilot study was described which involved determining the relative importance of four tangible and five intangible factors, and identifying a means of applying these measures of importance in selecting the most acceptable one of three hypothetical roadway solutions. The approach requires the assumption that frequency of citizen preference for one factor over another is directly related to importance of that factor. This approach could be used in the selection of an aquifer restoration strategy for meeting a given need, and public participation techniques are described in Chapter 8.

Nominal Group Process Technique

The Nominal Group Process technique (NGP), an interactive group technique, was developed in 1968. It was derived from social-psychological studies of decision conferences, management science studies of aggregating group judgments, and social work studies of problems surrounding citizen participation in program planning. The NGP has gained wide acceptance in health, social service, education, industry, and government organizations. For example, Voelker (1977) described the application of NGP to rank decision factors important in siting nuclear power plants. Siting decisions are routinely required for injection wells for treated ground water. The following four steps are involved in the use of NGP for importance weighting:

(1) Nominal (silent and independent) generation of ideas in writing by a panel of participants.
(2) Round-robin listing of ideas generated by participants on a flip chart in a serial discussion.
(3) Discussion of each recorded idea by the group for the purpose of clarification and evaluations.
(4) Independent voting on priority ideas, with group decision determined by mathematical rank ordering.

RATING TECHNIQUES FOR IMPORTANCE WEIGHTING

Rating techniques for importance weighting basically involve the assignment of importance numbers to a series of decision factors, and possibly, although not always, their subsequent normalization via a mathematical procedure. An example of a rating scale approach is shown in Figure 6.1 (Eckenrode, 1965). Four additional examples of rating techniques will be presented herein: (1) the use of a predefined importance scale (Linstone and Turoff, 1975); (2) the use of a group approach for the assignment of scales of importance (Rau, 1980); (3) the use of the multiattribute utility measurement technique (Edwards, 1976); (4) and the use of a computerized version of a multiattribute utility measurement technique (Rugg and Feldman, 1982). These four rating techniques could be used for importance weighting of decision factors associated with the selection of an aquifer restoration strategy for meeting a given need. It should be noted that the public preference approach (Falk, 1968) and the NGP technique (Voelker, 1977) described in the previous subsection could also be used for rating the importance of decision factors.

Predefined Importance Scale

Decision factors can be assigned numerical values based on predefined importance scales. Table 6.4 delineates five scales with definitions to be considered in the assignment of numerical values to decision factors (Linstone and Turoff, 1975). Usage of the predefined scales can aid in systematizing importance weight assignments.

Group Approach Using Scales of Importance

Rau (1980) suggested a group approach for importance weight assignments, with the approach following a procedure that will produce reliable results. Because these weights are essentially based on the judgmental values or attitudes of those surveyed, the selected procedure must be systematic and must be able to reduce all possible variation. The group of persons ultimately selected for the weighting should include a cross section of society such as individuals from governmental agencies, politicians and decision-makers, experts in the field of hydrogeology and environmental evaluation, representatives from special interest groups, and members of the general public. Groups of individuals representing this cross section must be sampled a number of times to obtain consistent estimates of the weights.

The procedure suggested by Rau (1980) for determining the relative importance of each decision factor (environmental impact area in this example) consists of ranking and pairwise comparisons. Each individual

Table 6.4   Importance Scale (Linstone and Turoff, 1975)

Scale Reference	Definitions
1. Very Important	A most relevant point First order priority Has direct bearing on major issues Must be resolved, dealt with or treated
2. Important	Is relevant to the issue Second order priority Significant impact but not until other items are treated Does not have to be fully resolved
3. Moderately Important	May be relevant to the issue Third order priority May have impact May be a determining factor to major issue
4. Unimportant	Insignificantly relevant Low priority Has little impact Not a determining factor to major issue
5. Most Unimportant	No priority No relevance No measurable effect Should be dropped as an item to consider

is required to rank the impact areas and to compare in pairwise fashion the degree of importance of highest rank with the one immediately following. If this procedure is followed in a systematic way, a weight will be developed for each area. The procedure is repeated a number of times for different groups in order to get the desired cross-sectional population representation and the reliability needed for an importance weighting. The basic steps for determining the importance weightings are as follows (Rau, 1980):

(1) Step 1: Select a group of individuals for evaluating and explain to them in detail the weighting concept and the use of rankings and weightings.
(2) Step 2: Prepare a table with columns corresponding to the range of values which can be assigned as a "score of importance" to each impact area – for example, if five values are possible, there would be five columns. The rows in the table would correspond to the impact areas being ranked as to importance. Table 6.5 contains an illustration of five columns for importance weighting (Rau, 1980).

Table 6.5 Example of the Development of Impact Area Importance Weightings (Rau, 1980)

Impact area	Low importance →1	2	Average importance →3	4	High importance →5	Total	Weighting
Park requirements		X				2	2/43
School age students generated			X			3	3/43
Trips generated		X				2	2/43
Police protection				X		4	4/43
Fire protection				X		4	4/43
Public service costs					X	5	5/43
Total revenues					X	5	5/43
Employment (long-term jobs)				X		4	4/43
Electricity consumption			X			3	3/43
Natural gas consumption			X			3	3/43
Solid waste generated		X				2	2/43
Sewage discharge			X			3	3/43
Water consumption			X			3	3/43
						43	1.0

(3) Step 3: Give a copy of the table developed in Step 2 to each individual evaluator and repeat steps 4–9 until no further changes in the table entries are desired.

(4) Step 4: Ask each individual to place an "X", or other signifying mark, in each column for each impact area. Thus, a value of importance is assigned to each impact area.

(5) Step 5: Ask all individuals to compare the marked columns on a pairwise basis to insure that the impact areas are ordered on the proper relative basis in their opinion. If not, they should reassign their scores so as to have the desired relative ordering of impact areas. (For example, on a scale from 1 to 10, if a value of 10 has been assigned to impact area A and it appears that A is twice as important as B, impact area B should be assigned a value of 5.)

(6) Step 6: Ask each individual to add the value (or importance score) selected for each of the impact areas to obtain a total.

(7) Step 7: The individual should then divide the value selected for each impact area by the total obtained in Step 6 to determine the desired weighting for each area.

(8) Step 8: Collect the tables from each individual evaluator and average the weightings determined for each impact area to obtain a "group or composite average."

(9) Step 9: Present the averages obtained to the individual evaluators and ask them to compare the group weightings with those derived by each of them individually in Step 7.

(10) Step 10: If any one or more individuals desires to change the assignment of scores based on what the group decided, go to Step 4 and repeat the entire process. If none desire to change their scores, stop the experiment, because the impact area relative weightings of importance will have been derived.

As referred to earlier, Table 6.5 contains an example of importance weight assignments to 13 impact areas of interest based on five possible importance scores (Rau, 1980). By adding the scores corresponding to each "X", a total of 43 points is obtained. Dividing each score by 43, the relative importance weightings given by the last column in Table 6.5 are determined.

Multiattribute Utility Measurement Technique

Edwards (1976) described the multiattribute utility measurement (MAUM) technique for use in decision-making involving different publics. The MAUM technique can be used to spell out explicitly what the values of each participant (decision-maker, expert, pressure group, government, etc.) are, show how much they differ, and in the process can frequently reduce the extent of such differences. The basic assumption is that the values of the participants are reflected in the importance weights assigned to individual factors. It should also be noted that the usage of the MAUM technique can be by an individual, a small group of persons, or multiple publics. The ten basic steps in the MAUM technique are (Edwards, 1976):

Step 1: Identify the person or organization whose utilities are to be maximized. If, as is often the case, several organizations have stakes and voices in the decision, they must all be identified. (Utilities refer to general goals or objectives in the terminology used herein.)

Step 2: Identify the issue or issues to which the utilities needed are relevant. (Issues refer to needs being addressed.)

Step 3: Identify the entities to be evaluated. Formally, they are outcomes of possible actions. But in a sense, the distinction between an outcome and the opportunity for further actions is usually fictitious. (Entities refer to alternatives in the terminology used herein; outcomes would reflect the evaluation of each entity relative to the decision factors.)

Step 4: Identify the relevant dimensions of value for evaluation of the entities. As has often been noted, goals ordinarily come in hierarchies. But it is often practical and useful to ignore their hierarchical structure, and instead to specify a simple list of goals that seem important for the purpose at hand. It is important not to be too expansive at this stage. The number of relevant dimensions of value should be modest, for reasons that will be apparent shortly. (It should be noted that dimensions of value refer to the decision factors for evaluation of the alternatives.)

Step 5: Rank the dimensions in order of importance. This ranking job, like Step 4, can be performed either by an individual or by representatives of conflicting values acting separately or by those representatives acting as a group.

Step 6: Rate dimensions in importance, preserving ratios. To do this, start by assigning the least important dimension an importance of 10. Now consider the next least important dimension. How much more important (if at all) is it than the least important? Assign it a number that reflects that ratio. Continue up the list, checking each set of implied ratios as each new judgment is made. Thus, if a dimension is assigned a weight of 20, while another is assigned a weight of 80, it means that the 20 dimension is $1/4$ as important as the 80 dimension, and so on. By the time you get to the most important dimensions, there will be many checks to perform; typically, respondents will want to review previous judgments to make them consistent with present ones.

Step 7: Sum the importance weights, divide each by the sum, and multiply by 100. This is a purely computational step which converts importance weights into numbers that, mathematically, are rather like probabilities. The choice of a 1 to 100 scale is, of course, completely arbitrary.

Step 8: Measure the location of each entity being evaluated on each dimension. The word "measure" is used rather loosely here. There are three classes of dimensions: purely subjective, partly subjective, and purely objective. The purely subjective dimensions are perhaps the easiest; you simply get an appropriate expert to estimate the position of the entity on that dimension on a 0 to 100 scale, where 0 is defined as the minimum plausible value and 100 is defined as the maximum plausible value. Note "minimum and maximum plausible" rather than "minimum and maximum possible." The minimum plausible value often is not total absence of the dimension. A partly subjective dimension is one in which the units of measurement are objective, but the locations of the entities must be subjectively estimated. A purely objective dimension is one that can be measured nonjudgmentally, in objective units, before the decision. For partly or purely objective dimensions, it is necessary to have the estimators provide not only values for each entity to be evaluated, but also minimum and maximum plausible values, in the natural units of each dimension.

Step 9: Calculate utilities for entities. The equation is:

$$U_i = \sum_j w_j u_{ij}$$

where U_i = aggregate utility for ith entity (overall evaluation score, or index, for the ith alternative)

    i = number of entities (number of alternatives)

    w_j = normalized importance weight of the jth dimension of value (importance weight for jth decision factor); the w_j values are the output from Step 7.

    j = number of dimensions of value (number of decision factors).

    u_{ij} = scaled position of the ith entity on the jth dimension (scaled position of ith alternative on jth decision factor); the u_{ij} measures are the output from Step 8.

Step 10: Decide. If a single act is to be chosen, the rule is simple: maximize u_i.

Computerized Version of MAUM Technique

A simple computer program, called DECIDE, for use in importance weighting and decision-making is available for the TRS-80 computer system (Rugg and Feldman, 1982). The program can aid in decision-making when the decision involves the selection of one alternative from several choices. The following information describes how the program can be used.

(1) The first thing the program does is ask you to categorize the decision at hand into one of three categories: (1) choosing an item (or thing), (2) choosing a course of action, or (3) making a yes or no decision. You simply press 1, 2, or 3 followed by the ENTER key to indicate which type of decision is facing you. If you are choosing an item, you will be asked what type of item it is. To illustrate the mechanisms of the computer program, Table 6.6 contains a sample run (Rugg and Feldman, 1982). This illustration could easily be adapted to decisions related to the selection of an aquifer restoration strategy for meeting a given need.

(2) If the decision is either of the first two types, you must next enter a list of all the possibilities under consideration. A question mark will prompt you for each one. When the list is complete, type "END" in response to the last question mark. You must, of course, enter at least two possibilities. After the list is finished, it will be redisplayed so that you can verify that it is correct. If not, you must reenter it.

(3) Now you must think of the different factors that are important to you in making your decision. Each factor is to be entered in one at a time with the word "END" used to terminate the list. When complete, the list will be redisplayed. You must now decide which single factor is the most important and input its number. (You can enter 0 if you wish to change the list of factors.)

(4) The program now asks you to rate the importance of each of the other factors relative to the most important one. This is done by first assigning a value of 10 to the main factor. Then you must assign a value from 0–10 to each of the other factors. These numbers reflect your assessment of each factor's relative impor-

Table 6.6   Illustration of the Use of DECIDE (Rugg and Feldman, 1982)

I CAN HELP YOU MAKE A DECISION.   ALL I NEED TO DO IS ASK SOME QUESTIONS
AND THEN ANALYZE YOUR RESPONSES.

- - - - - - - - - -

WHICH OF THESE BEST DESCRIBES THE DECISION FACING YOU?

    1)   CHOOSING AN ITEM FROM VARIOUS ALTERNATIVES.
    2)   CHOOSING A COURSE OF ACTION FROM VARIOUS ALTERNATIVES.
    3)   DECIDING "YES" OR "NO".

WHICH ONE (1, 2, OR 3)? 1

WHAT TYPE OF ITEM IS IT

?   VACATION

I NEED TO HAVE A LIST OF EACH VACATION UNDER CONSIDERATION.

INPUT THEM ONE AT A TIME IN RESPONSE TO EACH QUESTION MARK.

TYPE THE WORD "END" TO INDICATE THAT THE WHOLE LIST HAS BEEN ENTERED.

?   CAMPING
?   SAFARI
?   TRIP TO D.C.
?   END

OK. HERE'S YOUR LIST:

    1)   CAMPING
    2)   SAFARI
    3)   TRIP TO D.C.

IS THE LIST CORRECT (Y OR N)? Y

NOW, THINK OF THE FACTORS THAT ARE IMPORTANT IN CHOOSING THE BEST
VACATION.

INPUT THEM ONE AT A TIME IN RESPONSE TO EACH QUESTION MARK.

TYPE THE WORD "END" TO TERMINATE THE LIST.

?   RELAXATION
?   AFFORDABILITY
?   CHANGE OF PACE
?   END

HERE'S YOUR LIST OF FACTORS:

    1)   RELAXATION
    2)   AFFORDABILITY
    3)   CHANGE OF PACE

DECIDE WHICH FACTOR ON THE LIST IS THE MOST IMPORTANT AND INPUT ITS
NUMBER. (TYPE 0 IF THE LIST NEEDS CHANGING.)

?   2

Table 6.6, continued

NOW LET'S SUPPOSE WE HAVE A SCALE OF IMPORTANCE RANGING FROM 0-10. WE'LL GIVE AFFORDABILITY A VALUE OF 10 SINCE AFFORDABILITY WAS RATED THE MOST IMPORTANT.

ON THIS SCALE, WHAT VALUE OF IMPORTANCE WOULD THE OTHER FACTORS HAVE?

RELAXATION

?   5.5

CHANGE OF PACE

?   9

EACH VACATION MUST NOW BE COMPARED WITH RESPECT TO EACH IMPORTANCE FACTOR.

WE'LL CONSIDER EACH FACTOR SEPARATELY AND THEN RATE EACH VACATION IN TERMS OF THAT FACTOR ONLY.

*** (HIT ANY KEY TO CONTINUE)

(A key is pressed)

LET'S GIVE CAMPING A VALUE OF 10 ON EVERY SCALE.

EVERY OTHER VACATION WILL BE ASSIGNED A VALUE HIGHER OR LOWER THAN 10. THIS VALUE DEPENDS ON HOW MUCH YOU THINK IT IS BETTER OR WORSE THAN CAMPING.

- - - - - - - - - -

CONSIDERING ONLY RELAXATION AND ASSIGNING 10 TO CAMPING,

WHAT VALUE WOULD YOU ASSIGN TO

SAFARI? 3

TRIP TO D.C.? 9

- - - - - - - - - -

CONSIDERING ONLY AFFORDABILITY AND ASSIGNING 10 TO CAMPING,

WHAT VALUE WOULD YOU ASSIGN TO

SAFARI? 1

TRIP TO D.C.? 8

- - - - - - - - - -

CONSIDERING ONLY CHANGE OF PACE AND ASSIGNING 10 TO CAMPING,

WHAT VALUE WOULD YOU ASSIGN TO

SAFARI? 60

TRIP TO D.C.? 25

Table 6.6 , continued

```
TRIP TO D.C. IS BEST BUT IT'S VERY CLOSE.

HERE'S THE FINAL LIST IN ORDER.  TRIP DO D.C. HAS BEEN GIVEN A VALUE OF
100 AND THE OTHERS RATED ACCORDINGLY.

HIT ANY KEY TO SEE THE LIST.

(A key is pressed)

100       TRIP TO D.C.

 98.7     CAMPING

 78.8     SAFARI

OK
```

tance as compared to the main one. A value of 10 means it is just as important; lesser values indicate how much less importance you place on it.

(5) Now you must rate the decision possibilities (alternatives) with respect to each of the importance factors. Each importance factor will be treated separately. Considering only that importance factor, you must rate each decision possibility. The program first assigns a value of 10 to one of the decision possibilities. Then you must assign a relative number (lower, higher, or equal to 10) to each of the other decision possibilities.

(6) Armed with all this information, the program will now determine which choice is best. The various possibilities are listed in order of ranking. Alongside each one is a relative rating with the best choice being normalized to a value of 100.

The DECIDE program listing is in Table 6.7; the program can currently accept up to ten alternatives and ten decision factors. If more alternatives are to be evaluated, or if more decision factors are involved, the value of MD in line 160 in the program listing should be increased. The main routines in the DECIDE computer program are as follows (Rugg and Feldman, 1982):

150– 190 Initializes and dimensions variables.

200– 360 Determines category of decision.

400– 490 Gets or sets T$.

500– 810 Gets list of possible alternatives from user.

900–1200 Gets list of importance factors from user.

1300–1490 User rates each importance factor.

1500–1900 User rates the decision alternatives with respect to each importance factor.

2000–2110 Evaluates the various alternatives.

Table 6.7   Program Listing for DECIDE (Rugg and Feldman, 1982)

```
100    REM:  DECIDE - 16K
110    REM:  (C) 1981, PHIL FELDMAN AND TOM RUGG
150    CLEAR 500
160    MD=10
170    DIM L$(MD),F$(MD),V(MD)
180    DIM C(MD,MD),D(MD),Z(MD)
190    E$="END"
200    GOSUB 5000
210    PRINT" I CAN HELP YOU MAKE A"
220    PRINT"DECISION.  ALL I NEED TO DO IS"
230    PRINT"ASK SOME QUESTIONS AND THEN"
240    PRINT"ANALYZE YOUR RESPONSES."
250    FOR J=1 TO 30:PRINT"-";:NEXT
260    PRINT
270    PRINT"WHICH OF THESE BEST DESCRIBES"
280    PRINT"THE DECISION FACING YOU?"
290    PRINT" 1) CHOOSING AN ITEM FROM"
300    PRINT" VARIOUS ALTERNATIVES."
310    PRINT" 2) CHOOSING A COURSE OF ACTION"
320    PRINT" FROM VARIOUS ALTERNATIVES."
330    PRINT" 3) DECIDING 'YES' OR 'NO'."
340    PRINT
350    INPUT"WHICH ONE (1,2,OR 3)";T
360    IF T<1 OR T>3 THEN 200
400    GOSUB 5000
410    ON T GOTO 420,440,460
420    PRINT"WHAT TYPE OF ITEM IS IT"
430    INPUT T$:GOTO 500

440    T$="COURSE OF ACTION"
450    GOTO 500
460    T$="'YES' OR 'NO'":NI=2
470    L$(1)="DECIDING YES"
480    L$(2)="DECIDING NO"
490    GOTO 900
500    GOSUB 5000:NI=0
510    PRINT" I NEED TO HAVE A LIST OF EACH"
520    PRINT T$;" UNDER"
530    PRINT"CONSIDERATION.":PRINT
540    PRINT" INPUT THEM ONE AT A TIME IN"
550    PRINT"RESPONSE TO EACH QUESTION MARK."
560    PRINT
570    PRINT" TYPE THE WORD '";E$;"' TO"
580    PRINT"INDICATE THAT THE WHOLE LIST"
590    PRINT"HAS BEEN ENTERED.":PRINT
600    IF NI MD THEN 620
610    PRINT"-LIST FULL-":GOTO 650
620    NI=NI+1:INPUT L$(NI)
630    IF L$(NI) E$ THEN 600
640    NI=NI-1
650    IF NI =2 THEN 700
660    PRINT
670    PRINT"YOU NEED AT LEAST 2 CHOICES!"
680    PRINT:PRINT"TRY AGAIN"
690    GOSUB 5200:GOTO 500
700    GOSUB 5000
710    PRINT"OK. HERE'S YOUR LIST:"
720    PRINT:FOR J=1 TO NI
730    PRINT J;CHR$(8);") ";L$(J)
740    NEXT:PRINT
```

Table 6.7, continued

```
750      FOR J=1 TO 9:R$=INKEY$:NEXT
760      INPUT"IS THE LIST CORRECT (Y OR N)";R$
770      IF R$="Y" THEN 900
780      IF R$<>"N" GOTO 700
790      PRINT
800      PRINT"THE LIST MUST BE RE-ENTERED"
810      GOSUB 5200:GOTO 500
900      GOSUB 5000:R$=INKEY$
910      PRINT" NOW, THINK OF THE FACTORS THAT"
920      IF T<3 THEN PRINT"ARE IMPORTANT IN CHOOSING THE"
930      IF T<3 THEN PRINT"BEST ";T$;"."
940      IF T=3 THEN PRINT"ARE IMPORTANT TO YOU IN"
950      IF T=3 THEN PRINT"DECIDING ";T$;"."
960      PRINT:PRINT" INPUT THEM ONE AT A TIME IN"
970      PRINT"RESPONSE TO EACH QUESTION MARK."
980      PRINT:PRINT" TYPE THE WORD '";E$;"' TO"
990      PRINT"TERMINATE THE LIST."
1000     PRINT:NF=0
1010     IF NF>=MD THEN PRINT"-- LIST FULL --":PRINT
1020     IF NF>=MD THEN GOTO 1060
1030     NF=NF+1:INPUT F$(NF)
1040     IF F$(NF)<>E$ THEN 1010
1050     NF=NF-1:PRINT
1060     IF NF<1 THEN PRINT"YOU NEED AT LEAST 1 - REDO IT !"
1070     IF NF<1 THEN GOSUB 5200
1080     IF NF<1 THEN 900
1100     GOSUB 5000
1110     PRINT"HERE'S YOUR LIST OF FACTORS:"
1130     FOR J=1 TO NF
1140     PRINT J;CHR$(B);") ";F$(J)
1150     NEXT
1160     PRINT"DECIDE WHICH FACTOR ON THE"
1170     PRINT"LIST IS THE MOST IMPORTANT AND"
1180     PRINT"INPUT ITS NUMBER. (TYPE 0 IF"
1190     PRINT"THE LIST NEEDS CHANGING.)"
1200     INPUT A:A=INT(A)
1210     IF A=0 THEN 900
1220     IF A>NF OR A<0 THEN 1100
1300     GOSUB 5000
1310     IF NF=1 THEN 1500
1320     PRINT" NOW LET'S SUPPOSE WE HAVE A"
1330     PRINT"SCALE OF IMPORTANCE RANGING"
1340     PRINT"FROM 0-10. WE'LL GIVE"
1350     PRINT F$(A);" A VALUE OF"
1360     PRINT"10 SINCE ";F$(A)
1370     PRINT"WAS RATED THE MOST IMPORTANT."
1380     PRINT:PRINT" ON THIS SCALE, WHAT VALUE OF"
1390     PRINT"IMPORTANCE WOULD THE OTHER"
1400     PRINT"FACTORS HAVE?"
1410     FOR J=1 TO NF
1420     IF J=A THEN 1490
1430     PRINT:PRINT F$(J)
1440     INPUT V(J)
1450     IF V(J)<0 THEN 1480
1460     IF V(J)>10 THEN 1480
1470     GOTO 1490
1480     PRINT" IMPOSSIBLE VALUE - TRY AGAIN":GOTO 1430
1490     NEXT
1500     V(A)=10:Q=0:FOR J=1 TO NF
1510     Q=Q+V(J):NEXT:FOR J=1 TO NF
```

Table 6.7, continued

```
1520   V(J)=V(J)/Q:NEXT:GOSUB 5000
1530   IF T<>3 THEN PRINT" EACH ";T$;" MUST NOW"
1540   IF T=3 THEN PRINT" DECIDING 'YES' OR DECIDING"
1550   IF T=3 THEN PRINT"'NO' MUST NOW"
1560   PRINT"BE COMPARED WITH RESPECT TO"
1570   PRINT"EACH IMPORTANCE FACTOR."
1580   PRINT"WE'LL CONSIDER EACH FACTOR"
1590   PRINT"SEPARATELY AND THEN RATE"
1600   IF T<>3 THEN PRINT"EACH ";T$;" IN TERMS"
1610   IF T=3 THEN PRINT"DECIDING 'YES' OR DECIDING"
1620   IF T=3 THEN PRINT"'NO' IN TERMS"
1630   PRINT"OF THAT FACTOR ONLY."
1634   PRINT:PRINT"xxx (HIT ANY KEY TO CONTINUE)"
1638   R$=INKEY$:IF R$="" THEN1638
1640   PRINT:PRINT" LET'S GIVE ";L$(1)
1650   PRINT"A VALUE OF 10 ON EVERY SCALE."
1660   IF T<>3 THEN PRINT" EVERY OTHER ";T$
1670   IF T=3 THEN PRINT" THEN DECIDING 'NO'"
1680   PRINT"WILL BE ASSIGNED A VALUE HIGHER"
1690   PRINT"OR LOWER THAN 10.  THIS VALUE"
1700   PRINT"DEPENDS ON HOW MUCH YOU THINK"
1710   PRINT"IT IS BETTER OR WORSE THAN"
1720   PRINT L$(1);"."
1800   FOR J=1 TO NF
1810   PRINT" ---------------"
1820   PRINT" CONSIDERING ONLY ";F$(J)
1830   PRINT"AND ASSIGNING 10 TO"
1835   PRINT L$(1);","
1840   PRINT"WHAT VALUE WOULD YOU ASSIGN TO"
1850   FOR K=2 TO NI
1860   PRINT L$(K);::INPUT C(K,J)
1870   IF C(K,J)>=0 THEN 1900
1880   PRINT" -- NEGATIVE VALUES ILLEGAL --"
1890   GOTO 1860
1900   NEXT:PRINT:C(1,J)=10:NEXT
2000   FOR J=1 TO NF:Q=0
2010   FOR K=1 TO NI
2020   Q=Q+C(K,J):NEXT
2030   FOR K=1 TO NI
2040   C(K,J)=C(K,J)/Q:NEXT:NEXT
2050   FOR K=1 TO NI:D(K)=0
2060   FOR J=1 TO NF
2070   D(K)=D(K)+C(K,J)*V(J):NEXT
2080   NEXT:MX=0:FOR K=1 TO NI
2090   IF D(K)>MX THEN MX=D(K)
2100   NEXT:FOR K=1 TO NI
2110   D(K)=D(K)*100/MX:NEXT
2200   FOR K=1 TO NI:Z(K)=K:NEXT
2210   NM=NI-1:FOR K=1 TO NI
2220   FOR J=1 TO NM:N1=Z(J)
2230   N2=Z(J+1)
2240   IF D(N1)>D(N2) THEN 2260
2250   Z(J+1)=N1:Z(J)=N2
2260   NEXT:NEXT:J1=Z(1):J2=Z(2)
2270   DF=D(J1)-D(J2):GOSUB 5000
2300   PRINT L$(J1);" IS BEST"
2310   IF DF<5 THEN PRINT"BUT IT'S VERY CLOSE."
2320   IF DF<5 THEN 2380
2330   IF DF<10 THEN PRINT"BUT IT'S FAIRLY CLOSE."
2340   IF DF<10 THEN 2380
2350   IF DF<20 THEN PRINT"BY A FAIR AMOUNT."
```

Table 6.7, continued

```
2360    IF DF<20 THEN 2380
2370    PRINT"QUITE DECISIVELY."
2380    PRINT" HERE'S THE FINAL LIST IN"
2390    PRINT"ORDER. ";L$(J1)
2400    PRINT"HAS BEEN GIVEN A VALUE OF 100"
2410    PRINT"AND THE OTHERS RATED"
2420    PRINT"ACCORDINGLY."
2430    PRINT
2440    PRINT" HIT ANY KEY TO SEE THE LIST."
2450    R$=INKEY$
2460    IF R$="" THEN 2450
2470    PRINT
2400    FOR J=1 TO NI:Q=Z(J)

2490    PRINT D(Q),L$(Q):NEXT
3000    END
5000    FOR J=1 TO 500:NEXT
5010    CLS:PRINT@12,"DECIDE"
5020    PRINT:RETURN
5200    FOR J=1 TO 1500:NEXT:RETURN
```

2200–2270 Sorts alternatives into their relative ranking.

2300–3000 Displays results.

5000–5020 Subroutine to clear screen and display header.

5200   Time wasting subroutine.

Definitions for the main variables in the DECIDE computer program are as follows (Rugg and Feldmen, 1982):

MD Maximum number of decision alternatives.

NI Number of decision alternatives.

NM NI-1.

L$ String array of the decision alternatives.

NF Number of importance factors.

F$ String array of the importance factors.

V Array of the relative values of each importance factor.

A Index number of most important factor.

C Array of relative values of each alternative with respect to each importance factor.

T Decision category (1 = item, 2 = course of action, 3 = yes or no).

T$ String name of decision category.

E$ String to signal the end of an input data list.

J,K Loop indices.

R$ User reply string.

Q,N1,N2 Work variables.

D Array of each alternative's value.

MX Maximum value of all alternatives.

DF Rating difference between best two alternatives.

Z Array of the relative rankings of each alternative.

PAIRED COMPARISON TECHNIQUES FOR IMPORTANCE WEIGHTING

Paired comparison techniques for importance weighting basically involve a series of comparisons between decision factors, and a systematic tabulation of the numerical results of the comparisons. These types of techniques have been extensively used in decision-making efforts, including numerous examples related to environmental improvement projects. Presented herein will be two examples of an unranked paired comparison technique (Dean and Nishry, 1965; Canter, 1983; Ross, 1976; and Hyman, Moreau and Stiftel, 1982); and two examples of a ranked pairwise comparison technique (Dee et al., 1972; and School of Civil Engineering and Environmental Science and Oklahoma Biological Survey, 1974). Three examples of unranked paired comparison techniques were discussed earlier in relation to a systematic comparison of structured importance weighting techniques (Eckenrode, 1965). Both unranked and ranked pairwise comparison techniques can be used for weighting decision factors associated with the selection of an aquifer restoration strategy for meeting a given need.

Unranked Paired Comparison Techniques

One of the most useful techniques for importance weighting of a series of decision factors is the paired comparison technique developed by Dean and Nishry (1965). This technique, which can be used by an individual or group, involves the comparison of each decision factor to each other decision factor in a systematic manner. Suppose that there are four basic decision factors known as F1 through F4 (F1 could be health risks, F2 could be economic efficiency, F3 could be social concern, and F4 could be environmental impacts). The weighting technique consists of considering each factor relative to every other factor and assigning to the one of the pair considered to be the most important a value of 1, and to the lesser important of the pair a value of 0. The use of this paired comparison technique is shown in Table 6.8. It should be noted that the assignment of 0 to a member of a pair does not denote zero importance; it simply means that in the pair considered it is of lesser importance.

A dummy factor, called F5, is also included in Table 6.8. The dummy factor is included so as to preclude the net assignment of a value of 0 to any of the basic factors (F1 through F4) in the process of each paired comparison. The dummy factor is defined as the least important in each paired comparison within which it is included. If two factors are consid-

Table 6.8 Use of Paired-Comparison Technique for Importance Weight Assignments

Factor	Assignment of Weights	Sum	FIC
F1	1 1 1 1	4	0.40
F2	0 1 0 1	2	0.20
F3	0 0 0 1	1	0.10
F4	0 1 1 1	3	0.30
F5 (dummy)	0 0 0 0	0	0
		10	1.00

ered to be of equal importance, then a value of 0.5 is assigned to each factor within the pair (Canter, 1983).

Following the assignment of relative importance to each factor relative to each other factor, with this process only being completed following several iterations to be sure that each factor is being considered in a consistent manner with each other factor, the rationale for each decision should be documented. It cannot be over-stressed that the most important aspect of using this technique is the careful delineation of the rationale basic to each 1 and 0 assignment. Following this documentation then the individual weight assignments are summed, with the factor importance coefficient (FIC) being equal to the sum value for an individual factor divided by the sum for all of the factors. The total of the sum column should equal to $(N)(N-1)/2$ where N is equal to the number of factors included in the assignment of weights. In the example in Table 6.8, five factors were included, hence the sum column total should be equal to 10. The total of the FIC column in Table 6.8 should be equal to 1.00.

The FIC column in Table 6.8 indicates that F1 is most important, followed by F4, F2, and F3. Whether or not the actual FIC fractions are used in a trade-off analysis, this paired-comparison approach has enabled the rank ordering of the four decision factors from most important to least important. In addition, importance weighting of subfactors can also be done by the same method. For example, in Table 6.1 environmental impacts also includes biophysical, cultural, and socioeconomic impacts.

The paired-comparison technique based on Dean and Nishry (1965) has been used in a number of environmentally related studies. Examples include selection of a sanitary landfill site (Morrison, 1974), selection of a wastewater treatment system (Canter, 1976), presentation of an environmental impact methodology for water resources projects (Solomon et al., 1977), and evaluation of the environmental impacts of a wastewater treatment system (Canter and Reid, 1977).

Ross (1976) described a method to check to the consistency of importance weight assignments made via the paired comparison technique. The paired comparison technique and consistency procedure proposed by Ross (1976) are as follows:

(1) When using this technique, an individual who has been requested to assess a number (N) of decision factors is presented with every possible combination of these factors, and asked to make judgments as to which of each pair is more important. His decisions are recorded in a paired-comparison matrix as shown in Table 6.9. An entry (C_{ij}) of "1" in this matrix denotes that the row decision factor i (row stimulus) was judged as being better, or more desirable, than the column stimulus j. Once all possible pairs have been compared, and the decisions recorded in the matrix, the ranking of stimuli can be readily ascertained by summing the rows of the matrix. The stimuli are ranked in order of these row sums.

(2) For the information in Table 6.9, the resultant ordering of the stimuli is 5, 6, 1, 3, 2, and 4. When the paired-comparison matrix is permuted according to the ranking derived, a characteristic pattern appears in which the upper right portion of the matrix is observed to be composed of 1's, and the lower left portion of 0's (see Table 6.10).

In the foregoing example, the individual making the comparisons, hereafter termed the judge, has been perfectly consistent in his judgments. It reveals that he has a clear idea of the stimuli, and that he has a good decision rule to follow while making the individual paired comparisons. Such is often not the case. Inconsistent judgments are revealed by the presence of 1's below the diagonal in the permuted matrix. In Table 6.11, for example, the preference of the judge for 4 over 5 is revealed. It is clearly a case of inconsistent judgment and may indicate an unclear understanding of the stimuli, or a confused or poor decision rule. It might also indicate that one attribute of stimulus 4 was so far superior to that of stimulus 5 that it became the sole determinant of the choice made in that particular comparison. The paired-comparison technique permits and identifies inconsistencies that would be lost in more traditional ranking approaches (Ross, 1976).

Hyman, Moreau and Stiftel (1982) have indicated that substantial advances were made during the first decade of environmental impact studies relative to eliciting and displaying expert judgments about the relative importance of effects. In order to further these advances they developed a new environmental assessment method known as SAGE. SAGE stand for Social-judgment capturing, Adaptive, Goals-achievement, Environmental assessment, because it builds on the best characteristics of several existing methods. The concepts of SAGE can be used in the selection of an aquifer restoration strategy from a set of alternatives.

Table 6.9   Example Comparison Matrix (Ross, 1976)

Stimulus[a]	1	2	3	4	5	6	Row Sum
1		1	1	1	0	0	3
2	0		0	1	0	0	1
3	0	1		1	0	0	2
4	0	0	0		0	0	0
5	1	1	1	1		1	5
6	1	1	1	1	0		4

[a]Decision factors.

Table 6.10   Permuted Example Comparison Matrix (Ross, 1976)

Stimulus	5	6	1	3	2	4	Row Sum
5		1	1	1	1	1	5
6	0		1	1	1	1	4
1	0	0		1	1	1	3
3	0	0	0		1	1	2
2	0	0	0	0		1	1
4	0	0	0	0	0		0

Table 6.11   Permuted Example Comparison Matrix, Inconsistent (Ross, 1976)

Stimulus	5	6	1	3	2	4	Row Sum
5		1	1	1	1	0	4
6	0		1	1	1	1	4
1	0	0		1	1	1	3
3	0	0	0		1	1	2
2	0	0	0	0		1	1
4	1	0	0	0	0		1

SAGE must be understood within an overall framework for assessing the relative social worth of alternative projects or policies. The general framework is that of the rational planning model, which is shown in Figure 6.2 as consisting of two phases: (1) a design phase; and (2) an analytical phase. The design phase includes the tasks of (1) identifying objectives, (2) establishing planning guides and criteria, and (3) searching for and synthesizing alternative designs. The analytical phase comprises four tasks: (1) identifying and predicting operational measures of the physical, chemical, and biological attributes of each alternative;

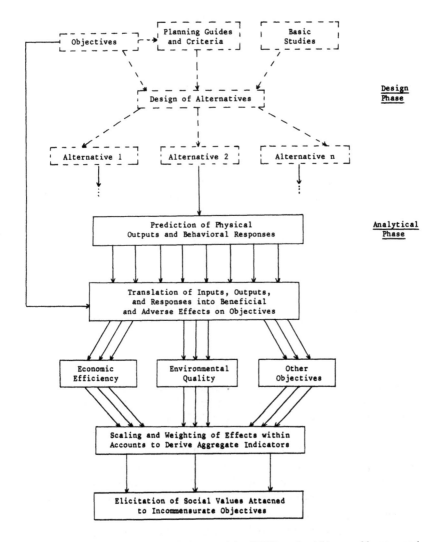

Figure 6.2. Design and analytical phases of the SAGE method (Hyman, Moreau, and Stiftel, 1982).

(2) scaling or translating those consequences into accounts of beneficial and adverse effects on each objective including behavioral consequences; (3) eliciting relative value weights that affected individuals or groups attach to the contributions to each objective when they rank alternatives preferentially; and (4) presenting the results in a form that is useful to decision-makers. This division of the rational planning model is conven-

ient because SAGE is concerned primarily with the analytical phase (Hyman, Moreau and Stiftel, 1982).

The basic concept of the SAGE method is that a system of accounts with characteristic attributes is used as the decision factors. A relative importance coefficient (RIC) must be derived for each attribute within each account. The RIC for each attribute is calculated using a scoring procedure of attributes in an account involving pairwise comparisons of all combinations (Dean and Nishry, 1965). The procedure ensures that the sum of the RIC within each account is unity. An aggregate index of the account score for an objective is obtained by taking a linear combination of the scaled effects of the attributes where each attribute is weighted by its RIC. This last step can be summarized mathematically as:

$$X_{ij} = \sum_k (RIC_{jk})(G_{ijk})$$

where (RIC_{jk}) = the relative importance coefficient for attribute k under the jth objective,

(G_{ijk}) = the effect of the ith alternative on each attribute,

and X_{ij} = the scaled score of the ith alternative on the jth objective.

As noted earlier, the importance weighting approach used by Hyman, Moreau and Stiftel (1982) was the same as that used by Dean and Nishry (1965). Hyman, Moreau and Stiftel (1982) noted that the approach has some limitations. For example, the largest RIC that can be assigned to any attribute is (2/n), where n is the number of attributes. Also, the entire importance weighting process relies heavily upon expert judgment. Nonetheless, when accompanied by an explanatory text and a full description of the actual effects, the method can be useful for summarizing and comparing features of alternatives.

The unique aspect of the SAGE method is that there is weighting of the accounts relative to each other. For example, Table 6.1 lists four major decision factors (health risk, economic efficiency, social concerns, and environmental impacts) which would be called accounts in the SAGE method. Attributes within an account listed in Table 6.1 include biophysical, cultural, and socioeconomic impacts. The SAGE method elicits weights between accounts by a technique based on social judgment theory (Hyman, Moreau and Stiftel, 1982). The technique involves presenting respondents with an identical set of cards. Each card in the deck contains a verbal description of the effects of an alternative on each of the decision factors, as well as the scaled numerical scores for all of the accounts. In this Q-sort procedure, the participants are asked to arrange the cards in order of their preferences and then to score the alternatives on a scale ranging from 0 to 100.

Next, the analyst infers the weights from these rankings through the use of regression analysis methods. To do so, it is necessary to hypothesize a model that relates each participant's rankings of an alternative on an interval scale to the contribution of the alternative to each of the objectives. In mathematical terms,

$$Y_i = f(X_{i1}, X_{i2}, X_{i3}, X_{i4}, \ldots \ldots X_{in})$$

where Y_i is the score assigned to the ith alternative and the variables X_{i1} to X_{in} refer to the contributions of the ith alternative to the n objectives. In estimating the model, an error term is added to reflect inconsistencies in the participant's application of the model and also failures of the model to replicate a participant's actual weighting process. The simplest form of the regression model is linear:

$$Y_i = a + b_1X_{i1} + b_2K_{i2} + \ldots b_nX_{in} + e_i$$

where a is a constant, the b variables are weights attached to each of the n objectives, and e_i is the error term. Given a set of n alternatives and data on the X variables and Y_i for the set of participants, the analyst can infer the weights and perform an analysis of variance. If there is reason to believe that the rankings include nonlinearities or interactions, a more complex, polynomial model can be used instead.

Ranked Paired Comparison Techniques

The key feature of ranked pairwise comparison techniques relative to unranked techniques is that an initial ranking of all decision factors is required. Two examples of the use of ranked pairwise comparisons in environmental impact studies will be cited. The first deals with importance weighting for water resources projects (Dee et al., 1972), and the second with importance weighting for a waterway navigation project (School of Civil Engineering and Environmental Science and Oklahoma Biological Survey, 1974). These examples could be adapted for usage in selecting an aquifer restoration strategy.

The general methodology developed for water resources projects is called the Environmental Evaluation System (EES). The relative importance of the 78 parameters in the EES was expressed in commensurate units, called parameter importance units (PIU), by quantifying several individuals' subjective value judgments. The weighting technique used by the method developers, Battelle-Columbus, was based on sociopsychological scaling techniques and the Delphi procedure (Dee et al., 1972). The Delphi procedure will be discussed later. The importance weighting technique in the EES is systematic, minimizes individual bias, produces consistent comparisons, and aids in the convergence of judgment.

The ranked pairwise comparison technique used by Battelle-Columbus is a commonly used sociopsychological scaling technique. In ranked pairwise comparison, the list of decision factors (or parameters) to be compared is ranked according to selected criteria and then successive pairwise comparisons are made between contiguous parameters to select for each parameter pair the degree of difference in importance. A weighted list of the parameters is the output from this procedure. The initial ranking was made based on considering the following three criteria relative to each parameter (decision factor):

(1) Inclusiveness of parameter
(2) Reliability of parameter measurements
(3) Sensitivity of parameter to changes in the environment

It should be noted that the 78 parameters are basically environmental parameters grouped into four categories (ecology, environmental pollution, aesthetics, and human interest) and 17 quantitative components (species and populations, habitats and communities, water pollution, air pollution, land pollution, noise pollution, land, air, water, biota, manufactured objects, composition, educational/scientific packages, historical packages, cultures, mood/atmosphere, and life patterns) as shown in Figure 6.3 (Dee et al., 1972). The following ten steps were used in the importance weighting technique:

Step 1: Select a group of individuals for conducting the evaluation and explain to them in detail the weighting concept the use of their rankings and weightings.

Step 2: Rank the categories, components, or parameters that are to be evaluated.

Step 3: Assign a value of 1 to the first category on the list. Then compare the second category with the first to determine how much the second is worth compared to the first. Express this value as a decimal ($0 < x \leq 1$).

Step 4: Continue with these pairwise comparisons until all in the list have been evaluated (Compare 3rd with 2nd, 4th with 3rd, etc.)

Step 5: Multiply out percentages and express over a common denominator, and average overall individuals in the experiment.

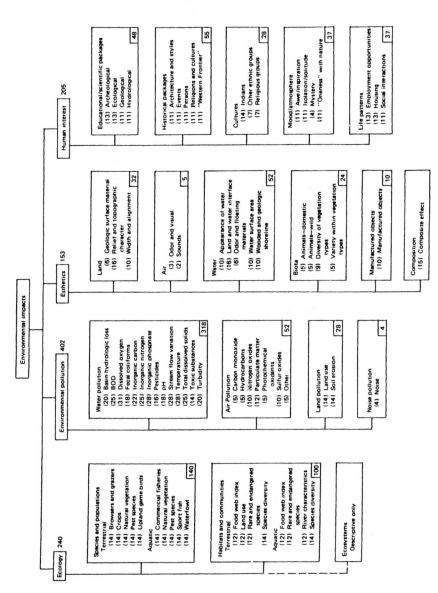

Figure 6.3.   Environmental evaluation system (Dee et al., 1972).

Step 6: In weighting the categories or components, adjust the decimal values from Step 5 if unequal numbers of parameters exist in the parameter groups being evaluated. Adjustment is made by proportioning these decimal values in proportion to the number of parameters included in that grouping.*

Step 7: Multiply these averages the the number of parameter importance units to be distributed to the respective grouping.

Step 8: Do Steps 2-7 for all categories, components, and parameters in the EES.

Step 9: Indicate to the individuals by controlled feedback the group results of the weighting procedure.

Step 10: Repeat the experiment with the same group of individuals or another group to increase the reliability of the results.

The following numerical example illustrates the use of the 10 steps (Dee et al., 1972). Consider three components (A, B, C) that have been selected in Steps 1 and 2, with these components consisting of 8 parameters, 4 in A, 2 in B, and 2 in C.

Step 2   Ranking of component—B, C, A

Steps 3,4 Assign weights

> B = 1
> C = ½ the importance of B
> A = ½ the importance of C

Step 5 Multiply out percentages and express over common denominator. Assume the average values of all individuals are given below.

> B = 1
> C = 0.5
> A = 0.25
> ‾‾‾‾
> 1.75

---

*The hierarchical system shown in Figure 6.3 has an unequal number of elements in each grouping. To be mathematically correct all levels of the EES hierarchy should have an equal number of elements. However, at the present time, there has not been sufficient knowledge in many of these areas to permit an equal number of elements at the same level of detail. This difference in the number of elements from group to group must be taken into consideration when assigning parameter importance units in the ranking and weighting. Therefore, in the ranking and weighting procedure (Steps 1-5), the researchers were asked to assume an equal number of elements in the groupings being compared. These value judgments were then "adjusted" in proportion to the number of elements in each group. Because the purpose of the weighting procedure was to assign weights to parameters, if an "adjustment" were not made the individual parameters grouped under water pollution would not receive sufficient weight as compared to those under noise because the total number of units available under water pollution would have to be distributed among 14 parameters whereas those for noise would all be assigned to the single noise parameter. For this reason, comparisons between components and categories should be based on average values of the PIU for the grouping not the sum of the values.

$$B = 1/1.75 = 0.57$$
$$C = 0.5/1.75 = 0.29$$
$$A = 0.25/1.75 = \underline{0.14}$$
$$\overline{1.00}$$

Step 6   Adjust for unequal number of parameters in each component.

$$B = 0.57 \times \frac{1}{4} = 0.14$$
$$C = 0.29 \times \frac{1}{4} = 0.07$$
$$A = 0.14 \times \frac{1}{2} = \underline{0.07}$$
$$\overline{.28}$$

Using the new total, the components values are

$$B = 0.14/0.28 = 0.50$$
$$C = 0.07/0.28 = 0.25$$
$$A = 0.07/0.28 = \underline{0.25}$$
$$\overline{1.00}$$

and the average values are

$$B = 0.50/2 = 0.25$$
$$C = 0.25/2 = 0.135$$
$$A = 0.25/4 = 0.0625$$

Step 7   Multiply these adjusted values by appropriate PIU, which is assumed in this case to be 20.

$$20 \times 0.5 = 10$$
$$20 \times 0.25 = 5$$
$$20 \times 0.25 = 5$$

Step 8   Continue until reliable estimates are obtained.

In the EES, a total of 1000 PIU are assigned to the parameters by first distributing to the four categories, then to the 17 quantitative components, and finally, to the 78 parameters. That is, the participating group specifies, for example, which is more important, aesthetics or environmental pollution, and then assigns appropriate weights. The process is continued until all the units are distributed among all parameters. Dee et al. (1972) noted that instead of using the initial weight resulting from the importance weighting procedure, an aggregate weight based on several iterations of the technique is preferred. After each iteration, the participants are given selected information about the group weights. This information can include the group mean and variance, or other pertinent information. In the weighting procedure employed in developing the EES, the participants' mean value was given in the feedback stage. All of the weighting and feedback was performed via formal feedback statements, thereby avoiding undesirable direct interchange of judgments of the individuals in the test.

The waterway navigation project included an evaluation of the environmental impacts of nine alternatives (eight waterway locational routes and the no-action alternatives) for extending waterway navigation from Tulsa, Oklahoma to Wichita, Kansas (School of Civil Engineering and Environmental Science and Oklahoma Biological Survey, 1974). Figure 6.4 presents a flow diagram of the environmental analysis approach used in the study. The multidisciplinary team consisted of an environmental engineer, botanist, zoologist, planner, and archaeologist. Six basic environmental categories were selected, and a total of 102 specific parameters were identified and grouped into the categories. The categories and parameters are listed in Table 6.12 (School of Civil Engineering and Environmental Science and Oklahoma Biological Survey, 1974).

The research team discussed the six categories and agreed to following rank order of decreasing importance: biology, physical-chemical, regional compatibility, archaeology, aesthetics, and climatology. A total of 1000 importance points (or weights) were distributed by the ranked pairwise comparison technique as used in the EES (Dee et al., 1972). In most cases within a category, the parameters were classified as to high, medium, or low importance, and the allotted points for that particular category were distributed accordingly. High importance parameters were allotted three times as many points as low importance parameters, and medium importance parameters were allotted twice as many points as the low importance parameters.

Evaluation of the nine alternatives in the waterway navigation project relative to the 78 parameters was made via the use of parameter function graphs yielding environmental quality (EQ) scores. Development and usage of the graphs will be discussed later. The overall impact evaluation was made via the development of Route Scores for each alternative, with the Route Scores defined as follows (School of Civil Engineering and Environmental Science and Oklahoma Biological Survey, 1974):

$$\text{Route Score}_j = \sum_{i=1}^{102} (EQ)_{ij}\, IW_i$$

where EQ_{ij} = environmental quality for jth alternative relative to ith parameter.
       IW_i = importance weight of ith parameter.

Since the importance weights for the six categories and 102 parameters involved considerable subjective judgment, the weight term in the above Route Score equation was allowed to vary randomly ($\pm$ 50%) in calculating Route Scores. A computer program was used to calculate an average route score following 25 iterations of the above equation.

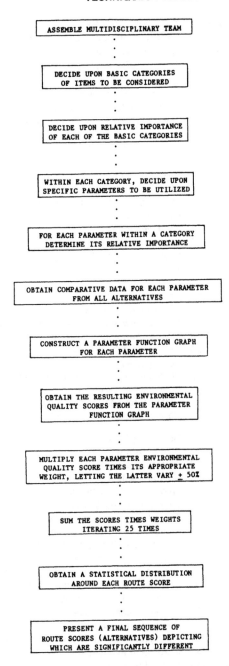

Figure 6.4.   Flow diagram of environmental analysis procedure (School of Civil Engineering and Environmental Science and Oklahoma Biological Survey, 1974).

Table 6.12   Environmental Categories and Parameters for Waterway Navigation Project (School of Civil Engineering and Environmental Science and Oklahoma Biological Survey, 1974)

Category	Parameter
Biology	1. Biological effect of water quality change
	2. Biological effect of salinization change
	3. Biological effect of eutrophication change
	4. Biological implications of siltation
	5. Browser populations (deer)
	6. Change from lotic to lentic habitat
	7. Change from lowland feeding and breeding habitats
	8. Changes in lowland nesting habitat
	9. Expansion of population ranges
	10. Fishing pressure
	11. Food web index
	12. Grazers (cattle)
	13. Interruption of wildlife refuges
	14. Forest land removed
	15. Cropland removed
	16. Grassland removed
	17. Migration of waterfowl
	18. Miles of new shoreline (channel)
	19. Miles of new shoreline (lake)
	20. Number of oxbow lakes
	21. Pest species
	22. Potential areas for fish and wildlife management
	23. Potential for recreation areas
	24. Potential for sport and commercial fishing
	25. Rare and endangered species (plants and animals)
Physical/Chemical	26. Recruitment from tributaries
	27. Size of oxbow lakes
	28. Terrestrial ecosystem stability or diversity
	29. Unique habitats
	30. Weedy aquatic vegetation
	31. Weedy terrestrial vegetation
	32. Air pollution from construction
	33. Alteration of waste assimilative capacity
	34. Blasting and drilling (noise and dust)
	35. Change in appearance of water
	36. Change in salt load carried by stream
	37. Change in sediment load carried by stream
	38. Commitment of resources
	39. Degree of change in natural surface drainage
	40. Depth of cuts and effect on natural setting
	41. Flow modification and reduction in flood damage
	42. Increase in evaporation
	43. Miles of new stream
	44. Number of acres of land lost
	45. Number and type of new bridges

Table 6.12, continued

Category	Parameter
	46. Number of lakes created
	47. Number of locks and route distance (travel time in route)
	48. Number of recreational areas developed and the resulting pollution
	49. Noise from operation
	50. Other noise from construction
	51. Pollution from shipping
	52. Potential for affecting municipal pollution
	53. Potential for shipping accident
	54. Relocations (utilities, railroads, highways)
	55. Size of work force and effects of temporary camps
	56. Susceptibility of cuts to erosion
	57. Total quantity of excavation
	58. Water and air pollution resulting from industrialization
Regional Compatibility	59. Accessibility to first order transportation (non-rail)
	60. Accessibility to railroad transportation
	61. Accessibility to second order transportation
	62. Conformance to existing and developing urban patterns
	63. Construction-intermediate urban to site distance (impact of construction phase)
	64. Construction-major urban to site distance (impact of construction phase)
	65. Construction-minor urban to site distance (impact of construction phase)
	66. Proximity to urban centers (intermediate)
	67. Proximity to urban centers (major)
	68. Proximity to urban centers (minor)
	69. Severance of surface transport routes
Archeology	70. Age or occupational period of the site
	71. Concern by local population
	72. Cost of conducting on foot archaeological survey
	73. Depth of the occupational area
	74. Ecological setting
	75. Eligibility of the site for state or national register
	76. Estimated number of sites to be obtained by the survey
	77. Importance in terms of local, state, and national level
	78. Minimum salvage costs for estimated sites
	79. Nature of the site
	80. Number of known sites to be damaged
	81. Presence of single or multiple occupations
	82. Preservation of archaeological data
	83. Previous knowledge of the area
	84. Site frequency within the area concerned
	85. Site importance in terms of geographic area and problems
	86. Site preservation or damages
	87. Size of occupational area
	88. Value of site for nonarchaeological fields

Table 6.12, continued

Category	Parameter
Aesthetics	89. Atmospheric turbidity
	90. Confinement of view
	91. Degree of alteration of natural setting
	92. Degree of urbanization
	93. Diversity of vegetation types
	94. Landform diversity
	95. Man-made visual obstructions
	96. River pattern (channelization)
	97. Turbidity
	98. Wildlife diversity
Climatology	99. Increase in fog frequency
	100. Changes in microclimates ($\pm$ 1 mile)
	101. Evaporative potential
	102. Inadvertent weather modification

DELPHI TECHNIQUE FOR IMPORTANCE WEIGHTING

The Delphi technique is a structured approach for achieving group concensus about common issues of concern. It was used in a modified form in the development of importance weights in the EES described earlier (Dee et al., 1972). The Delphi technique has been compared to other group communication techniques as shown in Table 6.13 (Linstone and Turoff, 1975). Conventional Delphi is the method discussed herein, while real time Delphi, that is, computer conferencing, refers to a variation of conventional Delphi. The technique has broad versatility and the results are easy to understand. Application of the method involves a few days to as much as three months which makes the cost low to medium compared to other methods. The expertise required to administer the study ranges from very little to average and the assistance required ranges from none to highly sophisticated computer data analysis (Stanford Research Institute, 1975). The Delphi technique could be used to weight decision factors associated with the selection of an aquifer restoration strategy for meeting a given need.

Key elements in the successful conduction of a Delphi study are the study director and the chosen panel of experts. The panel of experts chosen may come from academic, governmental, consulting or industrial backgrounds (Toussaint, 1975; Martino, 1972). There are three other

Table 6.13 Group Communication Techniques (Linstone and Turoff, 1975)

	Conference Telephone Call	Committee Meeting	Formal Conference or Seminar	Conventional Delphi	Real-Time Delphi
Effective Group Size	Small	Small to medium	Small to large	Small to large	Small to large
Occurrence of Interaction by Individual	Coincident with group	Coincident with group	Coincident with group	Random	Random
Length of Interaction	Short	Medium to long	Long	Short to medium	Short
Number of Interactions	Multiple, as required by group	Multiple, necessary time delays between	Single	Multiple, necessary time delays between	Multiple, as required by individual
Normal Mode Range	Equality to chairman control (flexible)	Equality to chairman control (flexible)	Presentation (directed)	Equality to monitor control (structured)	Equality to monitor control or group control and no monitor (structured)
Principal Costs	Communications	Travel; Individuals' time	Travel; Individuals' time; Fees	Monitor time; Clerical; Secretarial	Communications; Computer usage
Other Characteristics	Time-urgent considerations; Equal flow of information to and from all; Can maximize psychological effects	Forced delays; Equal flow of information to and from all; Can maximize psychological effects	Efficient flow of information from few to many	Forced delays; Equal flow of information to and from all; Can minimize psychological effects; Can minimize time demanded of conferees	Time-urgent considerations; Can minimize time demanded of respondents or conferees

types of panelists: stakeholders (those directly affected), experts who can supply experience and special application techniques, and facilitators who can clarify, organize or stimulate (Linstone and Turoff, 1975). The director nominates outside experts based on a combination of their performance, years of experience, number of publications, and status with peers (Helmer, 1966). Expertise can also be assessed by contrasting it with public views of whomever is willing to buy and pay for each individual opinion. Those selected may be asked to nominate other panel members. The availability and willingness to serve for the entire study is an important consideration. The number of experts chosen is usually seven to ten for environmental issues, although more panelists are used in forecasting social impacts (Toussaint, 1975). Studies have shown that when employing a group to arrive at a decision, the average group error decreases with an increase in the number of individuals. As the panel increases in size there is an increase in reliability up to 15 panel members (Gordon and Helmer, 1964).

Several rounds of questionnaires are used in the conventional Delphi technique. The questionnaires may be prepared before or during the selection of panel members. Once the questionnaires have been designed and the panel selected, the next step is the administration of the Delphi sequence which involves the mailing of questionnaires and their return to the director. The director compiles and edits the information received for successive rounds.

If the Delphi technique is being used for importance weighting, the objective of round one is to get an initial estimate of the ranking and relative importance weights of the decision factors. The objective could be to identify potential decision factors given information about the environmental setting and potential project. The information generated the first round must be carefully reviewed by the director and duplications eliminated. The director determines the median (point of concensus) and the spread of opinion which is expressed by the interquartile range (IQR) encompassing 25% of the responses above and below the median. The panelists receive this written feedback through the mail which eliminates the noise of an open discussion and the dominance of certain individuals in the panel.

In round two panelists weigh the strength of their own convictions against the group median and may even revise the ranking and relative importance of their round one estimates. Experts who give responses outside the IQR must write down their reasons. On the average, the questionnaire and written material for the second round will be five to ten times that of the first round (Linstone and Turoff, 1975). The developments are reordered by using a combination of probability, desirability, or feasibility scales. In many cases it may be desirable to keep track of certain subgroups making up the respondent group as a whole. This provides a mechanism to decide whether polarized views reflect the affiliations or the background of the respondents. The director compiles round two estimates and presents the new median, IQR, and reasons for their distribution in round three.

Round three is conducted similar to round two and again experts write down their reasons for disagreeing with estimates outside the new IQR. The initial disagreers and new ones have a chance to convince other panelists to change their estimates. The responses are summarized by median, IQR, disagreement reasons and counter arguments. If it is the last round, the director summarizes majority and minority opinions on projections for each development along with the above information before a revote is taken. If possible, the revote should be put off until a fourth round when everyone can see the additional remarks. In the fourth round a revote is taken and the director compiles the new median, IQR and reasons for consensus for each issue being addressed. Delphi studies could utilize two to five rounds, but it has been found that responses tend to stabilize after three rounds (Helmer, 1966). Four rounds may be necessary to give minority opinions a chance at shifting the median and IQR (Toussaint, 1975).

A number of major criticisms have been made against usage of the Delphi technique. Helmer (1966) identified the following criticisms:

(1) There is instability in the panel memberships. Many studies have shown that only a few of the panelists respond to questionnaires, many drop out on succeeding rounds, and convergence is impeded by too many panel substitutions.

(2) A big time lapse between succeeding rounds can cause shifting of opinions.

(3) Ambiguous questions confuse panelists.

(4) Panelists' competence is questionable. Helmer (1966) suggests asking experts questions only in their particular field and to leave blanks when unsure of their own judgment.

On the positive side, there are several things which can be done to improve the results of the use of the Delphi technique for importance weighting. Martino (1972) recommended the following eight steps to insure the smooth conduction and reliability of a Delphi study:

(1) Obtain agreement of experts to serve on a panel.

(2) Explain the Delphi procedure thoroughly.

(3) Avoid compound events in one analysis. If the event statement contains one part the panelist agrees with, and another part he disagrees with, it is difficult for him to know how to answer. Avoid compound questions such as: Capability A will be achieved through the use of Technology B in the year _____. The panelist may

find two distinct parts for a single event, and this is where feedback is available to help the director improve his questions.

(4) Avoid ambiguous statements of events, and the careless use of technical jargon. Avoid using the terms "everybody knows," "common," "widely used," "normal," and "in general use."

(5) Make questions easy to answer such as filling in the blank, or multiple choice. The arguments for and against each event should be summarized so the panelists can connect them to specific questions.

(6) Ask no more than 25 to 40 questions in any one Delphi round.

(7) The director should never inject his own opinion(s). If the director becomes convinced that the panelists are overlooking some significant element, he should recognize that he has picked an unqualified panel and should repeat the work with another panel.

(8) The director should be able to estimate the work load involved for each panel member in a Delphi sequence. As a planning factor, the director should allow two professional man-hours per panelist per questionnaire. When using 100 or more panelists it is advisable to use a computer to aid in the data analysis.

Linstone and Turoff (1975) have made other procedural recommendations which are stated below:

(1) Lay out the expected processing of the data throughout all the rounds of the Delphi before you finalize the design. You may be forced to later modify the procedure, but the process of planning ahead will usually turn up any large problems in your initial questionnaire design.

(2) Design the handling of data so that each response can be processed as it comes in. By doing this you will not have a frantic rush to analyze all the responses at once when the last return comes in.

(3) At least two professionals should work on monitoring a Delphi exercise, especially when the abstracting of comments is a good portion of the exercise. With two individuals one can always review what the other has done.

(4) Pretest your questionnaire on a group before using it for the rounds sequence.

(5) If you are covering a number of fields of expertise, make sure each field is adequately represented in your group.

(6) The criteria for retaining an item for further evaluation should be made clear at the beginning of the Delphi sequence.

(7) Interpersonal techniques such as interviews and seminars should be interspersed with the rounds of questionnaires and information feedback.

(8) When editing respondents' comments, try to preserve the intent of the originator. When editing from round to round, avoid changing a statement so that it has one meaning in one round and another meaning in another round.

(9) Standardized measures should be available to a respondent so he can self-rate his competence or familiarity to specific questions.

(10) If a multidisciplinary approach is desired respondents should be encouraged to consider all items, but to make estimates only on those items with which they feel comfortable. Respondents may indicate their familiarity with a specialized area or the importance of an item, without making probability estimates.

(11)The source of a suggested item should be identified taking care not to compromise the anonymity of specific inputs. Keep track of how different subgroups in your respondent group vote on specific items. This can be very useful in analyzing the results and will produce situations where you want the respondents to know the existence of polarizations or differences based on particular backgrounds.

To serve as an example for an environmentally related study, Toussaint (1975) used the Delphi technique to develop importance weights for 14 water pollution parameters related to the proposed Aubrey Reservoir project in northern Texas. Eighteen environmental experts from the region established regional values for the two judgment-based decisions currently used in environmental impact assessment. These two decisions were referred to as the weighting (determination of the relative emphasis or degree of importance each parameter is to be assigned in the assessment) and scaling (determination of the magnitude of effect resulting from a change in a parameter measure) process. The Delphi Procedure was used to reach a consensus of opinion.

SCALING/RATING/RANKING OF ALTERNATIVES

Scaling/rating/ranking of each alternative for each decision factor is the second major aspect in the use of the multicriteria decision-making approach. Several different techniques have been used for this evaluation of alternatives in a decision. Examples of techniques which have potential applicability to decisions related to the selection of an aquifer restoration strategy for meeting a given need include the use of (1) the alternative profile concept, (2) a reference alternative, (3) linear scaling based on the maximum change, (4) letter or number assignments designating impact categories, (5) evaluation guidelines, (6) functional curves, or (7) the paired-comparison technique. Examples of each of these techniques are as follows:

(1) Bishop et al. (1970) contains information on the alternative profile concept for impact scaling. This concept is represented by a graphical presentation of the effects of each alternative relative to each decision factor. Each profile scale is expressed on a percentage basis ranging from a negative to a positive 100%, with 100% being the maximum absolute value of the impact measure adopted for each decision factor. The impact measure represents the maximum change, either plus or minus, associated with a given alternative being evaluated. If the decision factors are displayed along with the impact scale from +100% to -100%, a dotted line can be used to connect the plotted points for each alternative and thus describe its "profile". The alternative profile concept is useful for visually displaying the relative impacts of a series of alternatives.

(2) Salomon (1974) describes a scaling technique for evaluation of cooling system alternatives for nuclear power plants. To determine scale values, a reference cooling system was used and each alternative system compared to it. The following scale values were assigned to the alternatives based on the reference alternative: very superior ($+8$), superior ($+4$), moderately superior ($+2$), marginally superior ($+1$), no difference (0), marginally inferior (-1), moderately inferior (-2), inferior (-4), and very inferior (-8).

(3) Odum et al. (1971) utilized a scaling technique in which the actual measures of the decision factor for each alternative plan are normalized and expressed as a decimal of the largest measure for that factor. This represents linear scaling based on the maximum change.

(4) A letter scaling system is used in Voorhees and Associates (1975). This methodology incorporates 80 environmental factors oriented to the types of projects conducted by the Department of Housing and Urban Development. The scaling system consists of the assignment of a letter grade from A+ to C- for the impacts, with A+ representing a major beneficial impact and C- an undesirable detrimental change.

(5) Duke et al. (1977) describe a scaling checklist for the EQ account for water resources projects. Scaling is accomplished following the establishment of an evaluation guideline for each environmental factor. An evaluation guideline is defined as the smallest change in the highest existing quality in the region that would be considered significant. For example, assuming that the highest existing quality for dissolved oxygen in a region is 8 mg/l, if a reduction of 1.5 mg/l is considered as significant, then the evaluation guideline is 1.5 mg/l irrespective of the existing quality in a given regional stream. Scaling is accomplished by quantifying the impact of each alternative relative to each environmental factor, and if the net change is less than the evaluation guideline it is insignificant. If the net change is greater and moves the environmental factor toward its highest quality, then it is considered to be a beneficial impact; the reverse exists for those impacts that move the measure of the environmental factor away from its highest existing quality.

(6) A paired-comparison technique can also be used for assigning scale values to alternatives based on their impact on environmental factors. The paired comparison technique for accomplishing scaling is described in Dean and Nishry (1965).

(7) Functional curves can also be used to accomplish impact scaling for environmental factors. The functional curve is used to relate the objective evaluation of an environmental factor to a subjective judgment regarding its quality based on a range from high to low quality (Dee et al., 1972).

Paired Comparison Technique for Scaling/Rating/Ranking

One of the most useful techniques for scaling/rating/ranking of alternatives relative to each decision factor is the paired-comparison technique described by Dean and Nishry (1965). This technique was also described earlier relative to its use for importance weighting of decision factors. Again, this technique can be used by an individual or group for the scaling/rating/ranking of alternatives. Suppose that the decision to be made involves four decision factors and that the importance weights have been assigned as shown in Table 6.8. Furthermore, suppose that there are three alternatives (A1, A2, and A3) to be evaluated relative to the four decision factors, and that Table 6.14 contains the relevant qualitative and quantitative information. The scaling/rating/ranking technique consists of considering each alternative relative to every other alternative and assigning to the one of the pair considered to be the most desirable a value of 1, and to the lesser desirable of the pair a value of 0. The use of this paired comparison technique for the three basic alternatives and four basic decision factors is shown in Tables 6.15 through 6.18, respectively. It should be noted that the assignment of 0 to a member of a pair does not denote zero desirability; it simply means that in the pair considered it is of lesser desirability.

A dummy alternative, called A4, is also included in Tables 6.15 through 6.18. The dummy alternative is included so as to preclude the net

Table 6.14   Information for Trade-Off Analysis

Decision Factor	Alternatives		
	A1	A2	A3
F1	Has lowest health risk	Has highest health risk	Has medium degree of health risk
F2	Medium economic efficiency	Low economic efficiency	High economic efficiency
F3	Undesirable social impacts expected	No social impacts expected	Beneficial social impacts expected
F4	Decrease overall environmental quality by 20%	Increase overall environmental quality by 10%	Increase overall environmental quality by 10%

Table 6.15   Scaling/Rating/Ranking of Alternatives Relative to F1

Alternative	Assignment of Desirability						Sum	ACC
A1	1	1	1				3	0.50
A2	0			0	1		1	0.17
A3		0		1		1	2	0.33
A4 (dummy)			0		0	0	0	0
							6	1.00

Table 6.16 Scaling/Rating/Ranking of Alternatives Relative to F2

Alternative	Assignment of Desirability						Sum	ACC
A1	1	0	1				2	0.33
A2	0			0	1		1	0.17
A3		1		1		1	3	0.50
A4 (dummy)			0		0	0	0	0
							6	1.00

Table 6.17 Scaling/Rating/Ranking of Alternatives Relative to F3

Alternative	Assignment of Desirability						Sum	ACC
A1	0	0	1				1	0.17
A2	1			0	1		2	0.33
A3		1		1		1	3	0.50
A4 (dummy)			0		0	0	0	0
							6	1.00

Table 6.18 Scaling/Rating/Ranking of Alternatives Relative to F4

Alternative	Assignment of Desirability						Sum	ACC
A1	0	0	1				1	0.17
A2	1			0.5	1		2.5	0.415
A3		1		0.5		1	2.5	0.415
A4ʻ(dummy)			0		0	0	0	0
							6.0	1.00

assignment of a value of 0 to any of the basic alternatives (A1 through A3) in the process of each paired comparison. The dummy alternative is defined as the least desirable in each paired comparison within which it is included. If two alternatives have the same desirability relative to a decision factor (one is not more desirable than another), then a value of 0.5 is assigned to each of the pair.

Following the assignment of the relative desirability of each alternative relative to each other alternative, with this process based on the qualitative and quantitative information in Table 6.14, then individual choice assignments are summed, with the alternative choice coefficient (ACC) being equal to the sum value for an individual alternative divided by the sum for all of the alternatives. The total of the sum column in Tables 6.15 through 6.18 should equal (M)(M-1)/2 where M is equal to the number of alternatives included in the assignments. In this example, four alternatives were included, hence the sum column total in Tables 6.15 through 6.18 should be equal to 6. The total of the ACC column should equal to 1.00.

The ACC column in Table 6.15 indicates that alternative A1 is the most desirable relative to decision factor F1, and is followed by A3 and A2. Similar types of comments could be made for the ACC values in Tables 6.16 through 6.18. Whether or not the actual ACC fractions are used in a trade-off analysis, this paired comparison approach has enabled the rank ordering of the desirability of each alternative relative to each decision factor. As noted earlier, the paired comparison technique based on Dean and Nishry (1965), including the development of both FIC and ACC values, has been used in a number of environmentally related studies (Morrison, 1974; Canter, 1976; Solomon et al., 1977; and Canter and Reid, 1977). This technique could be easily applied to decision-making needs associated with the selection of an aquifer restoration strategy for meeting a given need.

Functional Curves for Scaling/Rating/Ranking

Functional curves, also called value functions and parameter function graphs, have been used in a number of environmental impact studies for scaling/rating/ranking the impacts of alternatives relative to a series of decision factors. They have potential for application in aquifer restoration programs. Dee et al. (1972) described the following seven steps which could be used in developing a functional curve (relationship) for an environmental parameter (decision factor):

Step 1: Obtain scientific information when available on the relationship between the parameter and the quality of the environment. Also, obtain experts in the field to develop the value functions.

Step 2: Order the parameter scale so that the lowest value of the parameter is zero and it increases in the positive directions — no negative values.

Step 3: Divide the quality scale (0–1) into equal intervals and express the relationship between an interval and the parameter. Continue this procedure until a curve exists.

Step 4: Average the curves over all experts in the experiment to obtain a group curve. (For parameters based solely on judgment, value functions should be determined by a representative population cross section.)

Step 5: Indicate to the experts doing the value function estimation the group curve and expected results of using the curves in the EES. Decide if a modification is desired; if needed go to Step 3, if not continue.

Step 6: Do Steps 1–5 until a curve exists for all parameters.

Step 7: Repeat experiment with the same group or another group of persons to increase the reliability of the functions.

The same basic approach as used by Dee et al. (1972) was also described by Rau (1980). Parameter function graphs were also developed and used in the environmental impact study for the waterway navigation project (School of Civil Engineering and Environmental Science and Oklahoma Biological Survey, 1974). Examples of two functional curves are shown in Figures 6.5 and 6.6 (Dee et al., 1972). Usage of the curves would involve entering the x-axis with extant or predicted information and reading the resultant y-axis environmental quality scale.

DEVELOPMENT OF DECISION MATRIX

The final step in multicriteria decision-making is to develop a decision matrix displaying the products of the importance weights (or ranks) and the alternative scales (or ranks). To complete the example using the paired comparison technique (Canter, 1983), Table 6.19 summarizes the FIC values for the four decision factors (from Table 6.8), and the ACC values for the three alternatives (from Tables 6.15 through 6.18). The final product matrix is shown in Table 6.20. Summation of the products

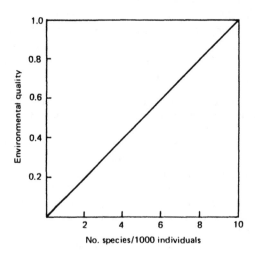

Figure 6.5. Functional curve for species diversity (Dee et al., 1972).

for each alternative indicates that alternative A3 would be the best choice following by A1 and A2. The bases for the differences in the three alternatives are indicated by the fractions shown in Table 6.20.

Rau (1980) provided an illustration of using importance weighting and impact rating to develop a total impact evaluation for four alternatives. As described earlier, importance weighting involved a group approach using scales of importance. Table 6.5 contains an example of the use of this approach (Rau, 1980). Impact rating was used based on predefined values as shown in Table 6.21 (Rau, 1980). The decision matrix is developed by multiplying each alternative's impact ratings by the importance weights from each of the decision factors.

In the Battelle EES an overall environmental impact index is calculated based on the following two steps (Dee et al., 1972):

(1) Obtain parameter data without the project for each of the 78 environmental factors. Convert these parameter data into EQ scale values for each of the 78 parameters. Multiply these scale values by the PIU for each of the individual parameters to develop a composite score for the environment without the project.

(2) For each alternative predict the change in the environmental parameters. Utilizing predicted changes in the parameter values, determine the environmental quality scale for each parameter and each alternative. Multiply the environmental quality values for each alternative by each PIU, and aggregate the information for a total composite score.

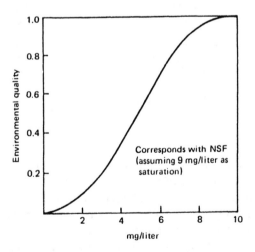

Figure 6.6.   Functional curve for dissolved oxygen in water (Dee et al., 1972).

Table 6.19   FIC and ACC Values for Example Decision Problems

Decision Factors	FIC Values	ACC Values		
		A1	A2	A3
F1	0.40	0.50	0.17	0.33
F2	0.20	0.33	0.17	0.50
F3	0.10	0.17	0.33	0.50
F4	0.30	0.17	0.415	0.415

Table 6.20   Product Matrix for Trade-Off Analysis for Example Decision Problem

Decision Factor	FIC × ACC		
	A1	A2	A3
F1	0.200	0.068	0.132
F2	0.066	0.034	0.100
F3	0.017	0.033	0.050
F4	0.051	0.124	0.124
	0.334	0.259	0.406

Table 6.21   Basis for Impact Rating (Rau, 1980)

Impact on Present Condition	Value
≥ 100% increase	+7
50 – 99.9% increase	+5
25 – 49.9% increase	+3
0 – 24.9% increase	+1
No change	0
0 – 24.9% decrease	−1
25 – 49.9% decrease	−3
50 – 99.9% decrease	−5
≥ 100% decrease	−7

The overall impact evaluations approach used in the waterway naviga-tion project was comprised of two parts (School of Civil Engineering and Environmental Science and Oklahoma Biological Survey, 1974). The first part, called Optimum Pathway Matrix Analysis (OPMA), uses a large number of data items from each route to derive a single value for each route on an arbitrary scale of 0 (low environmental quality) to 1000 (high environmental quality). The second part used is a multidimensional ordination of routes based on principal components analysis of parameter values (Jeffers, 1967). The ordination uses the data to construct a multidimensional ordering of the routes based on their similarity to each other rather than on some externally determined scale.

In the OPMA approach information from the importance weighting of 102 parameters was coupled with impact scaling information for the 102 parameters based on the use of parameter function graphs. As described earlier, Route Scores were developed for the nine alternatives. The OPMA procedure was converted to a Fortran computer program which builds an n times m matrix consisting of the scaled parameter values for n routes with m components for each route. Each component vector is then multiplied by its weight and each route vector is summed, giving a score for each route. Since the weights were subjectively determined, each route score was calculated 25 times with a different random error between $\pm 50\%$ introduced each time an importance weight was multiplied by a scaled parameter value. The random error was modified by a parabolic transformation of its frequency distribution such that the probability of an error greater than 45% is 0.02, while the probability of an error less than 5% is 0.2. Introduction of random errors and calculating each route score 25 times allows the calculation of confidence intervals and the use of statistical tests to determine whether differences between routes are significant.

The computer output for OPMA consisted of a list of scores for each route with the mean and confidence integral for that route, plus the results of a Student's T evaluation of the differences between routes, and a ranked list of the route scores. OPMA was done on each of the major categories (Biology, Physical/Chemical, Regional Compatibility, Archaeology, Aesthetics and Climatology) and on the total set of parameter values. In addition, OPMA was performed separately using parameters of high, medium and low importance. Finally, OPMA was done combining selected categories: Biology and Physical/Chemical; Biology, Physical/Chemical, Regional Compatibility, and Archaeology; and Biology, Physical/Chemical, Regional Compatibility, Archaeology and Aesthetics (School of Civil Engineering and Environmental Science and Oklahoma Biological Survey, 1974).

In the second part of the overall analysis of waterway navigation routes, a nonpolar principal components-based ordination was constructed for the routes by the method of Jeffers (1967). In this case (1) the data consisted of the matrix of scaled parameter values; (2) the basic data were not transformed; (3) a correlation matrix was calculated; and (4) parameters with scaled eigenvector values having absolute values greater than 0.7 were used in the calculation of route positions. A stan-

Table 6.22   Summary of Some Weighting-Scaling/Rating/Ranking Methods for Trade-Off Analyses

Author(s)	Salient Features
Canter (1976)	Weighting-scaling checklist using weighted rankings technique is described.
Canter and Reid (1977)	A weighting-scaling technique for evaluating the environmental impact of wastewater treatment process is described.
Crawford (1973)	Weighting-scaling checklist for evaluation of impacts of alternatives.
Dee et al. (1972)	Weighting-scaling checklist with a good listing of biological, physical-chemical, aesthetic and cultural variables.
Dee et al. (1973)	Weighting-scaling checklist based on relevant matrices and networks. Ranges of scale values. Concepts of "environmental assessment trees" to account for interrelationships among environmental factors.
Dunne (1977)	Weighting-scaling checklist for sanitary landfill site selection.
Gann (1975)	Weighting-scaling checklist using paired comparison technique is described.
Lower Mississippi Valley Division (1976)	Weighting-scaling checklist using habitat approach is presented.
Morrison (1974)	Weighting-scaling checklist using paired comparison technique is described.
Odum et al. (1971)	This weighting-scaling checklist includes an error term to allow for misjudgment in the assignment of importance weights. Computerization of the methodology enables the conduction of a sensitivity analysis.
Paul (1977)	Weighting-scaling checklist used to prioritize potential projects.
Raines (1972)	Weighting-scaling checklist; use of functional curves to translate environmental impact quantities into environmental cost units.
Salomon (1974)	Weighting-scaling checklist; relative scaling based on a reference alternative is used.
School of Civil Engineering and Environmental Science and Oklahoma Biological Survey (1974)	Weighting-scaling checklist, which is similar in concept to Dee et al. (1972); an error team is included to account for subjective misjudgments.
Smith (1974)	Weighting-scaling checklist for a rapid transit system.
Toussaint (1975)	Weighting and scaling of 14 water pollution parameters was established by the Delphi procedure using two separate groups of nine experts.
Wenger and Rhyner (1972)	A stochastic computer procedure is used to account for uncertainty in the weighting and scaling checklist procedure for evaluation of solid waste system alternatives.

Table 6.23 Delineation of a Methodology for Trade-Off Analysis Involving Environmental Impacts

Element	Delineation
A. Establish Interdisciplinary Team	1. Selection (a) Select members of interdisciplinary team. (b) Designate team leader. 2. Review and Familiarization (a) Review information on potential technologies. (b) Visit locations with technologies being applied.
B. Select Decision Factors and Assemble Basic Information	1. Selection (a) Assemble preliminary list of decision factors (b) Use technical questions and findings from A.2, along with professional judgment, to select additional relevant factors. (c) Identify any resulting interactive or cross-impact factors or categories. 2. Environmental Inventory (a) Assemble extant baseline data for selected factors. (b) Identify factors with data deficiencies, and plan data collection effort. (c) Conduct field studies or assemble information on data-deficient factors.
C. Evaluate Alternatives Relative to Decision Factors	1. Prediction and Delineation (a) Predict changes in each factor for each alternative using available techniques and/or professional judgment. (b) Delineate potential impacts of alternatives. (c) Highlight significant impacts and "red flag" any critical issues. 2. Weighting and Scaling (a) Use paired comparison technique, or some other importance weighting technique, to determine importance coefficients for each factor (FIC). (b) Scale/rate/rank predicted impacts through development of alternative choice coefficients, or use some other technique for evaluation of alternatives relative to decision factors (ACC). 3. Evaluation and Interpretation of Results (a) Multiply FIC by ACC to obtain final coefficient matrix. Sum coefficient values for each alternative. (b) Use values in final coefficient matrix as basis for description of impacts of alternatives and trade-offs between alternatives. (c) Discuss any critical issues and predicted impacts.
D. Document Results	1. Rationale (a) Describe rationale for selection of decision factors. (b) Describe procedure for impact identification and predication, and rationale for weighting, scaling/rating/ranking and interpreting results. 2. Provide Referencing of Sources of Information

dard library program (FACTO) was used to calculate the initial principal components analysis. Each eigenvector from the principal components analysis was used to calculate route positions on one axis of the ordination. Positions of the routes on an ordination axis were calculated by (1) scaling the elements of the corresponding eigenvector to a maximum value of $+1.0$; (2) deleting all component vectors having scaled eigenvectors with absolute values less than 0.7; (3) multiplying each of the remaining component vectors by its eigenvector; and (4) summing the route vectors to produce the route positions on the axis. The result of the ordination procedure is a multidimensional graphical representation of similarity between the routes (School of Civil Engineering and Environmental Science and Oklahoma Biological Survey, 1974).

The concepts of a weighting-scaling/rating/ranking approach as demonstrated in these example studies have been used in a number of other studies involving decisions which included environmental matters. Table 6.22 contains a summary listing of these examples along with some other weighting-scaling checklists which have been used in environmental decision-making and impact assessment studies (Canter, 1979). The general steps associated with this type of approach for doing a trade-off analysis are outlined in Table 6.23. These steps could be used to aid in decision-making associated with the selection of an aquifer restoration strategy for meeting a given need.

SELECTED REFERENCES

Bishop, A. B., et al., "Socio-Economic and Community Factors in Planning Urban Freeways", Sept. 1970, Department of Civil Engineering, Stanford University, Menlo Park, California.

Canter, L. W., "Supplement to Environmental Impact Assessment, Terrebonne Regional Sewerage Facilities", Aug. 1976, Report submitted to GST Engineers, Houma, Louisiana.

Canter, L. W., *Water Resources Assessment — Methodology and Technology Sourcebook*, 1979, Ann Arbor Science, Ann Arbor, Michigan.

Canter, L. W., "Evaluation of Social and Environmental Impacts of Emerging Technologies in Agricultural Production", Oct. 1983, University of Oklahoma, Norman, Oklahoma.

Canter, L. W. and Reid, G. W., "Environmental Factors Affecting Treatment Process Selection", Paper presented at Oklahoma Water Pollution Control Federation Annual Meeting, 1977, Stillwater, Oklahoma.

Crawford, A. B., "Impact Analysis Using Differential Weighted Evaluation Criteria", 1973, in J. L. Cochrane and M. Zeleny, editors, *Multiple Criteria*

Decision Making, University of South Carolina Press, Columbia, South Carolina.

Dalkey, N. C., "The Delphi Method: An Experimental Study of Group Opinion", Memorandum RM-5888-PR, June 1969, The Rank Corporation, Santa Monica, California.

Dean, B. V. and Nishry, J. J., "Scoring and Profitability Models for Evaluating and Selecting Engineering Products", *Journal Operations Research Society of America*, Vol. 13, No. 4, July–Aug. 1965, pp. 550–569.

Dee, N., et al., "Environmental Evaluation System for Water Resources Planning", Final Report, 1972, Battelle-Columbus Laboratories, Columbus, Ohio.

Dee, N., et al., "Planning Methodology for Water Quality Management: Environmental Evaluation System", July 1973, Battelle-Columbus Laboratories, Columbus, Ohio.

Duke, K. M., et al., "Environmental Quality Assessment in Multi-objective Planning", Nov. 1977, Final Report to U.S. Bureau of Reclamation, Denver, Colorado.

Dunne, N. G., "Successful Sanitary Landfill Siting: County of San Bernardino, California", SW-617, 1977, U.S. Environmental Protection Agency, Cincinnati, Ohio.

Eckenrode, R. T., "Weighting Multiple Criteria", *Management Science*, Vol. 12, No. 3, Nov. 1965, pp. 180–192.

Edwards, W., "How to Use Multi-attribute Utility Measurement for Social Decision Making", SSRI Research Report 76-3, Aug. 1976, Social Science Research Institute, University of Southern California, Los Angeles, California.

Falk, E. L., "Measurement of Community Values: The Spokane Experiment", *Highway Research Record*, 1968, No. 229, pp. 53–64.

Finsterbusch, K., "Methods for Evaluating Non-Market Impacts in Policy Decisions with Special Reference to Water Resources Development Projects", IWR Contract Report 77-78, Nov. 1977, U.S. Army Engineer Institute for Water Resources, Fort Belvoir, Virginia.

Gann, D. A., "Thermal Reduction of Municipal Solid Waste", Master's Thesis, 1975, School of Engineering and Environmental Science, University of Oklahoma, Norman, Oklahoma.

Gordon, T. J. and Helmer, O., "Report on a Long Range Forecasting Study", P-2982, Sept. 1964, Rand Corporation, Santa Monica, California.

Gum, R. L., Roefs, T. G. and Kimball, D. B., "Quantifying Societal Goals: Development of a Weighting Methodology", *Water Resources Research*, Vol. 12, No. 4, Aug. 1976, pp. 617–622.

Helmer, O., *Social Technology*, 1966, Basic Books, New York, New York.

Hyman, E. L., Moreau, D. H. and Stiftel, B., "SAGE: A New Participant-Value Method for Environmental Assessment", Feb. 1982, Environment and Policy Institute, East-West Center, Honolulu, Hawaii.

Jeffers, J., "Two Case Studies in the Application of Principal Component Analysis", *Journal of Applied Statistics*, Vol. 16, 1967, pp. 225–236.

Lower Mississippi Valley Division, "A Tentative Habitat Evaluation System (HES) for Water Resources Planning", Nov. 1976, U.S. Army Corps of Engineering, Waterways Experiment Station, Vicksburg, Mississippi.

Martino, J. P., *Technological Forecasting for Decision Making*, 1972, American Elsevier, New York, New York.

Morrison, T. H., "Sanitary Landfill Site Selection by the Weighted Rankings Method", Master's Thesis, 1974, School of Civil Engineering and Environmental Science, University of Oklahoma, Norman, Oklahoma.

O'Connor, M. F., "The Application of Multi-Attribute Scaling Procedures to the Development of Indices of Value", June 1972, Engineering Psychology Laboratory, University of Michigan, Ann Arbor, Michigan.

Odum, E. P., et al., "Optimum Pathway Matrix Analysis Approach to the Environmental Decision Making Process — Test Case: Relative Impact of Preposed Highway Alternates", 1971, Institute of Ecology, University of Georgia, Athens, Georgia.

Paul, B. W., "Subjective Prioritization of Energy Development Proposals Using Alternative Scenarios", (Paper presented at the Joint National ORSA/TIMS Meeting, San Francisco, California, May 1977), Engineering and Research Center, U.S. Bureau of Reclamation, Denver, Colorado.

Raines, G., "Environmental Impact Assessment of New Installations", (Paper presented at International Pollution Engineering Congress, Cleveland, Ohio, Dec. 4-6, 1972), Battelle Memorial Institute, Columbus, Ohio.

Rau, J. G., "Summarization of Environmental Impact", in *Environmental Impact Analysis Handbook*, Rau, J. G. and Wooten, D. C., editors, 1980, McGraw-Hill Book Company, Inc., New York, New York, pp. 8–17 to 8–25.

Ross, J. H., "The Numeric Weighting of Environmental Interactions", Occasional Paper No. 10, July 1976, Lands Directorate, Environment Canada, Ottawa, Canada.

Rugg, T. and Feldman, P., "TRS-80 Color Computer Programs", 1982, Dilithium Press, Beaverton, Oregon, pp. 25–36.

Salomon, S. N., "Cost-Benefit Methodology for the Selection of a Nuclear Power Plant Cooling System", Paper presented at the Energy Forum, 1974 Spring Meeting of the American Physical Society, Washington, D.C., Apr. 22, 1974.

School of Civil Engineering and Environmental Science and Oklahoma Biological survey, "Mid-Arkansas River Basin Study — Effects Assessment of Alternative Navigation Routes from Tulsa, Oklahoma to Vicinity of Wichita, Kansas", June 1974, University of Oklahoma, Norman, Oklahoma.

Smith, M. A., "Field Test of an Environmental Impact Assessment Methodology", Report ERC-1574, August 1974, Environmental Resources Center, Georgia Institute of technology, Atlanta, Georgia.

Solomon, R. C., et al., "Water Resources Assessment Methodology (WRAM): Impact Assessment and Alternatives Evaluation", Report 77-1, 1977, U.S. Army Engineer Waterways Experiment Station, Vicksburg, Mississippi.

Stanford Research Institute, "Handbook of Forecasting Techniques", IWR Contract at Report 75-7, Dec. 1975, A report submitted to U.S. Army Engineers Institute for Water Resources, Menlo Park, California.

Toussaint, C. R., "A Method for the Determination of Regional Values Associated with the Assessment of Environmental Impacts", Ph.D. Dissertation, 1975, School of Civil Engineering and Environmental Science, University of Oklahoma, Norman, Oklahoma.

Voelker, A. H., "Power Plant Siting, An Application of the Nominal Group Process Technique", ORNL/NUREG/TM-81, Feb. 1977, Oak Ridge National Laboratory, Oak Ridge, Tennessee.

Voorhees, A. M. and Associates, "Interim Guide for Environmental Assessment: HUD Field Office Edition", June 1975, Washington, D.C.

Wenger, R. B. and Rhyner, C. R., "Evaluation of Alternatives for Solid Waste Systems", *Journal of Environmental Systems*, Vol. 2, No. 2, June 1972, pp. 89–108.

CHAPTER **7**

Risk Assessment as Related to Aquifer Restoration Planning

Risk assessment is a frequently used term in the 1980s, with its meaning encompassing a wide range of concerns. In the early 1970s a popular term was "environmentalist," with its meaning encompassing a wide spectrum of disciplines and perspectives related to environmental management. In like manner, risk assessment also includes a wide spectrum of disciplines and perspectives. Persons engaged in risk assessment include, but are not limited to, engineers, hydrogeologists, environmental scientists, chemists, microbiologists, geologists, geographers, and public administrators. Perspectives on risk assessment range from concerns about ground water pollution to human health to financial decision-making. Approaches for doing a risk assessment may range from considering selected chemical properties of materials to calculation of numerical indices to presentation of information on the probabilities and likely consequences of catastrophic events. Therefore, the growing field of risk assessment is very broad. The purpose of this chapter is to summarize some information related to this field, particularly as focused on ground water pollution concerns. Background information on, and elements of, risk assessment will be presented along with a discussion of environmental risk analysis and risk assessment as related to examples of hazardous waste disposal sites.

BACKGROUND INFORMATION

Terminology associated with risk assessment is illustrated by the following definitions:

Risk: (1) function of the probability of an event occurring and the magnitude or severity of the event should it occur (Berger, 1982); (2) measure of the probability and severity of adverse effects (Conway, 1982); (3) possibility of harm, loss, or injury (Lee and Nair, 1979); and (4) collective chance or probability of accidents and disease resulting in injury or death (Inhaber, 1982).

Risk Assessment: (1) the identification of hazards, the allocation of cause, the estimation of probability that harm will result, and the balancing of harm with benefit (Berger, 1982); process of determining the probability that certain activities will produce certain adverse effects (Cornaby, et al., 1982); process including both risk analysis and safety analysis, where risk analysis is a quantitative assessment of the consequences of decisions, and safety analysis is the assessment of the level of risk acceptable to society (Jutro and Nerode, 1981).

Environmental Risk Analysis: process of predicting where in the environment a chemical will be transported, the rate and extent of its transformation, and its effect on organisms and environmental processes at expected ambient levels (Conway, 1982).

Environmental Risk Assessment: evaluation of both scientific data and the social, economic, and political factors that must be considered in reaching an ultimate decision on the prohibition, control, or management of chemicals in the environment; the final decision-making involves the scientific measurement of risk and then a social judgment of whether the benefits of the product outweigh the risk (Conway, 1982).

Risk assessments are being used to satisfy a wide range of purposes. The chief purpose of risk assessment should be to aid decision-making, and this focus should be maintained throughout any risk assessment (Lee and Nair, 1979). Risk assessments may be used to determine whether a particular site is suitable for waste disposal. Relative to hazardous waste sites, risk assessments are intended to clarify the degree of hazard and thereby help determine: (1) the priority for conducting detailed environmental evaluations and for undertaking preventive or remedial action or (2) the extent and character of the preventive or remedial action (Schweitzer, 1981). The decision as to whether or not to clean up an uncontrolled hazardous waste site that is leaking involves weighing the costs of cleanup against the risks resulting from continued leakage (Nisbet, 1982).

New information and various societal actions have increased the need for conduction of risk assessments. Conway (1982) cited the following as increasing the needs for environmental risk analyses:

(1) Discovery that certain chemicals persist in the environment and could have adverse effects at higher levels, e.g., polychlorinated biphenyls (PCB), dioxins, and fluorocarbons.
(2) Rapid growth of the chemical industry and the many other human activities (e.g., fuel production and use) that introduce chemicals into the environment.

(3) The federal Toxic Substances Control Act (TSCA), which requires information on ecological fate and effects for new products, significant-new-use products, priority commercial materials, and chemicals suspected of having substantial risks.

(4) The federal Resource Conservation and Recovery Act (RCRA), which requires information regarding the fate and effects of leachates from land disposal of hazardous solid wastes.

(5) The federal Clean Water Act Amendments, which require that receiving water quality criteria be developed for toxic pollutants.

(6) Studies of hazardous air pollutants under the federal Clean Air Act.

(7) The need for managers of manufacturing and distribution operations to know the environmental impact of materials they handle so that they are able to make decisions, especially in emergencies.

(8) Customers' demand for environmental fate-and-effects information from their suppliers.

(9) The need for product development to be guided by a combination of information on efficacy, economics, health effects, and environmental effects.

Required information for risk assessment must be assembled from a variety of sources (Schweitzer, 1981). For example, clinical and epidemiological studies to investigate human health effects as the pivotal concern in risk assessments have been triggered by reports of worker exposures in certain manufacturing facilities. Modeling techniques, using limited data on chemical properties, have been employed to predict future risks that might result from the manufacture and use of newly developed chemicals. Materials balance studies have been important in estimating environmental discharges from certain production processes and certain uses as a basis for risk assessments of industrial chemicals such as vinyl chloride and formaldehyde. While monitoring data have often been used in such assessments, only on a few occasions were the monitoring data collected in a systematic manner designed to support authoritative risk assessments. The most widely used methodologies for assessing chemical risks have emphasized a chemical-by-chemical analysis. One approach has been to estimate the toxicity of the chemical as a function of dose, usually drawing on the results of laboratory experiments. A second approach relies on direct observations of the biological effects of the chemical exposure, effects either on human populations or more frequently on biological surrogate populations. In either approach the dose levels or the effects to critical receptor populations are estimated, with the risk then determined to be the probability that a person within the exposed population of concern will suffer an adverse effect from the likely exposure. The uncertainties in these methodologies as illustrated in Figure 7.1 are well known. They include the problems of extrapolating effects from laboratory animals to man, in determining the critical receptor populations and their activity patterns, and in estimating dose levels.

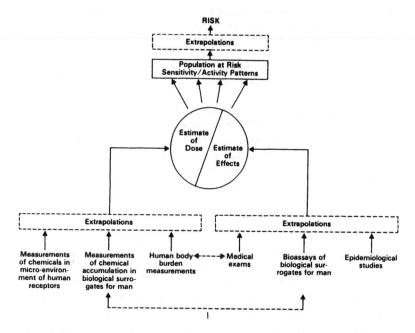

Figure 7.1 Extrapolations required in determining risk (Schweitzer, 1981).

Additional uncertainties or limits related to risk assessments have been delineated by Ess and Shih (1982), and by Nisbet (1982). Ess and Shih (1982) presented uncertainties or limits in terms of the characteristics of risk. They noted that risk can be modeled as a chain or series of events as shown in Figure 7.2. As indicated in Figure 7.2, the events that occur in this path are termed hazard, outcome, exposure, and consequence. Interconnected hazards, outcomes, etc., exist in the real world, thus risk assessment must address complicated and overlapping risk paths. The second characteristic of risk is that many of the events of concern are relatively rare occurrences. This means that a good base of historical statistical data on event occurrence frequencies is limited or nonexistent. This problem is further complicated by the final characteristic which needs to be considered. Many of the elements of risk pathways, such as the specific toxic substance involved, are new and therefore unknown.

Special difficulties are associated with risk assessments for toxic chemicals and hazardous waste sites. Nisbet (1982) indicated that risk assessments for toxic chemicals are difficult because exposure is usually variable and poorly characterized, and because toxicity information is difficult to extrapolate from animals to humans. The problems are much

Figure 7.2   Pathway concept (Ess and Shih, 1982).

greater for hazardous waste sites, for several reasons. First, the wastes
are often poorly characterized chemically, especially when they result
from disposal in the "bad old days" of the past. Even when the chemicals
present in the wastes are reasonably well known, they are usually present
as complex and variable mixtures, which makes it very difficult to assess
either exposure or toxicity. In uncontrolled landfills, for example, where
wastes are leaking into subsoil and ground water, the variability of
underground media makes it very difficult to calculate rates of dispersion, and the variability of underground concentrations makes it difficult
to measure rates of dispersion without elaborate and expensive monitoring programs. The toxicity of mixtures is very rarely measured, and is
difficult to assess theoretically because so little is known about interactions and synergisms. Finally, without knowledge of the rate of dispersion, even the size of the population at risk is difficult to estimate reli-
ably.

ELEMENTS OF RISK ASSESSMENT

Lee and Nair (1979) identified two basic elements associated with risk
assessment: (1) risk quantification or estimation; and (2) risk evaluation.
Risk quantification includes defining the hazard, identifying the initiating event(s) which would cause the hazard, determining the response and
consequences to the receptor system, and assigning probabilities of
occurrence. One tool which provides a structure for risk quantification is
the event tree. Event trees such as the one in Figure 7.3 provide a structured approach for risk quantification. For example, assume that the
initiating event in Figure 7.3 is the placement of hazardous materials at a
site with two liners. System 1 represents the primary (top) liner and
System 2 the secondary (bottom) liner. If System 1 fails and System 2
works, there is no damage except expenses to repair System 1. However,
if both systems fail there is the possible release of the hazardous materials to the subsurface environment. Figure 7.3 shows how the probabilities of each branch of the event tree can be calculated. While not related

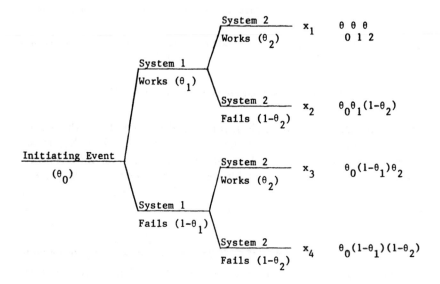

Figure 7.3   Risk quantification using an event tree (Lee and Nair, 1979).

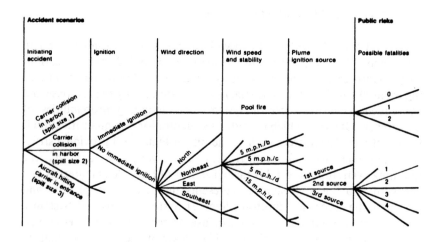

Figure 7.4   An event tree analysis of LNG spills (Lee and Nair, 1979).

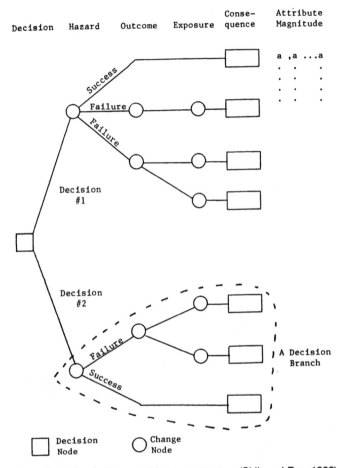

Figure 7.5 Generalized decision tree structure (Shih and Ess, 1982).

to aquifer restoration, Figure 7.4 is an example of an event tree used in a risk assessment of liquefied natural gas (LNG) spills in a coastal harbor (Lee and Nair, 1979). Decision trees can also be developed as a part of the risk quantification process. Figure 7.5 displays a generalized decision tree structure for problems which largely involve risk (Shih and Ess, 1982). Attribute magnitude refers to the probability of occurrence for a given consequence.

Risk evaluation involves determining what risks are socially acceptable. One approach for risk evaluation is to compare the quantified risk with various existing risks encountered in everyday life (Lee and Nair,

Table 7.1   Actions Increasing Risk of Death by One in a Million (Conway, 1982)

Action	Nature of Risk
Smoking 1.4 cigarettes	Cancer, heart disease
Drinking 0.5 liter of wine	Cirrhosis of the liver
Spending 1 hour in a coal mine	Black lung disease
Spending 3 hours in a coal mine	Accident
Living 2 days in New York or Boston	Air pollution/heart disease
Traveling 6 minutes by canoe	Accident
Traveling 10 miles by bicycle	Accident
Traveling 30 miles by car	Accident
Flying 1000 miles by jet	Accident
Flying 6000 miles by jet	Cancer caused by cosmic radiation
Living 2 months in Denver on vacation from New York	Cancer caused by cosmic radiation
Living 2 months in average stone or brick building	Cancer caused by natural radioactivity
One chest x-ray taken in a good hospital	Cancer caused by radiation
Living 2 months with a cigarette smoker	Cancer, heart disease
Eating 40 tablespoons of improperly stored peanut butter	Liver cancer caused by aflatoxin B
Drinking heavily chlorinated water (e.g., Miami) for 1 year	Cancer caused by chloroform
Drinking 30 12-oz. cans of diet soda	Cancer caused by saccharin
Living 5 years at site boundary of a typical nuclear power plant in the open	Cancer caused by radiation
Drinking 1000 24-oz. soft drinks from recently banned plastic bottles	Cancer from acrylonitrile monomer
Living 20 years near PVC plant	Cancer caused by vinyl chloride (1976 standard)
Living 150 years within 20 miles of a nuclear power plant	Cancer caused by radiation
Eating 100 charcoal broiled steaks	Cancer from benzopyrene
Risk of accident by living within 5 miles of a nuclear reactor for 50 yr	Cancer caused by radiation

1979). Table 7.1 summarizes some of these daily risks (Conway, 1982). This evaluation approach is known as the "revealed societal preferences" approach. Risk evaluation is the most difficult and potentially controversial part of risk assessment.

ENVIRONMENTAL RISK ANALYSIS

Moving from general risk assessment to consideration of environmental risk analysis, Conway (1982) suggested the following four major approaches:

(1) The "stochastic-statistical" approach, in which large amounts of data are obtained under a variety of conditions, and the correlations are established between the input of a certain material and its observed concentrations and its effects in various environmental compartments.

(2) The model ecosystem ("microcosm") approach, in which a physical model of a given environmental situation is constructed, dosed with a chemical, and the fate and effects of the chemical are observed.

(3) The "deterministic" approach, which uses a simple mathematical model to describe the rates of individual transformations and transports of the chemical in the environment. Only the dominant mechanisms, grouped as much as possible, are studied to arrive at an estimate of Expected Environmental Concentration (EEC). Separate determinations of toxicity with indicator organisms are made. This is the approach proposed in the TSCA premanufacture-notice testing guidelines. Figure 7.6 depicts this approach.

(4) The "baseline chemical" approach, in which transformations, transports, and effects are measured as in the deterministic approach, but the results then are compared with data on chemicals of known degrees of risk.

Cornaby et al. (1982) suggested a deterministic approach for hazardous waste sites when they identified the three elements shown in Figure 7.7. The environmental fate and potential receptors near a waste site are

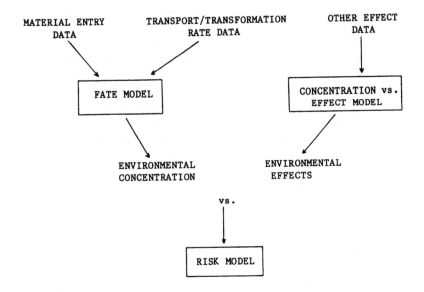

Note: Models can be simple mathematical expressions or yes–no comparisons with pre-established levels and ratios that trigger certain actions.

Figure 7.6 Deterministic approach to environmental risk analysis (Conway, 1982).

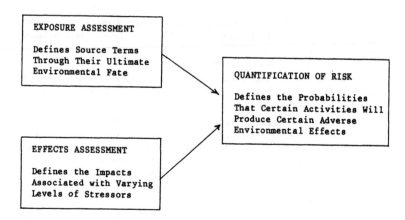

Figure 7.7   Overview of the three basic elements of environmental risk assessment (Cornaby et al., 1982).

depicted in Figure 7.8 (Schweitzer, 1981). The approach suggested by Cornaby et al. (1982) integrates human and nonhuman or ecological aspects of environmental concerns as shown in Figure 7.9.

One of the key aspects of environmental risk analysis is associated with determining the transport and fate of chemicals in surface and ground water, air, and soil. Table 7.2 provides a listing of mechanisms affecting the environmental fate of chemicals (Conway, 1982). It is beyond the scope of this presentation to discuss in detail various environmental transport and fate mathematical models. Another key aspect of environmental risk analysis is related to determining the ecological effects of chemicals to be released. Laboratory testing for both chemical fate and ecological effects may be necessary, and Table 7.3 lists some candidate tests (Conway, 1982).

RISK ASSESSMENT FOR HAZARDOUS WASTE DISPOSAL SITES

Existing and planned hazardous waste disposal sites are receiving increased attention as a result of the Resource Conservation and Recovery Act of 1976, and the "Superfund" Act of 1980. Environmental contamination has been documented at many existing sites. For example, the U.S. Environmental Protection Agency recently sponsored a study of damage case histories at 929 sites nationwide (Tusa and Gillen, 1982).

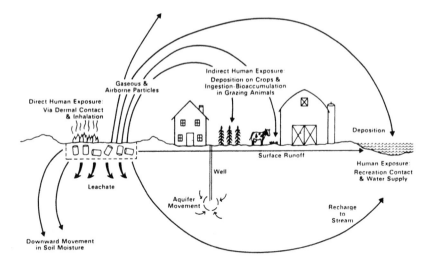

Figure 7.8 Environmental pathways from a waste site (Schweitzer, 1981).

Table 7.4 contains a summary of guidelines used in rating the severity of damage at the evaluated sites. Many of the sites contained multiple facilities. A total of 1,722 individual facility types were used in describing the 929 sites in the ten regions. Of the 1,722 facility types evaluated, 23% were landfills, 22% were containers, 16% were surface impoundments and 11% were tanks. The remaining 28% of the facilities were described by various other categories. Contamination, either documented or suspected, was identified in 834 sites, or at 90% of the sites evaluated. At 555 of the sites, or 60%, contamination was documented. Most of the contamination occurred in ground water 32%, with the remaining incidents occurring to soil (31%), surface water (29%), and air (8%). The events most often associated with contamination included leachate, leaks, spills, fire/explosion, toxic gas emission and erosion. The most commonly identified contaminants included metals, volatile halogenated organics, and volatile nonhalogenated organics.

A conceptual framework for preliminary site screening is shown in Figure 7.10 (Unites, Possidento and Housman, 1980). The guidelines in Table 7.4 could serve as a basis for ascertaining the severity of damage at a site. Nisbet (1982) identified the following seven steps for a qualitative risk assessment at a hazardous waste site:

(1) An engineering survey of the site, including an assessment of the propensity for scheduled and unscheduled release;

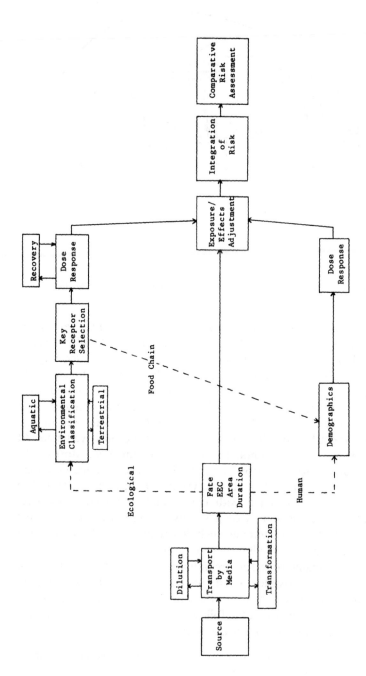

Figure 7.9 Overall environmental risk assessment scheme (Cornaby et al.,1982).

Table 7.2 Mechanisms Affecting Environmental Fate of Chemicals (Conway, 1982)

I. Original Input to Environment During Manufacturing, Distribution, Use, and Disposal

    A. To Water
        1. Treatment plant effluents at maufacturing and/or formulating plants
        2. Spills during manufacturing (original and formulating) and distribution
        3. Disposal after use

    B. To Soil
        1. Direct applications as an agricultural chemical or for vegetation control
        2. Land disposal, e.g., landfill or land cultivation operations
        3. Spills

    C. To Atmosphere
        1. Stack emission during manufacture
        2. Fugitive volatilization losses, e.g., from leaks, storage tank vents, and waste-water treatment
        3. Losses during use and subsequent disposal

II. Mechanisms for Transformation and Transport Within and Between Environmental Compartments

    A. Water
        1. Transport from water to atmosphere, sediments, or organisms
            a. Volatilization
            b. Adsorption onto sediments; desorption
            c. Absorption into cells (protista, plants, animals); desorption
        2. Transformations
            a. Biodegradation (effected by living organisms)
            b. Photochemical degradation (nonmetabolic degradation requiring light energy), direct or via a sensitizer
            c. Chemical degradation (effected by chemical agents), e.g., hydrolysis, free-radical oxidation

    B. Soil
        1. Transport to water, sediments, atmosphere, or cells
            a. Dissolution in rain water
            b. Absorbed on particles carried by runoff
            c. Volatilized from leaf or soil surfaces
            d. Taken up by protista, plants, and animals
        2. Transformation
            a. Biodegradation
            b. Photodegradation on plant and soil surfaces

    C. Atmosphere
        1. Transport from atmosphere to land or water
            a. Adsorption to particulate matter followed by gravitational settling or rain washout
            b. Washout by being dissolved in rain
            c. Dry deposition (direct absorption in water bodies)

Table 7.2, continued

---

    2. Transport within atmosphere
       a. Turbulent mixing and diffusion within troposphere
       b. Diffusion to stratosphere
    3. Atmospheric transformations
       a. Photochemical degradation by direct absorption of light, or by accepting energy from an excited donor molecule (sensitizer), or by reacting with another chemical that has reached an excited state
       b. Oxidation by ozone
       c. Reaction with free radicals
       d. Reactions with other chemical contaminants

---

(2) An inventory of the materials stored or disposed of at the site;

(3) An assessment of geological, hydrogeological and meteorological data, to assess the propensity for transport of materials away from the site; see Figure 7.11 for information on the migration of hazardous wastes in the environment (Murphy, 1982).

(4) A monitoring program for ground water, air, and other media, to measure the ambient concentrations of chemicals being transported away from the site;

(5) A survey of the distribution of the human population and other sensitive targets subject to exposure;

(6) A review of the toxicity of each of the major components of the material subject to release;

(7) Finally, an assessment of potential risks resulting from the exposure, including characterization of the uncertainty in this assessment.

Unless a very detailed monitoring program is carried out, it is rarely possible for the above 7-step assessment to be quantitatively reliable. In many cases, the risk assessor has to settle for much less—often a ranking or scoring procedure for factors controlling inherent hazard, release, environmental transport, and population at risk. At best, such a procedure can yield a ranking of risk on a qualitative or semiquantitative scale; for example, on a scale from 1–10, or from "low" to "high." This is sufficient for many purposes—e.g., for priority setting, for permitting, or for decisions on loss control programs or insurance (Nisbet, 1982).

Two numerical-based scoring procedures and one health effects criteria-based system will be described as examples for assessing the risks of hazardous waste disposal sites, including the Hazard Ranking System (Caldwell, Barrett and Chang, 1981), the Site Rating Methodology (Kufs et al., 1980), and the Air Quality Risk Assessment System (Walsh and Jones, 1982).

Table 7.3   Candidate Tests for Screening Ecological Impact of New Products (Conway, 1982)

I. Chemical Fate (Transport, Persistence)

  A. Transport
    1. Adsorption isotherm (soil)
    2. Partition coefficient (water/octanol)
    3. Water solubility
    4. Vapor pressure

  B. Other Physical-Chemical Properties
    1. Boiling/melting/sublimation points
    2. Density
    3. Dissociation constant
    4. Flammability/explodability
    5. Particle size
    6. pH
    7. Chemical incompatibility
    8. Vapor-phase UV spectrum for halocarbons
    9. UV and visible absorption spectra in aqueous solution

  C. Persistence
    1. Biodegradation
      a. Shake flask procedure following carbon loss
      b. Respirometric method following oxygen (BOD) and/or carbon dioxide
      c. Activated sludge test (simulation of treatment plant)
      d. Methane and CO_2 productions in anaerobic digestion
    2. Chemical degradation
      a. Oxidation (free-radical)
      b. Hydrolysis (25°C, pH 5.0 and 9.0)
    3. Photochemical transformation in water

II. Ecological Effects

  A. Microbial Effects
    1. Cellulose decomposition
    2. Ammonification of urea
    3. Sulfate reduction

  B. Plant Effects
    1. Algae inhibition (fresh and seawater, growth, nitrogen fixation)
    2. Duck weed inhibition (increase in fronds or dry weight)
    3. Seed germination and early growth

  C. Animal Effects Testing
    1. Aquatic invertebrates (*Daphnia*) acute toxicity (first instar)
    2. Fish acute toxicity (96 hr)
    3. Quail dietary LC_{50}
    4. Terrestrial mammal test
    5. *Daphnia* life cycle test
    6. *Mysidopsis bahia* life cycle
    7. Fish embryo-juvenile test
    8. Fish bioconcentration test

Table 7.4   Summary of Guidelines Used in Rating Severity of Damage at Evaluated Site (Tusa and Gillen, 1982)

Category	Severity		
	High	**Medium**	**Low**
Human Health	Damage incident to at least one person resulting in . . .		
	death	severe injury	minor injury
		Contamination of groundwater resulting in closure or restriction of drinking water in a . . .	
		community water supply	single private well
Environmental			
groundwater, surface water & air	Contamination incident where sampling indicates the presence of pollutants in concentrations . . .		
	at levels greater than 10 times applicable standards	at levels equal to applicable standards	at detectable levels, but less than applicable standards
food chain, flora	Contamination incident resulting in stress to vegetated or food crop area . . .		
	greater than 1 acre	greater than ½ acre	in limited areas only
fauna	Damage incident confirmed by . . .		
	massive kills	limited kills	bioassay studies confirming tissue contamination
soil	[a]	[a]	contamination incident confirmed by sampling data

[a]Higher levels of damage were typically identified via use of evidence in the other categories.

Hazard Ranking System

The U.S. Environmental Protection Agency, through a contract with MITRE Corporation, developed the Hazard Ranking System (HRS) as a method for ranking facilities for remedial action according to risks to health and the environment (Caldwell, Barrett and Chang, 1981). The HRS is a scoring system designed to address the full range of problems resulting from releases of hazardous substances, including surface water,

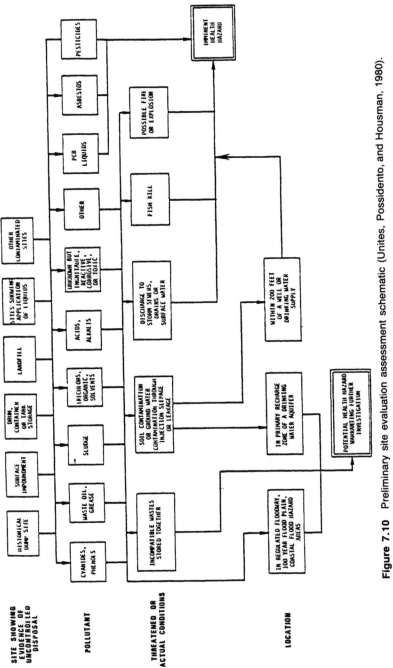

Figure 7.10 Preliminary site evaluation assessment schematic (Unites, Possidento, and Housman, 1980).

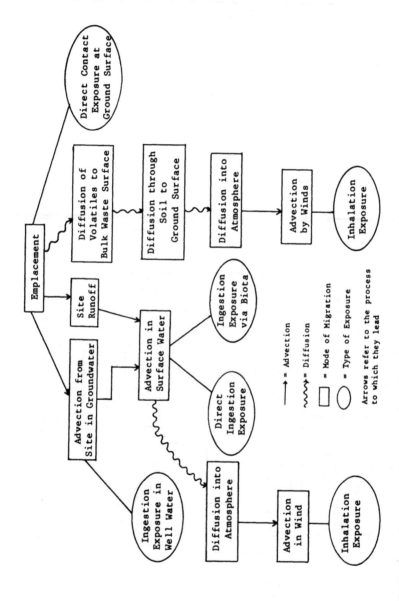

Figure 7.11 Sample diagram for migration of hazardous wastes in the environment (Murphy, 1982).

air, fire and explosion, and direct contact, in addition to ground water contamination. The HRS applies a structured value analysis approach. Three migration routes of exposure, ground water, surface water and air are evaluated and their scores are combined to derive a score representing the relative risk posed by the facility. Two additional routes of exposure — (1) fire and explosion and (2) direct contact — are measures of the need for emergency response.

Application of the HRS results in three scores for a hazardous waste facility. One score, S_M, reflects the potential for harm to humans or the environment as a result of migration of a hazardous substance away from the facility by routes involving ground water, surface water or air. S_M is a composite of separate scores for each of the three routes. Another score, S_{FE}, reflects the potential for harm from materials that can explode or cause fires. The third, S_{DC}, reflects the potential for harm as a result of direct human contact with hazardous materials at the facility (i.e., no migration need be involved). The score for each hazard mode (migration, fire and explosion, and direct contact) or route is obtained by considering a set of factors that characterize the hazardous potential of the facility. A comprehensive listing of factors for all of the hazard modes is given in Table 7.5 (Caldwell, Barrett and Chang, 1981). Each factor is assigned a numerical value (generally on a scale of 0 to 3, 5 or 8) according to prescribed guidelines. This value is then multiplied by a weighting factor to yield the factor score. The factor scores are then combined by following established guidelines: scores within a factor category are additive but the factor category scores are multiplicative.

In computing S_{FE} or S_{DC}, or an individual migration route score, the product of its factor category scores is divided by the maximum value the product can have and the resulting ratio is multiplied by 100, thus normalizing scores to a 100-point scale. Computation of S_M is slightly more complex since S_M is a composite of the scores for the three possible routes: ground water (S_{gw}), surface water (S_{sw}) and air (S_a). S_M is obtained from the equation:

$$S_M = \frac{1}{1.73} \sqrt{S_{gw}^2 + S_{sw}^2 + S_a^2}$$

The factor $1/1.73$ arises from the vector addition of the three route scores after the individual scores are normalized to a common denominator. This means of combining them gives added weight to routes with higher scores.

Table 7.5 Comprehensive List of Rating Factors in Hazard Ranking System (Caldwell, Barrett, and Chang, 1981)

Hazard Mode	Factor Category	Factors		
		Ground Water Route	Surface Water Route	Air Route
Migration	Containment	Containment	Containment	
	Route Characteristics	Depth to aquifer of concern	Site slope and intervening terrain	
		Net precipitation	1 Year 24 hour rainfall	
		Permeability of unsaturated zone	Distance to nearest surface water	
			Flood potential	Volatility/Physical State
	Waste Characteristics	Physical State	Physical State	Reactivity
		Persistence	Persistence	Incompatibility
		Toxicity	Toxicity	Toxicity
	Hazardous Waste Quantity	Total waste quantity	Total waste quantity	Total waste quantity
	Targets	Ground water use	Surface water use	Distance to nearest human population
		Distance to nearest well down gradient	Distance to a sensitive environment	Population within 1 mile radius
		Population served by ground water drawn within 3 mile radius	Population served by surface water drawn within 3 mile radius	Distance to sensitive environment
				Land use

Fire and Explosion	Containment	
	Waste Characteristics	Direct evidence of ignitibility or explosivity
		Ignitibility
		Reactivity
		Incompatibility
	Hazardous Waste Quantity	Total waste quantity
	Targets	Distance to nearest human population
		Distance to nearest building
		Distance to sensitive environment
		Land use
		Population within 2 mile radius
		Number of buildings within 2 mile radius
Direct Contact	Accessibility	
	Containment	
	Waste Characteristics	Toxicity
	Targets	Population within a 1 mile radius
		Distance to a critical habitat

The HRS does not result in quantitative estimates of the probability of harm from a facility or the magnitude of the harm that could result. Rather, it is a device for rank-ordering facilities in terms of the potential hazard they present. Risks are generally considered to be a function of the probability of an event occurring and the magnitude or severity should the event occur. Applying this approach to hazardous substance facilities, the probability and magnitude of a release are generally functions of the following areas:

(1) The manner in which the hazardous material is contained
(2) The route by which its release would occur
(3) The characteristics of the harmful substance
(4) The amount of hazardous substance
(5) The likely targets

These areas have been examined and representative factors were chosen to address each in the HRS (Caldwell, Barrett and Chang, 1981).

Site Rating Methodology

The Site Rating Methodology (SRM) has been used for the prioritization of Superfund sites in terms of remedial actions. The SRM has three elements: (1) a system for rating the general hazard potential of a site, (2) a system for modifying the general rating based on site-specific problems, and (3) a system for interpreting the ratings in meaningful terms. The first element is called the Rating Factor System, the second is called the Additional Points System, and the third is called the Scoring System (Kufs et al., 1980). The Rating Factor System is used for the initial rating of a waste disposal site based on a set of 31 generally applicable rating factors. As shown in Table 7.6, the 31 factors have been divided into four categories based on their focus. The four categories are: (1) receptors, (2) pathways, (3) waste characteristics, and (4) waste management practices. For each of the factors, a four-level rating scale has been developed which provides factor-specific levels ranging from 0 (indicating no potential hazard) to 3 (indicating a high potential hazard). The rating scales are also listed in Table 7.6. These scales have been defined so that the rating factors can typically be evaluated on the basis of readily available information from published materials, public and private records, contacts with knowledgeable parties or site visits.

The rating factors do not all assess the same magnitude of potential environmental impact. Consequently, a numerical value called a multiplier has been assigned to each factor in accordance with the relative magnitude of impact that it does assess. Multipliers were originally defined based on the judgment of individuals from several technical disciplines and were review based on the results of field testing. These values are multiplied by the appropriate factor ratings to result in factor scores for each of the rating factors.

The Additional Points System addresses special features of a facility's location, design or operation that cannot be handled satisfactorily by rating factors alone (Kufs et al., 1980). These features might present hazards that are unusually serious, unique to the site, or not assessable by rating scales. For example, an extremely high population density near a site should be considered even more hazardous than the rating factor for "population within 1,000 ft" indicates. Power lines running through sites that contain explosive or flammable wastes, though not typical of waste disposal sites in general, should be considered a potential hazard. Finally, the function of the nearest off-site building might indicate a serious threat of human exposure exists, even though types of functions cannot be quantitatively evaluated by rating scales the way distance can be. In such cases, raters can assign a greater hazard potential score to a site than it might otherwise receive by using the Additional Points System. Guidance on suggested point assignments are in Table 7.7 (Kufs et al., 1980). In order to maintain the objectivity of the rating methodology while allowing the assignment of additional points, the following limits have been placed on the number of additional points that can be assigned in each rating factor category: receptors, 50 points; pathways, 25 points; waste characteristics, 20 points; and waste management practices, 30 points.

The scoring system is based on all rating factors and additional points that are used to rate a site. Each subscore is based on those rating factors and additional points in that factor category which are used to rate a site. All of these scores are normalized so that they are on a scale of 0 to 100. Associated with every hazard potential score is a percentage of missing and assumed data. These percentages highlight scores that are based on large amounts of missing data and, in a general way, measure the reliability of the scores (Kufs et al., 1980).

Once a site has been rated using the SRM, the scores must be interpreted. Assessing the meaning of a score can be approached by either relative or absolute interpretation (Kufs et al., 1980). Relative interpretations are made by means of rankings. When a number of sites all require attention, rankings can be used to determine a preferential ordering. For example, when there are too many facilities to be addressed with available resources, rankings can be used to help determine the order of sites for: collection of additional background information, surveys of sites,

Table 7.6   Rating Factors and Scales for the Site Rating Methodology (Kufs, et al., 1980)

Rating Factors	Rating Scale Levels			
	0	1	2	3
Receptors				
Population within 1000 ft	0	1-25	26-100	>100
Distance to nearest drinking-water well	>3 mi	1-3 mi	3001 ft to 1 mi	0-3000 ft
Distance to nearest off-site building	>2 mi	1-2 mi	1001 ft to 1 mi	0-1000 ft
Land use/zoning	completely remote (zoning not applicable)	agricultural	commercial or industrial	residential
Critical environments	not a critical environment	pristine natural areas	wetlands, floodplains, and preserved areas	major habitat of an endangered or threatened species
Pathways				
Evidence of contamination	no contamination	indirect evidence	positive proof from direct observation	positive proof from laboratory analyses
Level of contamination	no contamination	low levels, trace levels, or unknown levels	moderate levels or levels that cannot be sensed during a site visit but which can be confirmed by a laboratory analysis	high levels or levels that can be sensed easily by investigators during a site visit
Type of contamination	no contamination	soil contamination only	biota contamination	air, water, or foodstuff contamination
Distance to nearest surface water	>5 mi	1-5 mi	1001 ft to 1 mi	0-1000 ft
Depth to groundwater	>100 ft	51-100 ft	21-50 ft	0-20 ft
Net precipitation	<-10 in.	-10 to +5 in.	+5 to +20 in.	>+20 in.
Soil permeability	>50% clay	30-50% clay	15-30% clay	0-15% clay
Bedrock permeability	impermeable	relatively impermeable	relatively permeable	very permeable
Depth to bedrock	>60 ft	31-60 ft	11-30 ft	0-10 ft

Waste Characteristics

Toxicity	SAX's level 0 or NFPA's level 0	SAX's level 1 or NFPA's level 1	SAX's level 2 or NFPA's level 2	SAX's level 3 or NFPA's levels 3 or 4
Radioactivity	At or below background levels	1 – 3 times background levels	3 – 5 times background levels	>5 times background levels
Persistence	easily biodegradable compounds	straight chain hydrocarbons	substituted and other ring compounds	metals, polycyclic compounds, and halogenated hydrocarbons
Ignitability	flash point >200° or NFPA's level 0	flash point of 140 – 200°F, or NFPA's level 1	flash point of 80 – 140°F, or NFPA's level 2	flash point <80°F, or NFPA's levels 3 or 4
Reactivity	NFPA's level 0	NFPA's level 1	NFPA's level 2	NFPA's levels 3 or 4
Corrosiveness	pH of 6 – 9	pH of 5 – 6 or 9 – 10	pH of 3 – 5 or 10 – 12	pH of 1 – 3 or 12 – 14
Solubility	insoluble	slightly soluble	soluble	very soluble
Volatility	vapor pressure <0.1 mm Hg	vapor pressure of 0.1 – 25 mm Hg	vapor pressure of 78 – 25 mm Hg	vapor pressure >78 mm Hg
Physical state	solid	sludge	liquid	gas

Waste Management Practices

Site security	secure fence with lock	security guard but no fence	remote location or breachable fence	no barriers
Hazardous waste quantity	0 – 250 tons	251 – 1000 tons	1001 – 2000 tons	>2000 tons
Total waste quantity	0 – 10 acre-ft	11 – 100 acre-ft	101 – 250 acre-ft	>250 acre-ft
Waste incompatibility	no incompatible wastes are present	present, but does not pose a hazard	present and may pose a future hazard	present and posing an immediate hazard
Use of liners	clay or other liner resistent to organic compounds	synthetic or concrete liner	asphalt-base liner	no liner used
Use of leachate collection systems	adequate collection and treatment	inadequate collection or treatment	inadequate collection and treatment	no collection or treatment
Use of gas collection systems	adequate collection and treatment	collection and controlled flaring	venting or inadequate treatment	no collection or treatment
Use and condition of containers	containers are used and appear to be in good condition	containers are used but a few are leaking	containers are used but many are leaking	no containers are used

Table 7.7   **Guidance for Additional Points System in the Site Rating Methodology (Kufs et al., 1980)**

Example Situation	Suggested Point Allotment
Receptors (50 Points Maximum)	
Use of site by nearby residents, especially children (for example, a site may be remote and/or fenced, but may still be used frequently by children as a play area or by adults with recreational vehicles).	0 to 4 points if used sparingly by adults; 4 to 10 points if used regularly by adults; 10 to 20 points if used regularly by children.
Type of building nearby (for example, a school versus a warehouse).	0 to 6 points for public use buildings (e.g., shopping centers); 5 to 15 points for schools and hospitals.
Presence of major surface water supplies, aquifers, or aquifer recharge areas near the site.	0 to 30 points depending on the proximity of the drinking water supply and the extent to which it is used.
Type of adjacent land use (for example, dairy farms, meat packing plants, orchards, and municipal water treatment plants would cause extreme concern).	0 to 10 points for recreational uses; 10 to 30 points for food or water-related uses.
Presence of economically important natural resources (e.g., shellfish beds, agricultural lands).	0 to 20 points depending on the number of people affected.
Presence of major transportation routes.	0 to 2 points for railways; 2 to 6 points for roads; and 4 to 10 points for foot paths or bicycle trails.
Residential population within 1000 ft over 100 people.	1 point per 25 people up to 10 points.
Pathways (25 Points Maximum)	
Erosion and runoff; susceptibility to washout from a flood; and slope instability.	0 to 4 points if a potential problem; 4 to 8 points if a moderate problem; 8 to 12 points if a severe problem.
Seismic activity.	0 to 10 points depending on the most likely adverse effects.
Waste Characteristics (20 Points Maximum)	
Substances that are carcinogenic, teratogenic, or mutagenic.	4 points per substance.
Any high-level radioactive wastes.	5 points if in minute quantities; 15 points if in significant quantities.
Substances with a high bioaccumulation potential.	2 points per substance.
Substances that are infectious.	0 to 5 points for wastes containing known transmittable pathogens depending on the potency of the wastes.

Table 7.7, continued

Example Situation	Suggested Point Allotment
Waste Management Practices (30 Points Maximum)	
No training or safety measures for personnel (active sites).	0 to 4 points depending on the number of people and their responsibilities.
Open burning (active sites).	0 to 10 points depending on regularity of the practice and the type of waste burned.
Site abandonment.	0 to 5 points depending on the reasons for the abandonment.
No waste mapping or records.	0 to 8 points depending on the presence of hazardous or incompatible wastes.
Heat sources or power lines near areas having explosives or flammable wastes.	0 to 8 points depending on the proximity and potential for ignition.
Less than 18 in. of cover over inactive landfills.	0 to 4 points for no apparent problems; 4 to 8 points for blowing trash; 6 to 12 points for hazardous vapors.
Less than 6 in. of daily cover on active landfills.	0 to 2 points for no apparent problems; 2 to 4 points for blowing trash; 3 to 6 points for hazardous vapors.
Total quantity of waste over 250 acre-ft.	1 point per 10 acre-ft up to 15 points.
Quantity of hazardous waste over 2000 tons (2370 yd³).	1 point per 4000 tons (4750 yd³) up to 25 points.

complete investigations of sites, implementation of remedial actions, and preparation of enforcement cases. Sites can be ranked in several ways including by overall scores, subscores, combinations of scores, and percentages of missing data. Overall scores are likely to be the most useful basis for ranking waste sites because all rating factors are included in the score. The subscores, however, may also provide valuable rankings. A scale developed for absolute interpretations can be used as a convenient point of reference for deciding "how bad is bad" at a site, and as a tool for deciding how urgent it is to respond to the problems at a site (Kufs et al., 1980). For example, Figure 7.12 is a flexible scale for overall hazard potential scores that can be used as a guide for defining absolute levels of hazard. Based on this use of hazard potential scores, a rater might decide to pursue in-depth investigations of sites classified high and very high before making initial surveys of sites classified very low and low.

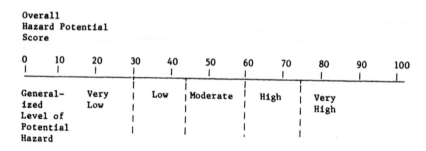

Figure 7.12   Guidelines for interpreting the level of potential hazard based on the site rating methodology (Kufs et al., 1980).

Air Quality Risk Assessment System

There are three steps in remedial action planning in which air exposure risks must be considered, both from a health impact and a liability point of view (Walsh and Jones, 1982). The first is an assessment of the present off-site risk before any type of action is considered since air risk considerations may be a remedial action design factor. This assessment would include an analysis of any existing air quality impact data. The second major task is associated with the potential air quality impacts of remedial action alternatives. An air risk assessment of the candidate alternatives must be undertaken to predict the ramifications of whichever cleanup alternative is chosen. Technical alternatives which have potential air risks may include: physical removal operations, air or steam stripping operations, chemical stabilization, and on-site combustion. Most remedial action alternatives are likely to increase the exposure risk in the short term, so this increase must be monitored and controlled by instituting operational practices to minimize off-site exposures. The final phase of a complete air quality risk assessment deals with postclosure considerations. This would primarily consist of continued area monitoring over a short period of time where air risks were clearly identified as a remedial action problem. The purpose of this monitoring is to document the fact that air risks have been reduced to acceptable levels. This is an additional liability prevention step.

A basic aspect of the Air Quality Risk Assessment System is the use of criteria to determine when a risk level is, or is not, present (Walsh and Jones, 1982). There is a good deal of uncertainty as to which effects criteria should be used especially when it relates to cancer risk. Examples of criteria used to evaluate the range of risk levels include:

Cancer Assessment Group Values (CAGs). These are recommended lifetime exposure limits to known carcinogens which have been developed by the U.S. Environmental Protection Agency for a limited number of toxic compounds. The CAG number represents maximum allowable concentrations that may result in incremental risk of human health over the short term or long term at an assumed risk. This assumed risk is the expected number of increased incidences of cancer in the effected population when the concentration over a lifetime equals the specified value.

Multimedia Environmental Goals (MEGs). The MEGs were developed in recent years by the Research Triangle Institute for the U.S. Environmental Protection Agency to meet the need for a workable system of evaluating and ranking pollutants for the purpose of multimedia environmental impact assessment. Consideration in arriving at these ambient level goals was given to existing federal standards or criteria, established or estimated human threshold levels, and acceptable risk levels for lifetime human exposure to suspected carcinogens or teratogens, among others.

Threshold Limit Value (TLV). TLVs are established by the American Conference of Governmental Industrial Hygienists as guidelines for prevention of adverse occupational exposures and are based on both animal studies and epidemiological findings and inferences. They refer to airborne concentrations of substances and represent conditions under which it is believed that nearly all workers may be repeatedly exposed for eight hours a day, five days a week, without adverse effects. Many TLVs also protect against short-term aggravations such as eye irritation, odor impacts, headache, etc. However, TLVs do not represent the hypersusceptible minority of individuals. Additionally, by definition, they assume a daily period (nonworking hours) of nonexposure time. Many states have developed or are currently developing ambient air hazardous waste guidelines and regulations based on these TLVs. A common practice is to adjust the TLV for a 24-hr exposure and then divide by some large uncertainty factor. Common limits are TLV/300–TLV/420.

SUMMARY

Risk assessment is a currently popular term being used to describe both risk quantification followed by risk evaluation. A confusing issue is the nonstandardization of terminology. Completion of risk assessments is often hampered by lack of understanding of risk paths and by lack of historical data. Environmental risk assessment encompasses the transport and fate of pollutants in the environment and their subsequent effects on human and nonhuman receptors. Due to limitations of data, relative risk approaches are often used. Examples have been presented for risk assessments for hazardous waste sites.

SELECTED REFERENCES

Berger, I. S., "Determination of Risk for Uncontrolled Hazardous Waste Sites", *Proceedings of the National Conference on Management of Uncontrolled*

Hazardous Waste Sites, 1982, Hazardous Materials Control Research Institute, Silver Spring, Maryland, pp. 23–26.

Caldwell, S., Barrett, K. W. and Chang, S. S., "Ranking System for Releases of Hazardous Substances", *Proceedings of the National Conference on Management of Uncontrolled Hazardous Waste Sites*, 1981, Hazardous Materials Control Research Institute, Silver Spring, Maryland, pp. 14–20.

Conway, R. A., "Introduction to Environmental Risk Analysis", Ch. 1 in *Environmental Risk Analysis for Chemicals*, R. A. Conway, editor, 1982, Van Nostrand Reinhold Company, New York, New York, pp. 1–30.

Cornaby, B. W., et al., "Application of Environmental Risk Techniques to Uncontrolled Hazardous Waste Sites", *Proceedings of the National Conference on Management of Uncontrolled Hazardous Waste Sites*, 1982, Hazardous Materials Control Research Institute, Silver Spring, Maryland, pp. 380–384.

Ess, T. H. and Shih, C. S., "Perspectives of Risk Assessment for Uncontrolled Hazardous Waste Sites", *Proceedings of the National Conference on Management of Uncontrolled Hazardous Waste Sites*, 1982, Hazardous Materials Control Research Institute, Silver Spring, Maryland, pp. 390–395.

Inhaber, H., *Energy Risk Assessment*, 1982, Gordon and Breach Science Publishers, Inc., New York, New York, pp. 1–54.

Jutro, P. R. and Nerode, A., "Development of Methodology for Determining Risk Assessment When Sludge is Applied to Land", EPA-600/S1-81-058, Aug. 1981, U.S. Environmental Protection Agency, Cincinnati, Ohio.

Kufs, C., et al., "Rating the Hazard Potential of Waste Disposal Facilities," *Proceedings of the National Conference on Management of Uncontrolled Hazardous Waste Sites*, 1980, Hazardous Materials Control Research Institute, Silver Spring, Maryland, pp. 30–41.

Lee, W. W. and Nair, K., "Risk Quantification and Risk Evaluation", *Proceedings of National Conference on Hazardous Material Risk Assessment, Disposal and Management*, 1979, Information Transfer, Inc., Silver Spring, Maryland, pp. 44–48.

Murphy, B. L., "Abandoned Site Risk Assessment Modeling and Sensitivity Analysis", *Proceedings of the National Conference on Management of Uncontrolled Hazardous Waste Sites*, 1982, Hazardous Materials Control Research Institute, Silver Spring, Maryland, pp. 396–398.

Nisbet, I. C., "Uses and Limitations of Risk Assessments in Decision-Making on Hazardous Waste Sites", *Proceedings of the National Conference on Management of Uncontrolled Hazardous Waste Sites*, 1982, Hazardous Materials Control Research Institute, Silver Spring, Maryland, pp. 406–407.

Schweitzer, G. E., "Risk Assessment Near Uncontrolled Hazardous Waste Sites: Role of Monitoring Data", *Proceedings of the National Conference on Management of Uncontrolled Hazardous Waste Sites*, 1981, Hazardous Materials Control Research Institute, Silver Spring, Maryland, pp. 238–247.

Shih, C. S. and Ess, T. H., "Multiattribute Decision-Making Imbedded with Risk Assessment for Uncontrolled Hazardous Waste Sites", *Proceedings of the National Conference on Management of Uncontrolled Hazardous Waste*

Sites, 1982, Hazardous Materials Control Research Institute, Silver Spring, Maryland, pp. 408–413.

Tusa, W. K. and Gillen, B. D., "Assessment of Hazardous Waste Mismanagement Cases", *Proceedings of the National Conference on Management of Uncontrolled Hazardous Waste Sites*, 1982, Hazardous Materials Control Research Institute, Silver Spring, Maryland, pp. 27–30.

Unites, D., Possidento, M. and Housman, J., "Preliminary Risk Evaluation for Suspected Hazardous Waste Disposal Sites in Connecticut", *Proceedings of the National Conference on Management of Uncontrolled Hazardous Waste Sites*, 1980, Hazardous Materials Control Research Institute, Silver Spring, Maryland, pp. 25–29.

Walsh, J. F. and Jones, K. H., "The Air Quality Impact Risk Assessment Aspects of Remedial Action Planning", *Proceedings of the National Conference on Management of Uncontrolled Hazardous Waste Sites*, 1982, Hazardous Materials Control Research Institute, Silver Spring, Maryland, pp. 63–66.

CHAPTER **8**

Public Participation in Aquifer Restoration Decision-Making

The basic purpose of public participation is to promote productive use of inputs and perceptions from private citizens and public interest groups in order to improve the quality of environmental decision-making (Canter, 1977). This type of decision-making can be associated with the selection of aquifer restoration strategies in a given geographical area. These citizen-oriented activities are variously referred to as citizen participation, public participation, public involvement, and citizen involvement. Interest groups include those representative of industry, government, conservation, and preservation. This chapter has been prepared to highlight issues related to the planning and implementation of public participation programs.

The introductory section contains some basic definitions, and delineates general advantages and disadvantages of public participation. Additional issues to be addressed in this chapter include background information on public participation problems and constraints, delineating objectives for public participation, identification of publics, techniques of public participation, conflict management and resolution, and practical suggestions for implementation of a public participation program.

BASIC DEFINITIONS

Public participation can be defined as a continuous, two-way communication process, which involves promoting full public understanding of the processes and mechanisms through which environmental problems and needs are investigated and solved by the responsible agency; keeping the public fully informed about the status and progress of studies and findings and implications of plan formulation and evaluation activities; and actively soliciting from all concerned citizens their opinions and perceptions of objectives and needs and their preferences regarding resource use and alternative development or management strategies and any other information and assistance relative to plan formulation and evaluation (Canter, 1977).

In essence, public participation involves both information feed-forward and feedback. Feedforward is the process whereby information is communicated from public officials to citizens concerning public policy. Feedback is the communication of information from citizens to public officials regarding public policy. Feedback information should be useful to decision-makers in making timing and content decisions (Canter, 1977).

ADVANTAGES AND DISADVANTAGES OF PUBLIC PARTICIPATION

There are advantages and disadvantages associated with public participation in environmental decision-making. Benefits accrue when affected persons likely to be unrepresented in decision-making processes are provided an opportunity to present their views (Canter, 1977). Creighton, Chalmers and Branch (1981) indicated that public participation can perform three vital functions: (1) serve as a mechanism for exchange of information, (2) provide a source of information on local values, and (3) aid in establishing the credibility of the planning and decision-making process. Additional comments related to these three functions are (Creighton, Chalmers and Branch, 1981):

(1) Mechanism for Exchange of Information—In order to provide the public with up-to-date information with which to assess study progress and options, it is necessary to summarize available planning and assessment information into short, succinct documents written in a language understandable to the public. Thus each public involvement activity becomes a natural point of synthesis for all the information generated during the study. The need to get information ready for public involvement provides a focus which can be otherwise missing, coming together only in a report at the end of the process. Public involvement is also a mechanism for obtaining information from the public. Much of this is in the form of values information, discussed below. But public involvement is also an indispensible component in the attempt to identify reliable information relevant to the planning and assessment problem at hand. The public (defined as everybody external to the agency conducting the study) is a major resource in identifying existing sources of information within the study area, thereby reducing search costs and the accidental duplication of previous work.

(2) Source of Information on Local Values — Increasingly, environmental, economic-demographic and social assessment methodologies have been recognizing the need for a values context by distinguishing between "Impacts" and "Effects." "Impacts" are the events that will occur if an action is not taken, for example, contamination of the local ground water resource. "Effects" are an assessment of what that impact means in the impacted community, in light of community conditions and attitudes. Public involvement is a necessary methodology by which to get from the assessment that an impact will occur, to an understanding of the effect of that impact upon the community. Again, public involvement provides a framework that integrates the other forms of assessment.

(3) Establish Credibility of Planning and Decision-Making Process — The ultimate synergy of well-integrated assessment and decision-making processes depends on their credibility in the eyes of potentially affected publics and in the eyes of the institution or group responsible for planning and implementing the proposed action. The credibility of any planning effort rests on the perception that the relevant issues were identified and addressed, appropriate information was obtained and correctly interpreted, and the significance of projected impacts assessed in the context of local values. Each of these criteria centers on the public involvement process. If the public perceives that it had access to the decision-making, if the role that public concerns and values had at each stage of the process is well documented, then the credibility of the process is enhanced. Effective public involvement often provides a kind of credibility within the planning agency as well. If the agency is confident that it is fully acquainted with public concerns and that a maximum effort has been made to incorporate public values into the planning process, it can commit to the implementation of the plan with a substantial increased confidence and security.

An additional benefit of public participation is that the agency, by constructing a record of decision-making (one to several reports), provides for both judicial and public examination of the factors and considerations in the decision-making process. Thus an added accountability is placed on political and administrative decision-makers since the process is open to public view. Openness exerts pressure on administrators to adhere to required procedures in decision-making. Finally, through public participation the agency is forced to be responsive to issues beyond those of the immediate project (Canter, 1977).

Disadvantages, or costs, of public participation include the potential for confusion of the issues since many new perspectives may be introduced. It is possible to receive erroneous information that results from the lack of knowledge on the part of the participants. Additional disadvantages include uncertainty of the results of the process of public participation, as well as potential project delay and increased project costs. A properly planned public participation program need not represent a major funding item, and it need not cause an extensive period of delay in the process of decision-making (Canter, 1977).

PUBLIC PARTICIPATION IN THE
DECISION-MAKING PROCESS

The decision-making process can be considered to consist of six stages: (1) issues identification; (2) conduction of baseline study of the environment; (3) prediction and evaluation of impacts; (4) mitigation planning; (5) comparison of alternatives; and (6) decision-making relative to the proposed action. Public participation should be associated with all six stages (Canter, Miller and Fairchild, 1982).

The initial identification of issues is done to establish the scope of the decision-making study. Public involvement at this stage is primarily devoted to informing the public about the project and getting to know what citizens feel about the problem being addressed. At this early stage, it will be possible to start identifying which groups see themselves as "winners" and which as "losers." The effort by the project planner should be to establish rapport and an air of cooperation.

The baseline study records the environmental status quo in the project area. At this stage, the flow of information to the public takes the form of what is being surveyed and why. Feedback to this information is often helpful in identifying existing data bases. Thus, the public's response can reduce the time and cost of the baseline survey.

Impact evaluation consists of the prediction and interpretation of changes due to project alternatives. The public can assist in this process in several ways. By reviewing the project alternatives being considered, they can ensure that no viable alternative is inadvertently omitted. Where legal standards are not in force, comments from the public can be useful in establishing project-specific criteria or maximum tolerable levels of change. Finally, the information-feedback cycle must be maintained to hold the public's interest and prevent alienation.

Mitigation measures are planned to reduce undesirable project effects. One of the major public inputs at this stage is ensuring that the mitigation measure is itself acceptable. Consider, for example, an aquifer restoration project that involves pumping of the polluted ground water and its treatment in an activated carbon plant located at the surface of the site. One mitigation measure might be to recharge the treated ground water. In many areas, particularly in the southwestern United States, this measure, though technically feasible, is culturally or legally unacceptable. As before, public review will ensure that all reasonable measures are considered.

The comparison of alternatives is done to determine the one or several preferred actions. Local values should be used to weight the importance of environmental factors at this stage. Some of the techniques which can be used to incorporate local social and cultural values into decision-making methods will be discussed in a later section. It is very important at this stage that the public have an input into what is recommended to decision-makers.

It is at the comparison of alternatives stage that the preferred project alternative is identified. It is therefore at this stage that conflicts will come clearly into focus. Methods of conflict resolution will be discussed in a later section of this chapter. If the public involvement program has been effective to this point, it would be possible to resolve conflicts in a spirit of cooperation.

The final step in a decision-focused study for aquifer restoration is the actual decision on which alternative will be implemented. At this stage, public involvement has three objectives. The public should be informed what the decision is and why. Ideally, the decision should be based on the recommendations arising out of the comparison of alternatives. However, this is not always the case. The second objective is final resolution of conflicts. It may be necessary to compensate certain publics in order to even out the distribution of benefits. Finally, if the decision-makers are responsible to the public, there will be feedback concerning the final decision.

If public participation is to be effective in the various stages of the decision-making process, the public participation program must be carefully planned. A good public participation program does not occur by accident (Canter, 1977). Planning for public participation should address the following elements:

(1) Delineation of objectives of public participation during the different stages.
(2) Identification of publics anticipated to be involved in the different stages.
(3) Selection of public participation techniques which are most appropriate for meeting the objectives and communicating with the publics. It may be necessary to delineate techniques for conflict management and resolution.
(4) Development of a practical plan for implementing the public participation program.

OBJECTIVES OF PUBLIC PARTICIPATION

The delineation of objectives of public participation during different stages within the decision-making process is an important element in developing a public participation plan. The objectives could be general or specific. Hanchey (1981) suggested three general objectives which should be considered in the design of a public participation program for a specific planning situation, including an aquifer restoration program.

These are referred to as: (1) the public relations objective, (2) the information objective, and (3) the conflict resolution objective. As shown in Figure 8.1, these general objectives can be disaggregated into eight second-order objectives which serve to clarify and to provide workable concepts for both the design and evaluation of such programs (Hanchey, 1981). Additional comments on each of these objectives and subobjectives are as follows:

(1) Public Relations Objective—The public relations objective is based on the premise that in order for the planning agency to develop plans which have broad public support and acceptance, the public must view the agency's role in the planning process as legitimate, and must have confidence and trust in the agency and its planning procedure. The need for legitimizing the agency's role in the planning process results from the fact that the public is frequently uninformed about the responsibilities and the authorities of the planning agency. Another important factor is the development of confidence and trust by the public toward the planning agency.

(2) Information Objective—The information objective deals with the stage of the planning process in which the planner determines the problems to be solved during the planning effort and searches for solutions which are acceptable to the public. There are three separate concepts making up this objective: (1) the diagnosis of community problems and needs relative to aquifer restoration, (2) development of alternative solutions, and (3) the evaluation of the consequences of solutions. Quite frequently projects have been rejected by the public because the planner and the public had a different view of the problems which needed solution. This is partly because people do not have the same values and thus do not perceive the same problems, even when viewing the same situation.

The need for involving the public during the development of alternative solutions stage of the study is based on the advantages to the planner of being able to test the social and political feasibility of alternatives early in the study. The purpose of public involvement at this early stage should be to allow the planner to begin to bracket the range of social and political feasibility early in the study, in order that more of the planning effort can be confined to plans more likely to be feasible and acceptable with the result that the planning process will more likely lead to a productive outcome. The planner should be careful, however, that he does not prematurely discard alternatives. This may happen for two reasons. First, it is very likely that the "public" as it is first encountered does not represent the full range of interests which will be affected by the ultimate plan and thus initial feasibility limits may not accurately reflect actual community feelings. Second, social and political feasibilities do not have fixed predetermined limits. They depend to a significant extent upon a clear understanding of the possibilities and the significance of choice. Thus, these limits are subject to change as the planning process progresses and increased information is exchanged between the participants.

(3) Conflict Resolution Objective—Conflicts among the participants in a study may arise from differences in opinions or beliefs; it may reflect differences in interests, desires, or values; or it may occur as a result of a scarcity of some resource such as ground water. Conflict can occur in a cooperative or competitive context and will be strongly influenced by the processes of conflict resolution employed

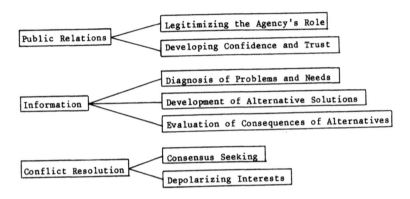

Figure 8.1. Objectives of public participation (Hanchey, 1981).

by the planner. There are two concepts which are useful in describing a favorable approach to conflict resolution, consensus seeking and the avoidance of extreme positions. It should be noted that these components of the conflict resolution objectives are not independent of the other two objectives; rather they are influenced to a great extent by the degree to which the planner has been successful in achieving the other objectives. Consensus seeking can be described as cooperative problem solving in which the conflicting parties have the joint interest of reaching a mutually satisfactory solution. Quite frequently, conflicts over issues have been perceived by participants as situations where a party to the conflict can take only one of two positions: for or against. This is unfortunate in that it implies that what is good for one party is necessarily bad for the other. Anyone who perceives it as such, must of course, align himself with one of the two positions.

Bishop (1975) delineated six objectives for public participation and tied them to the various stages of the decision-making process. The six general objectives are:

(1) Information, education, and liaison.
(2) Identification of problems, needs, and important values.
(3) Idea generation and problem solving.
(4) Reaction and feedback on proposals.
(5) Evaluation of alternatives.
(6) Conflict resolution consensus.

The relationship of these objectives to the six stages in the decision-making process are shown in Table 8.1 (Bishop, 1975). The first of these objectives is directed toward education of the citizenry on the decision-making process, its purpose, and the process of citizen participation. In

Table 8.1 Public Involvement Objectives at Various Decision-Making Process Stages (Bishop, 1975)

Objective	Assessment Stage					
	Issues Identification	Baseline Study	Impact Evaluation	Mitigation Planning	Comparison of Alternatives	Decision-Making
Inform/Educate	X	X	X	X	X	X
Identify Problems/Needs/Values	X	X	X	X	X	
Approaches to Problem Solving			X	X		
Feedback		X	X	X	X	X
Evaluate Alternatives			X	X	X	
Resolve Conflicts					X	X

addition, this objective includes dissemination of information on the study progress and findings, as well as data on potential environmental issues.

Identification of problems, needs, and important values is related to the determination of the environmental resources, including surface and ground water resources, important to various segments of the public in an area. In addition, this objective is focused on defining areas of ground water pollution and the relation of potential solutions being addressed in the project study.

The third objective is directed toward identification of alternatives that may not have been considered in normal planning processes for aquifer restoration. In addition to specific alternatives for identified needs, it is possible also to enumerate mitigating measures for various alternatives so as to minimize adverse environmental effects. The fourth objective attempts to probe public perceptions of the actions and resource interrelations. In addition, this objective can be used to assess significance of various types of impacts. The alternatives evaluation objective is closely related to reaction and feedback on proposals for aquifer restoration. In the process of evaluation of alternatives, valuable information can be received about the significance of unquantified and quantified environmental amenities. Public reaction to value trade-offs in the process of selection can also be assessed.

The final objective is related to resolving conflicts that exist over the proposed program for aquifer restoration. This objective may involve mediation of differences among various interest groups, development of mechanisms for environmental costs compensation, and effort directed toward arriving at a consensus opinion on a preferred program. Successful accomplishment of this objective can avoid unnecessary and costly litigation.

Extensive communication will be necessary to achieve the objectives of public participation. Bishop (1981) identified the following four communications process models which seem appropriate in meeting the basic objectives of public participation: (1) diffusion process, (2) collection process, (3) interaction process, and (4) diffusion-collection process. Explanations for these four models are as follows (Bishop, 1981):

(1) Diffusion Process—There is the possibility of multiple access to target groups or publics through the communications system. An operational example of this is illustrated in Figure 8.2. In this process, the agency sends a message via different media to various target groups, who in turn transmit the message to still other groups or individuals. The net result enables the agency to reach a broader segment of the public in terms of the total impact than just the initial target group. The diagram in Figure 8.2 brings out three important points. First, communication is not just a single, but a multistep process where target groups

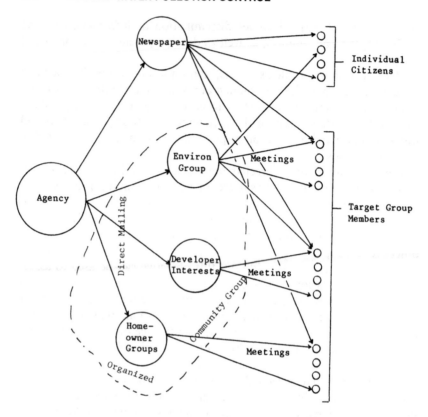

Figure 8.2 Example of a diffusion process (Bishop, 1985).

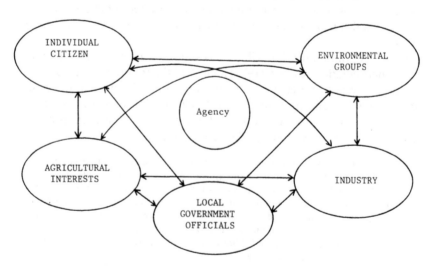

Figure 8.3 An interaction process (Bishop, 1981).

become senders in transferring a message to others through media which they can access. Corollary to this is the fact that the sender cannot completely control the communication process since intermediaries are present to influence or interrupt the process. Second, a target public can be contacted through several media, thus giving opportunity for reinforcing and clarifying the message. Third, if some media are inoperative due to frame of reference or noise problems, the diffusion process can still get the message to target groups through other media types.

(2) Collection Process—The collection process can be seen as diffusion in reverse. It may serve to obtain feedback to complete a communication loop or to collect information. The messages may or may not return by the same media.

(3) Interaction Process—Interaction describes the situation where communication is an interchange among several groups, as illustrated in Figure 8.3. The agency may assume the central role in acting as a moderator and facilitator in the communication exchange among other groups, or may simply take the role of one of the communicators in the interaction. The interactive processes generally imply communications media which involve meetings, work groups, committees, advisory panels, and the like.

(4) Diffusion-Collection Process—This process describes the situation where information is disseminated with the specific intent of eliciting some desired information in response. Usually, in addition, the mechanism or medium for response will be specified or provided in order to facilitate information collection. A simple example is a questionnaire that is sent to some public groups and to a newspaper (see Figure 8.2). Target publics are asked to send their responses by individual letter to the agency as the originator of the questionnaire.

The communications process models can be matched with the key objectives of public participation. These cross-comparisons as shown in Figure 8.4 (Bishop, 1981) can aid in selecting an appropriate communications approach for meeting a particular objective. For example, the inform, educate, and liaison objectives are all dependent on dissemination of information. The diffusion model describes this process. Identification, assessment, and feedback are objectives that are described by the reverse, the collection model. Idea generation, problem solving, conflict resolution, and consensus are generally best accomplished by interaction processes. Review, reaction, and evaluation objectives require a two-step process. An information "stimulus" is first directed to the "publics," then the publics respond with their reactions or evaluations. A total communication process will usually require all of these processes (Bishop, 1981).

IDENTIFICATION OF PUBLICS

The identification of publics which would potentially be involved in the various stages of the decision-making process is another basic element in the development of a public participation program. Also of importance is the delineation of public participation techniques which

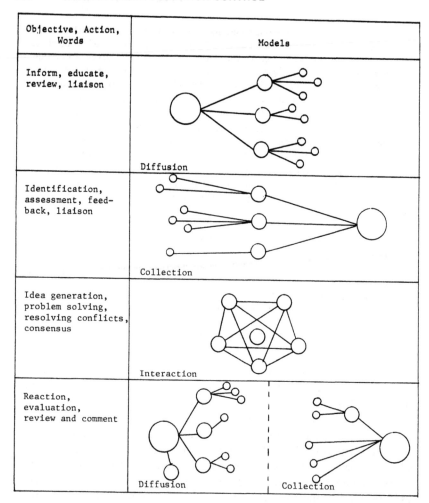

Objective, Action, Words	Models
Inform, educate, review, liaison	Diffusion
Identification, assessment, feedback, liaison	Collection
Idea generation, problem solving, resolving conflicts, consensus	Interaction
Reaction, evaluation, review and comment	Diffusion Collection

Figure 8.4 Correspondence of communications objectives and models (Bishop, 1981).

are most effective in communicating with different publics. This section of this chapter will address the recognition of types of publics, pragmatic approaches for identifying publics, and selected techniques for communicating with identified publics.

Recognition of Types of Publics

The general public cannot be considered as one body. The public is diffuse, but at the same time highly segmented into interest groups, geographic communities, and individuals. There are sets or groups of

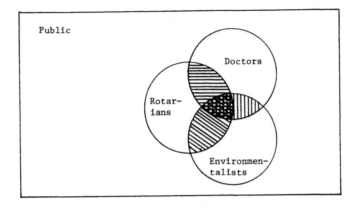

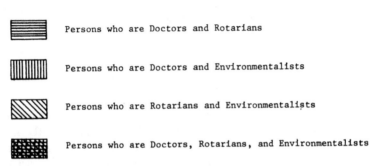

Persons who are Doctors and Rotarians

Persons who are Doctors and Environmentalists

Persons who are Rotarians and Environmentalists

Persons who are Doctors, Rotarians, and Environmentalists

Figure 8.5 Example of multiple public associations (Bishop, 1981).

"publics" that have common goals, ideals, and values. Any one person may belong to several different sets of these publics since they may be professionally, socially, or politically oriented. The Venn diagram shown in Figure 8.5 illustrates the overlapping of some of these groups, and the fact that an individual may identify with one, a combination of two, or all three of the groups (Bishop, 1981). Two significant points may be drawn from Figure 8.5 in terms of communication:

(1) Individuals are likely associated with various social, economic, and cultural orientations from which they draw their information and structure their values.

(2) Multiple association thus allows the opportunity for multiple access to individuals as participants, clients, or critics in a planning process.

Schwertz (1979) suggested that the purpose of identifying the various publics is twofold. The first is to ensure that no group of persons is excluded from participating. Such exclusion, whether intentional or not, will be resented. Those who are left out will often seek redress outside of the formal public involvement program—for example, by court action. The second purpose relates to the public involvement technique to be used. Some techniques are more effective with certain groups than others.

There are several ways of categorizing various publics that might be involved in a public participation program for an aquifer restoration project. One group of publics consists of four separate categories (Canter, 1977):

(1) Persons who are immediately affected by the project and live in the vicinity of the project.

(2) Ecologists ranging from preservationists to those who want to ensure that development is as effectively integrated into the needs of the environment as possible. Persons in this group are willing to incur substantial financial costs for environmental protection.

(3) Business and commercial developers who would benefit from initiation of the proposed action.

(4) The part of the general public who enjoy a high standard of living and who do not want to sacrifice this standard.

The U.S. Army Corps of Engineers has defined the following publics in conjunction with water resources development projects (U.S. Army Corps of Engineers, 1971). While oriented to surface water projects, many, if not all, of the following publics might be associated with aquifer restoration projects:

(1) Individual citizens, including the general public and key individuals who do not express their preferences through, or participate in, any groups or organizations.

(2) Sporting groups.

(3) Conservation/environmental groups.

(4) Farm organizations.

(5) Property owners and users, representing those persons who will be or might be displaced or affected by any alternative under study.

(6) Business and industrial groups, including Chambers of Commerce and selected trade and industrial associations.

(7) Professional groups and organizations, such as the American Institute of Planners, American Society of Civil Engineers, and others.

(8) Educational institutions, including universities, high schools, and vocational schools. General participation is by a few key faculty members and students or student groups and organizations.

(9) Service clubs and civic organizations, including service clubs in a community such as Rotary Club, Lions Club, League of Women Voters, and others.

(10) Labor unions.

(11) State and local governmental agencies, including planning commissions, councils of government, and individual agencies.

(12) State and local elected officials.

(13) Federal agencies.

(14) Other groups and organizations, possibly including various urban groups, economic opportunity groups, political clubs and associations, minority groups, religious groups and organizations, and many others.

(15) Media, including the staff of newspapers, radio, television, and various trade media.

In targeting publics an attempt is made to identify those persons who believe themselves to be affected by the aquifer restoration project outcome. The difficulty is that the degree to which people feel affected by a project is a result of their subjective perception; people the agency feels are most directly impacted may not be as concerned as someone that the agency perceives as only peripherally involved. However, the starting point always remains some effort to objectively analyze the likelihood that someone will feel affected by the study. Some of the bases on which people are most likely to feel affected are (Creighton, 1981c):

(1) Proximity: People who live in the immediate area of a project and who are likely to be affected by noise, dust, or possibly even threat of dislocation, are the most obvious publics to be included in the study.

(2) Economic: Groups that have jobs to gain or competitive advantages to win, e.g., industrial developers requiring water, are again an obvious starting point in any analysis of possible publics.

(3) Use: Those people whose use of the area is likely to be affected in any way by the outcome of the project are also likely to be interested in participating.

(4) Social: Increasingly, people who see projects as a threat to the tradition and culture of the local community are likely to be interested in projects. They may perceive that a large influx of construction workers into an area may produce either a positive or negative effect on the community. Or they may perceive that the project will allow for a substantial population growth in the area which they may again view either positively or negatively.

(5) Values: Some groups may be only peripherally affected by the first four criteria but find that some of the issues raised in the study directly affect their values, their "sense of the way things ought to be." Any time a study touches on such issues as free enterprise versus government control, or jobs versus environmental enhancement, there may be a number of individuals who participate primarily because of the values issues involved.

Pragmatic Approaches for Identifying Publics

Recognizing that there are different types of publics based on different classification schemes, the issue then turns to pragmatic approaches for identifying the potential publics which might be associated with the decision-making process during its various stages. Creighton (1981c) suggested three broad approaches to targeting the public: (1) self-identification, (2) third party identification, and (3) staff identification.

Self-identification simply means that individuals or groups step forward and indicate an interest in participating in the decision-making process. The use of the news media, the preparation of brochures and newsletters, and holding of well-publicized public meetings are all means of encouraging self-identification. Anyone who participates by attending a meeting or writing a letter or phoning on a hot line has clearly indicated an interest in being an active public in the decision-making process. As a result it is critical that anyone who expresses an interest in the process in any way quickly is placed on the mailing list and is continually informed of progress in the process (Creighton, 1981c).

One of the best ways to obtain information about other interests or individuals which should be included in the study is to ask an existing advisory committee, or representatives of known interests, who else should be involved. One variation on this theme is to enclose a response form in any mailings inviting people to suggest other groups that should be included. These simple techniques of consulting with known representatives to recommend others who should be involved often prove to be one of the most effective means of targeting the public (Creighton, 1981c).

There is a wide range of techniques by which internal staff can systematically approach targeting the public. These include (Creighton, 1981c):

(1) Intuitive/experimental information: Most planning staff that have worked in an area for some period of time can, if asked, immediately begin to identify individuals and groups that are likely to be involved in any new study. One of the richest sources of information for possible individuals or interests to be involved would be internal staff who have worked in the area for some period of time.

(2) Geographic analysis: In many cases just by looking at a map it is possible to identify publics who reside in a project site area and adjacent to it.

(3) Historical analysis: In many cases there is considerable information in old files. This includes:

    (a) Lists of previous participants in earlier studies included in reports.

    (b) Correspondence files.

    (c) Newspaper clippings regarding similar studies.

    (d) Library files on past projects.

(4) Consultation with other agencies: Since numerous agencies have held public involvement programs it can often be useful to explore their files or consult with them concerning possible publics.

Schwertz (1979) noted that many communities have existing organizations or interest groups which can and should be made a part of a public participation program. Many of these organizations have been created to address specific community problems. Other organizations, such as the

Table 8.2 Citizen Involvement Checklist (Schwertz, 1979)

____ Bankers' Associations	____ Retail Trade Associations
____ Business Associations	____ Service Organizations
____ Chamber of Commerce	____ Rotary Club
____ Civic Organizations	____ Others
____ Jaycees	____ Special Groups and/or Organizations
____ League of Women Voters	____ Audubon Society
____ Others	____ Conservation Societies
____ Developer Associations	____ Historic Preservation Societies
____ Elder Citizens' Organizations	
____ Entrepreneurs	____ Sierra Clubs
____ Executive and Professional Associations (local)	____ Others
____ Farmers' Organizations	____ Social Clubs
____ Fraternal Organizations	____ Trade Organizations
____ Labor Organizations	____ Utility Companies
____ Neighborhood Associations	____ Youth Organizations
____ Neighboring Communities (where appropriate)	____ Veterans' Organizations
____ Newspaper Editors and Owners	____ Women's Auxiliaries
____ Philanthropic Organizations and/or Foundations	____ _____
____ Real Estate Brokers' Associations	____ _____
____ Religious Organizations (By Denomination)	____ _____
	____ _____

local Chamber of Commerce, the Jaycees, or the League of Women Voters, are also active in many phases of community life. Table 8.2 contains a citizen involvement checklist which can be useful for the public participation program planning staff in identifying publics which would be potentially associated with a project.

While the mailing lists may be a useful tool for notifying parties of public meetings, there are problems in using them as the primary basis for identifying people for more intensive public participation (Ragan,

Jr., 1981a). First, because public meetings are "official" sessions, fully 75–80% of most mailing lists comprise public officials and agencies (as many as 45% are federal, with many of these in national or regional offices). Private organizations and individuals are not more strongly represented on the lists simply because they are much harder to identify. Second, mailing lists are hard to maintain; a study manager may not have the time. Third, many mailing lists may overlap and duplicate names of individuals or organizations. Fourth, mailing lists categorized by public organization, media, and all others make it difficult to identify potential interests to be contacted for special sessions. The study manager has no easy way to identify such interests; he must peruse the list and try to associate interests with organizational titles. This may be possible for organizations, but it is impossible for individuals – unless they and their interests are well-known.

Bishop (1981) has developed a basic scheme for pragmatically identifying publics, with the concept shown in Table 8.3. The list of interests/ issues groups coincides with the earlier cited list of publics associated with water resources development projects (U.S. Army Corps of Engineers, 1971). Use of Table 8.3 requires the identification of the relation of each group (public) to the decision-making process for an aquifer restoration project.

The planner must also be aware that identification of publics has the dimension of participation through time (Bishop, 1981). At the onset of planning, a certain segment of the public will have an interest in participating. These are usually people or groups that: (1) have participated in the past, (2) are affected by a problem, or (3) will be affected by a possible solution to the problem. Circle A in Figure 8.6 (a) indicates this identified portion of the public. As planning progresses, some of those identified do not participate, while some previously unidentified publics will identify themselves. Circle B in Figure 8.6 (b) illustrates those who are participating after the process has progressed for some time. Looking forward into time, there will always be those who may not be identified who may come into the process. This is shown by Circle C in Figure 8.6 (c). Hence, the planner must be prepared to communicate with three sets of publics: (1) those that can be identified and will participate, (2) those that become identified as the process progresses, and (3) those that will be identified in the future. Thus, of the publics initially identified by the agency, some will follow through, others will drop out, and some previously unidentified interests will enter the arena of participation. Indeed, controversies in aquifer restoration projects have often occurred as a result of new participants entering at the end of the process in opposition to proposed actions. Three approaches can be used in dealing with the

Table 8.3 Scheme for Identifying Publics (Bishop, 1981)

Interests/Issues Groups	Affected by the Problem			Relation to the Study		Not Affected
	Directly	Indirectly	Etc.	Affected by the Proposed Solution		
	Beneficial/Adverse			Users	Non-Users	
	⌐—	⌐—		…	…	—
Individuals						
Property Owners/Users						
Conservation/Environmental Groups						
Sportsman's Groups						
Farm Organization						
Business/Industrial						
Professional						
Education Institutions						
Labor Unions						
Service Clubs						
State/Local Agencies						
Elected Officials						
News Media						

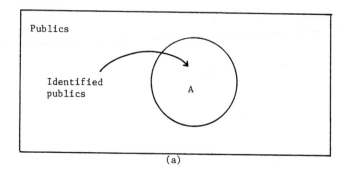

(a)

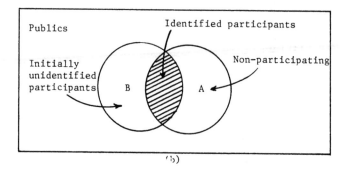

(b)

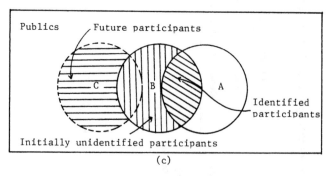

(c)

Figure 8.6 A temporal perspective of identification of publics (Bishop, 1981).

problem of publics changing over time: (1) actively seek out and engage at the outset of a study a broad and representative range of public interests; (2) keep as much flexibility for as long in the process as possible, insofar as selecting a plan or recommending action; and (3) document the process and the public inputs relating to alternatives and issues studied (Bishop, 1981).

Selected Techniques for Communicating with Identified Publics

It was noted earlier that some techniques are better than others for communicating with different publics. Table 8.4 summarizes the effectiveness of certain techniques for communicating with the publics identified earlier as associated with water resources projects (U.S. Army Corps of Engineers, 1971). Publics of the type listed in Table 8.4 could also be expected to be associated with an aquifer restoration project.

Table 8.4   Effectiveness of Different Media on Various "Publics" (Bishop, 1975)

Publics	Public Hearings and Meetings	Printed Brochures	Radio Programs and News	TV Programs and News	Newspaper Articles	Magazine Articles	Direct Mail and Newsletters	Motion Picture Film	Slide-Tape Presentation	Telelecture
Individual Citizens	M	L	H	H	H	L	L	M	M	L
Sportsmen Groups	M	M	M	M	M	H	H	H	H	M
Conservation-Environment Groups	M	M	M	M	M	H	H	H	H	M
Farm Organizations	M	M	M	M	M	H	H	M	M	M
Property Owners and Users	M	L	H	H	H	L	L	M	M	L
Business-Industrial	L	L	M	M	M	M	H	M	M	L
Professional Groups and Organizations	L	L	M	M	M	M	H	M	M	L
Educational Institutions	M	L	L	L	M	M	H	M	M	M
Service Clubs and Civic Organizations	L	L	M	M	M	M	L	H	H	M
Labor Unions	L	L	M	M	M	L	L	M	M	L
State-Local Agencies	H	M	L	L	L	M	H	H	H	H
State-Local Elected Officials	H	M	L	L	L	L	H	H	H	H
Federal Agencies	H	M	L	L	L	L	H	M	M	M
Other Groups and Organizations	H	M	M	M	M	M	H	H	H	M

H = Highly Effective
M = Moderately Effective
L = Least Effective

SELECTION OF PUBLIC PARTICIPATION
TECHNIQUES

A critical element in planning a public participation program is associated with the selection of public participation techniques to meet identified objectives and publics. It should be noted that there are numerous techniques, and a well-planned public participation program will probably involve the use of multiple techniques over the lifetime of an aquifer restoration study. The most traditional public participation technique is the public hearing, which is a formal meeting for which written statements are received and a transcript is kept. The public hearing is generally not the best forum for public participation in an aquifer restoration project. Several classification schemes have been developed for public participation techniques in accordance with their function (Schwertz, 1979) and communication characteristics and potential for meeting stated objectives (Bishop, 1975). In addition, Creighton (1981b) developed a short catalog of the advantages and disadvantages of several additional techniques. This section of this chapter will highlight the features of the classification schemes and the catalog.

Techniques Classification According to Function

Schwertz (1979) summarized a U.S. Department of Transportation (DOT) classification scheme for 37 public participation techniques grouped into six functional classes: (1) information dissemination; (2) information collection; (3) initiative planning; (4) reactive planning; (5) decision-making; and (6) participation process support. Table 8.5 lists the 37 citizen participation techniques by the functional classes (Schwertz, 1979). Volume II of the DOT report describes the 37 techniques using the following format: description, positive features, negative features, potential for resolving issues, program utilization, costs, and bibliography (U.S. Department of Transportation, 1976). Summary comments relative to the six functional classes are as follows (Schwertz, 1979):

(1) Information Dissemination techniques are used to inform the public of actions being taken in an area, the opportunities for citizen input, and any proposed plans that have been submitted. The flow of information is characteristically one-way. Accordingly, mass media techniques are most frequently used for this purpose. Many local governments have established Public Information Programs for coordinating information dissemination. The most frequently used mass media technique is the local daily newspaper. Where a local daily newspaper is not published, many communities will use the services of neighboring larger com-

Table 8.5 Participation Techniques Classified by Function (Schwertz, 1979)

1. **Information Dissemination**
   Public Information Programs
   Drop-In Centers
   Hot Lines
   Meetings—Open Information

2. **Information Collection**
   Surveys
   Focused Group Discussions
   Delphi
   Community-Sponsored Meetings
   Public Hearings
   Ombudsman

3. **Initiative Planning**
   Advocacy Planning
   Charettes
   Community Planning Centers
   Computer-Based Techniques
   Design-In and Color Mapping
   Plural Planning
   Task Force
   Workshops

4. **Reactive Planning**
   Citizen's Advisory Committees
   Citizen Representatives on Policy-
     Making Boards
   Fishbowl Planning
   Interactive Cable TV-Based
     Participation
   Meetings—Neighborhood
   Neighborhood Planning Councils
   Policy Capturing
   Value Analysis

5. **Decision-making**
   Arbitrative and Mediative Planning
   Citizen Referendum
   Citizen Review Board
   Media-Based Issue Balloting

6. **Participation Process Support**
   Citizen Employment
   Citizen Honoraria
   Citizen Training
   Community Technical Assistance
   Coordinator or Coordinator/Catalyst
   Game Simulation
   Group Dynamics

munities. Many smaller communities prepare newsletters and posters for disseminating information on meeting dates and proposed plans to supplement out-of-town newspapers. The use of radio and television is also common in Public Information Programs. In larger communities, television documentary films are often found. Radio talk shows are utilized in all communities to assist in "getting the word out" about local growth management programs. Another useful technique for disseminating information is Open Information Meetings. They are particularly effective when used early in the planning phases of the local growth management program. Widely publicized, and open to all interested parties, the Open Information Meetings present technical information, planning proposals, or details about a specific work program or aquifer restoration project.

(2) Information Collection techniques are used to gather both technical data as well as opinion-related information. The two most frequently used techniques are the Public Hearing and the Survey. The Public Hearing has been used by almost every community at one time or another. Although the Public Hearing does permit the parties-at-interest to be heard, it does not permit a two-way communication between the agency representatives holding the hearing and the individual parties-at-interest. The normal sequence of a Public Hearing begins with a presentation by the sponsoring public agency followed by testimony of individuals recognized by the presiding agency officer. Therefore, the Public Hearing can, at times, become heated as parties-at-interest seek to assert their position. A less

frequently utilized information collection technique is the Survey. However, as communities strive harder to ascertain the needs, attitudes, and opinions of its citizens, the use of the Survey becomes more prevalent. An advantage of the Survey over the Public Hearing is the ability to reach all segments of the population if it is properly administered. Public Hearings attract only the "vocal minority" and the parties-at-interest. Accordingly, a properly administered Survey can provide the community with the information it will require on needs, attitudes, and opinions to develop a responsible program. Proper administration includes the use of mailed survey questionnaires, telephone interviews, and personal interviews. Sole reliance on mailed questionnaires will not yield results representative of all segments of the population. Telephone and personal interviews may be needed in addition to using mailed surveys.

(3) Initiative Planning techniques assign the responsibility for producing conceptual proposals and plans to citizen representatives. The most frequently used techniques have been Task Force committees and Workshops. Although the citizen committee is responsible for producing the conceptual proposals and/or plans, it often requires significant financial resources and local staff assistance to complete its tasks. Therefore, Initiative Planning techniques should be used by communities which possess sufficient financial resources and the trained staff to provide technical assistance as needed. A Task Force is an ad hoc citizen committee in which the parties-at-interest are actively engaged in a well-defined problem solving or specific task effort in the planning process. In most larger communities, particularly those over 50,000 in population, Task Force techniques have been used to obtain citizen involvement. Workshops are working sessions lasting anywhere from several hours to several days. They are held to enable parties-at-interest to engage in mutual education by thoroughly discussing a technical issue or idea and by trying to reach some basic understanding about its nature, relevance, and role in the particular proposal, project, or plan. The workshops are frequently directed by trained professionals and involve citizens alone or together with local staff and elected officials. Although workshops promote the achievement of a consensus and the introduction of alternative viewpoints among those citizens, when used as the sole citizen participation technique the problem of nonrepresentation of all citizen segments arises. (It should be also noted that the same is true of using Task Force committees as the sole citizen participation technique in a community.)

(4) Reactive Planning techniques assign the responsibility for producing conceptual proposals and plans to the local planning agency, and citizens are asked to react to those proposals and plans. Based on citizen input, the conceptual proposals or plans are revised and resubmitted for reaction. Two of the most frequently utilized Reactive Planning techniques are the Citizens' Advisory Committees and the Neighborhood Planning Council. Citizens' Advisory Committees consist of a group of citizens called together by a local agency to represent their community and advise the local agency on specific projects or programs. They have limited advisory powers, are often large in membership (50 to 100), and tend to meet infrequently. Often the life of the committee is tied to the specific project for which they are to provide advice. Citizens' Advisory Committees are prevalent in communities of every size. If used properly, and if the members are representative of the community, this technique can serve as a community listening post and also serve to relay technical proposals and plans to the public.

(5) Decision-making techniques are not feasible in program areas where the responsibility for the decision is mandated to elected officials. However, the Citizen Referendum is the decision-making technique which has received the greatest use in the past.

(6) Participation Process Support techniques are not citizen participation techniques in themselves, but mechanisms to strengthen the effectiveness of the other techniques. For example, Citizen Training can be provided to assist citizen members of boards of committees with the technical aspects of the issues with which they are confronted. Citizen Honoraria can assist citizen members with the expenses incurred in serving on boards or committees. Group Dynamics can facilitate group interaction and achievement of group goals among Task Force members or Workshop participants.

Techniques Classification According to Communication Characteristics and Potential for Meeting Stated Objectives

Bishop (1975) developed a structured classification scheme for 24 public participation techniques. Table 8.6 displays the techniques in three groups: (1) the first six listed techniques represent public forums; (2) the following eleven listed techniques represent community contacts; and (3) the final seven listed techniques represent interactive group methods. The communication characteristics of the 24 techniques are displayed in Table 8.6 in terms of the level of public contact achieved, ability to handle specific interest, and degree of two-way communication. A relative scale of low (L), medium (M), and high (H) effectiveness is used to delineate the communication characteristics of the 24 techniques. Table 8.6 also has information on the potential usefulness of each technique relative to meeting one or more of six objectives for public participation in the decision-making process. Specific comments on the advantages and limitations of certain listed techniques are as follows (Bishop, 1975):

(1) Public hearings—Public hearings still tend to be formal and rather highly structured. However, there seems to be a trend away from this formality while still maintaining appropriate records, that is, the transcript and written statements for the hearing record. Because of the cost and delay in developing the hearing record, public hearings should generally be used only at stages in the decision-making process where a formal record or transcript is required. This is not to say that public hearings do not serve an important function. Although the hearing is more costly and formalized it provides an opportunity for any agency, group or individual to be on permanent public record as having put forth a certain view. In fact, in the view of many publics, the hearing, because it has a high degree of legitimacy, is the one place their voice must be heard. This single fact should not be overlooked when considering the virtues of the public hearing. Although hearings have a major advantage in public acceptance there are disadvantages;

Table 8.6 Capabilities of Public Participation Techniques (Bishop, 1975)

Public Participation Techniques	Level of Public Contact Achieved	Ability to Handle Specific Interest	Degree of Two-Way Communication	Inform/ Educate	Identify Problems/ Values	Get Ideas/ Solve Problems	Feedback	Evaluate	Resolve Conflict/ Consensus
Public hearings	M	L	L	X	X		X		
Public meetings	M	L	M	X	X		X		
Informal small group meetings	L	M	H	X	X	X	X	X	X
General public information meetings	M	L	M	X					
Presentations to community organization	L	M	M	X	X		X		
Information coordination seminars	L	H	H	X			X		
Operating field offices	L	M	L		X	X	X	X	
Local planning visits	L	H	H		X		X	X	
Planning brochures and workbooks	L	M	L	X		X	X	X	
Information brochures and pamphlets	M	M	L	X					
Field trips and site visits	L	H	H	X	X				
Public displays	H	L	M	X		X	X		
Model demonstration projects	M	L	M	X			X	X	X
Material for mass media	H	L	L	X					
Response to public inquiries	L	H	M	X					
Press releases inviting comments	H	L	L	X			X		
Letter requests for comments	L	H	L	X		X	X		
Workshops	L	H	H		X	X	X	X	X
Charettes	L	H	H			X		X	X
Advisory committees	L	H	H		X	X	X	X	
Task forces	L	H	H		X	X			
Employment of community residents	L	H	H		X	X			X
Community interest advocates	H	H	H			X	X	X	X
Ombudsman or representative	L	H	H		X	X	X	X	X

Communication Characteristics (first three columns) — *Impact Assessment Objectives* (last six columns)

[a] L, low; M, medium; H, high

for example, public hearings provide no guarantee of representativeness; and thus there is a high potential for bias. The chairman, being from the agency, may also strongly bias the hearing. Open ended statements presented are often hard to interpret and use in planning, and often persons testifying do not completely understand the issue or the plan on which they are speaking.

(2) Public meetings – The public meeting is less formalized than are hearings and do not require a transcript. However, detailed notes should be kept on file and recorded. These meetings, like the public hearing, provide the opportunity for participation by a wide cross section of the public. Although the tone here is informality, a discussion leader or chairman must maintain control over the meeting to prevent polarization of positions and to keep the meeting on the main issues and topics of discussion. Public meetings seem to have most of the advantages of the hearing without the rigidity and formality and the problems of cost of permanent records. This point becomes significant when a number of meetings are necessary in the decision-making process. In keeping with the public meeting's less formal and rigid structure, physical details such as placing the agency staff on the audience level and less formal seating of the audience itself should be used. This applies to meeting organization as well. After the opening statements by the chairman, the meeting can be turned over to the study manager. This tends to give the public the feeling that they are more intimately involved with the study as opposed to upper level staff. For purposes of efficiency this type of forum could often supplant the original hearing concept.

(3) Informal small group meetings – While this type of meeting may take several forms and serve several purposes, the overall format is much the same as a public meeting. In this respect, small group meetings may function as a series of small scale public meetings to allow more intimate contact with publics from various geographic or interest group areas. General community meetings may be of this sort. Meetings of general interest may be advertised by public notice while others will perhaps be invitational if a particular specialized discussion with key individuals or community leaders is to be held. The basic idea of this meeting, as with large-scale forums, is to present information and to ascertain the needs, desires, and opinions of the affected or interested public. The format should emphasize informality to the point of a round table type of discussion if feasible. Again, a discussion leader, preferably the project manager, should exercise a controlling function to prevent community or local factions from polarizing the meeting.

(4) Information and coordination seminars – This technique is not used to inform the general public directly, but functions to inform and coordinate with special interest groups, specific individuals and groups representing segments of the public. Often public interests and needs are voiced through key individuals, elected officials and non-elected leaders, rather than by involvement of the general public. Seminars could be effectively used with community and group leaders, public agencies or officials, and special interest groups. Individual citizens in an aquifer restoration project area might be a special interest group. Seminars are an excellent way of keeping elected officials up-to-date on a regular basis, providing specialized information to interest groups and clarifying policy and plans to any group or agency. These seminars could also aid in developing coordination between cooperating agencies. Seminars can be used as one technique for advanced preparation for workshops and special committees. This is an efficient method of providing select personnel with information neces-

sary to perform a prearranged future function. A major advantage to the seminars is that they have a low time budget. They can be organized on a regularly scheduled basis or only when needed.

(5) Forum of other agencies or groups—Forums of other agencies and groups such as civic group meetings, organization meetings, etc., can also be used for pertinent presentations or statements. These regularly scheduled meetings are an effective avenue for broadening contact during the planning process. Equally important is the opportunity to clarify policy and positions not only to the agency or group, but often directly to their constituency or organization as well. This approach should be utilized to the maximum extent possible by notifying potential audiences of the agencies' availability to provide pertinent information and discussion programs.

(6) Operating field offices—This serves a more or less specialized liaison function between the agency office and the public. In studies necessitating close local contact and coordination it may be used efficiently, particularly if well publicized. The field office can serve a planning study function and also be geared to providing public information. Duration of operation will depend on study planning and information needs. The field office could be of particular value in a controversial study in which the planning office, by virtue of distance or other limiting factors, is not readily accessible.

(7) Local planning visits—Community or on-site visits are oriented toward increased understanding and coordination with cooperating agencies, knowledgeable community interest groups, and individuals. The technique should serve a professional function rather than public relations. These visits can have the secondary function of providing advice on community or area problems related to the study. Visits should be tuned to community needs or specific aspects of the study. A practical example of the value of local visits would be liaison work or advance preparation for workshops. The press can often be usefully included in planning visits.

(8) Field trips and site visits—These differ from planning visits in that they are primarily nonprofessional "show-me" trips. These visits can be used to accurately inform public groups, local officials, and the media about the specifics of an aquifer restoration plan. Trips can be combined with or considered as field press conferences. In this fashion, the public is informed indirectly through their representatives—either groups or media. This is also excellent advance preparation for an open workshop or a later hearing.

(9) Public displays and model demonstration—Under appropriate conditions, displays and demonstrations can provide an overview of an aquifer restoration project, quick appraisal of alternatives, description of impacts of the project, and information on any number of project-related issues. The critical factors in good use of public displays seem to be their location and manning. Certainly the best locations are areas of high public use or areas where particular affected publics may be contacted. Displays may take several forms ranging from a manned project model to an automated slide show. Model demonstrations may be used as a public display or may be considered as an instructional aid at any type of public forum, workshop, or seminar. Displays or demonstrations, to be effective, must be clear, concise and directly related to areas of experience and familiarity of the publics.

(10) Workshops—The success of workshops depends, in large part, on the degree of advance preparation; therefore, this should be as comprehensive as possible.

Advance preparation for workshops might include distribution of the various types of brochures, planning visits, coverage by the media and direct contact of interested parties. Workshops can be several types depending on the planning activity and stage of the decision making process, the publics to be contacted, and the subject matter for discussion.

(a) Open public workshop—This type of workshop, in practice, is the most common. However, one major disadvantage of the open workshops is the uncertainty in the number of people that will attend, and their interests. With large numbers there is more limited opportunity for discussion and the high degree of interaction that is desired in a workshop.

(b) Invitational workshop—As implied by the name, invitational workshops are geared toward particular individuals and groups and around issues or alternatives that are somewhat specific in nature. This type of workshop has the advantage of being highly interactive, involving only interested publics on specific, critical issues.

(c) Invitational/Open—This workshop approach, a combination of the first two, provides a means of bringing all concerned publics into the planning process and providing a productive interchange. The workshop is structured to focus the beginning discussion with an invited group of interests, for example, a panel, then opening the meeting up to the general public.

There are, of course, several modifications to these three workshop types that can be introduced. Several varieties of mini-workshops are proving to be effective in stimulating interaction. Publics attending a workshop can be divided into small discussion groups, each with a leader, to exchange ideas on different subjects. Under certain circumstances revolving groups can be instituted, where individuals spend a set amount of time on one issue or subject, and then break up, with each individual going to a different group. As with all workshops, but especially the revolving mini-type, adequate preworkshop preparation is a necessity.

(11) Charettes—The charette functions as a highly intense, resolution-oriented meeting. It can be thought of as a mini-workshop or small select group meeting with the express purpose of reaching a decision or resolving a conflict. The charette goes beyond the interaction levels of an ordinary workshop and is problem solving and decision oriented in its subject matter. Hence, it presupposes a certain amount of advance preparation to assure a thorough understanding of the subject and a common ground on which to begin. Charettes can function at the interagency level, or community level, with special interest groups. In this setting the planner is often a negotiator of community interests. The intensity of charette sessions are certainly not necessary in all aquifer restoration studies, but in certain cases resolution and/or decision comes only through this type of interactive situation.

(12) Special committees—Several types of citizens committees have been used in planning studies. Committees, as representative public bodies, can serve a very useful purpose in aquifer restoration studies, but the overall concept must be well considered before the planner can effectively utilize the committee approach. Types of committees that have functioned in planning studies are the following:

(a) Citizens advisory committee—The concept of an advisory committee has recently been receiving criticism concerning its role in the planning process. The major question with the role of the CAC is that perhaps it was never advisory and indeed it should not be. It is extremely difficult to determine the extent of the public interests that a committee's advice or actions may represent. Since a committee with a true advisory capacity is difficult to staff with people representing a broad range of community interests as well as expertise pertinent to the aquifer restoration study, this type of committee often ends up serving a limited or issue specific function. However, a committee that is broadly representative can be extremely useful as a sounding board for study proposals.

(b) Ad hoc task force/committee — Aquifer restoration problems of a technical or local nature can often be effectively approached by a committee or task force which works towards solutions and advises the planning agency of local preferences on those particular study problems. A committee or task force should be limited to consideration of special problems. When controversial aspects of a plan are involved, a group representing all sides of the issue is necessary for lasting conflict resolution or problem solution. Since both ad hoc committees and special task forces are set up to work on a particular problem area, they should be dissolved once a solution has been rendered.

(c) Citizens committees — A new trend in committee roles in the planning process is that of the nonadvisory, nondebating forum. The task of a citizens committee is fivefold: (1) to provide fact-supported suggestions or arguments on various problems or issues that might arise; (2) to act as a sounding board to reflect community or subregional interests and preferences in regard to issues, alternatives or problems which will arise during the course of the decision-making process; (3) to act as a catalyst for the expansion of public participation by utilizing other techniques (workshops, small group meetings, etc.); (4) to involve the committee member's constituency by hosting and participating in various public forums; and (5) to bring other participants into the planning process.

The success or failure of citizens' committees seems to hinge on selection of committee members and timing. Selection of members often becomes the responsibility of the agency, but organizations or local officials should be invited to designate members, or at least suggest names. Representatives from certain major groups must be included from the very beginning, with additions or changes being a function of the committee or the supporting organizations. Committees are often unproductive because they are initiated too late in the decision-making process. In this situation, members feel they are little more than a token gesture and can contribute nothing that will influence what has already been determined. On the other extreme, beginning too early when the members have nothing on which to work may result in apathy and dissolution of the committee. The citizens' committee can be an effective tool for public involvement in decision-making. Unfortunately, however, there are some major difficulties that have prevented its widespread acceptance. First and foremost of these is the time commitment required of the planners and the participants. The planner usually must spend a great deal of his time organizing and participating in committee functions. The committee members, if they take their task seriously, can also devote a considerable amount of time to the committee. Considering these factors individual citizens committees, on the average, may not have a long life expectancy. In studies where there is not a considerable degree of opposition and interest, apathy takes its toll. This is not to discourage the planner from utilizing citizens committees. On the contrary, they can be one of the better participation tools available. However, it is only realistic to be aware of the disadvantages.

Due to the growing importance of citizen committees in public participation programs, some additional relevant information from Creighton and Delli Priscoli (1981) will be highlighted. They indicated that a citizens' committee can provide a number of helpful functions in the decision-making process, including:

(1) Help set planning priorities.
(2) Review technical data and make recommendations on its adequacy.
(3) Help resolve conflicts among various interests.
(4) Help in the design and evaluation of the public involvement program.
(5) Serve as a communication link to other groups and agencies and bring reactions back to the agency.
(6) Review and make recommendations on the decision-making process.
(7) Assist in developing and evaluating alternatives.
(8) Help select consultants and review contracts.
(9) Review and make recommendations on the program budget.
(10) Review written material prior to release to the general public.
(11) Help host and participate in public meetings.
(12) Assist in educating the public about the project and the decision-making process.

Creighton and Delli Priscoli (1981) delineated the following four general principles for establishing citizens' committees:

(1) Clearly define the limits of authority of the citizen's committee. It is extremely important that the authority of the citizens' committee be defined as there is frequent confusion as to the difference between a group that is an advisory and a decision-making body. It is easier for a citizens' committee to cope with limits to their authorities if they are clearly defined at the beginning of the study. If expectations are created of greater authority than actually exists, the sense of betrayal is often greater than if there had been clearly defined limits in the first place.
(2) Citizens' committees must be representative of the full range of values within the community. A citizens' committee that represents only a few limited viewpoints may mislead the agency and embitter those publics who are not included in the committee. Typically, citizens' committees are large enough so that it is possible to have direct representation for all the different viewpoints. Every effort must be made to insure that the citizens' committee represents the full spectrum of values within the community.
(3) The life of the committee should be limited. The longer that a committee is in existence, the more likely it is that the members of the committee become unrepresentative of their constituencies and instead become a new kind of elite. As a result it is important to establish from the beginning what the life of the commit-

tee will be. Typically, the life of the committee coincides with some major function such as the selection of the aquifer restoration strategy.

(4) Efforts should be made to insure that members of citizens' committees maintain
regular communication with the constituencies they are supposed to represent. As
suggested above, citizens' committees tend over time to become a new kind of
elite, and unless the expectation is established from the beginning that one of the
duties of citizen committee members is to maintain communication with their
constituencies, then the membership may become increasingly unrepresentative
of the public at large. This communication with their constituencies could take
the form of briefings of the groups they represent on study progress, informing
their constituencies through their own organizational newsletters, or occasional
interviews with other leaders from their constituencies.

The biggest single problem in establishing a citizens' committee is to
select members in such a way that the public believes the committee
represents the community (Creighton and Delli Priscoli, 1981). Because
an attitude of suspicion often exists towards the agency, there are frequent accusations that agencies have established citizens' committees in
such a way that they were "stacked" toward the desired ends of the
agency. There are five basic strategies by which members of a citizens'
committee can be selected:

(1) Members are selected by the agency with an effort to balance the different interests. As mentioned above, this runs the risk of the public believing that the agency
has established the committee to serve the agency's purposes. This danger can be
reduced somewhat if the agency has consulted thoroughly with various other
governmental agencies and interest groups prior to making these selections and
the selections clearly encompass many of these recommendations.

(2) The agency may turn over the selection of the citizens' committee to some third
party or group. One approach is to have some local elected body such as a city
council or board of supervisors select the membership. An alternative approach is
for the agency to select a small committee and permit the committee to select a
predetermined number of additional members. In either of these cases, it is
extremely important that the agency communicate its expectations that the membership of the committee should reflect the entire range of values within the
community.

(3) An alternative method is for the agency to identify the interests it wishes to have
represented and allow the various groups within those interests to select their own
representatives. This can create administrative problems as volunteer groups
sometimes have difficulty coordinating between themselves to select a representative, but it does eliminate the risk that the agency will be seen as "stacking the
deck."

(4) It is also possible to use any of the three methods above and then augment the
membership with the addition of volunteers. This in effect allows the different
interests to adjust the membership of the group by obtaining volunteers from
their own ranks. But if votes are being taken, it does lead to the risk that various
groups will "stack the deck" by trying to add a large number of additional
volunteers.

(5) In a few cases, membership on a citizens' committee has been determined by popular election. This last technique has been utilized only on projects where the target publics are clearly identified and limited.

Catalog of Additional Techniques

Creighton (1981b) prepared a short catalog of public involvement techniques. Within the catalog a short description is provided for each technique plus a discussion of relevant advantages and disadvantages. Techniques not previously discussed will be presented herein, including interviews, hotlines, newspaper inserts, reports, brochures, information bulletins, surveys and questionnaires, participatory television, cumulative brochures, mediation, Delphi, and simulation games. Summary information related to these techniques is as follows (Creighton, 1981b):

(1) Interviews—One technique for quickly assessing public sentiment is to conduct a series of interviews with key individuals representing the range of publics most likely to be interested or affected by the decision-making process. The kinds of information which might be discussed in an interview would include the amount of interest in the aquifer restoration study, the goals and values of the interest group the individual represents, the manner in which the interest group would like to participate in the study, political climate and relationship between the various interest groups. Interviews can either be nonstructured, or the interviewer can prepare a list of questions or topics to be discussed in each interview, so that responses can be easily compared and summarized. Since there are skills involved in effective interviewing, interviews should be conducted by persons with experience or training in interviewing. Federal agencies are required to get approval from the Office of Management and Budget for all surveys and questionnaires. Structured interviews may fall under these approval requirements, so federal agencies may find it preferable to stay with unstructured interviews.

  (a) Advantages of Interviews: Interviews can provide a quick picture of the political situation in which an aquifer restoration study will be conducted. Interviews can provide important information about how various interests wish to participate. Personal relationships can be built with key individuals and more direct communication links established with the publics. Once communication has been established through an interview, individuals and groups are more likely to participate.

  (b) Disadvantages of Interviews: Poor interviewing can create a negative impression of the individual and the agency. Interviews may not be entirely representative of public sentiment.

(2) Hotline—A hotline is an "easy to remember" telephone number which is publicized through repetition in brochures, reports, news stories, paid advertising, etc., as a single telephone number that citizens can call to ask questions or make comments about project-related issues. If the public which the agency wishes to reach is large geographically, the hotline is usually established so that the call is toll-free to the public regardless of where the call is placed. The hotline is manned with staff who will take responsibility for finding answers to questions

from the public, or for relaying comments or complaints from the public to appropriate staff persons. Hotlines have been used as a method of handling public complaints, and as coordination points for individuals requiring information about the progress of a study. Comments received over a hotline can be incorporated as part of the record of a public meeting or hearing.

(a) Advantages of the Hotline: The hotline provides a convenient means by which citizens can participate in the study. The hotline assists citizens in locating the staff most likely to be able to answer their questions or receive their comments. The hotline may be a useful means of providing information about meetings or other public involvement activities. The hotline is a communication to the public of the sponsoring agency's interest in their comments or questions.

(b) Disadvantages of the Hotline: Defensive or insensitive comments may produce a negative reaction from the public. The hotline must be staffed by people able and willing to deal with public comments effectively.

(3) Newspaper inserts — One technique which has been used to provide information to the broad general public and, at the same time solicit comment back from the public, is a newspaper insert including a response form distributed through the local newspaper. Most newspapers are able to handle the distribution of inserts for a modest cost, and are often able to print the insert at considerably less cost than other commercial printers. The newspaper insert can describe the aquifer restoration study and decision-making process and the various means by which the public can be involved; it can also include a response form which will allow people to express opinions or indicate their willingness to be involved in other public involvement techniques. Most urban newspapers are able to distribute inserts to selected geographical areas, rather than their entire readership, so that it is possible to target the insert at those areas which will have the highest interest in the study. On a percentage basis, the return of response forms is not likely to be very high, although on a total quantity basis, it may provide a means of participation for the largest number of citizens compared with other public involvement techniques. Because respondents are self-selecting, a statistical bias is introduced into the responses, so that they cannot be represented as statistically valid like a survey.

(a) Advantages of a Newspaper Insert: Newspaper inserts reach a much greater percentage of the population than most other public information techniques. The inserts provide an opportunity for a large number of citizens to participate. Newspaper insert response forms provide a means for identifying other individuals and groups interested in participating in public involvement activities.

(b) Disadvantages of Newspaper Inserts: Newspaper inserts are relatively expensive to produce and distribute in large numbers. The response rate from newspaper inserts is relatively low, and it cannot be represented as statistically valid.

(4) Reports, Brochures, and Information Bulletins — Reports, brochures, and information bulletins are an essential part of every public involvement effort. They are an essential vehicle for informing the public of the opportunities for participation, the progress of the study to date, and any decisions that have been made. There are three times at which reports are typically published in a public involvement program. These include: after problem definition and issues identification, including initial data collection; upon identification of a set of general alterna-

tives; and upon identification of specific detailed alternatives and their environmental consequences and costs. Because reports contain technical information, one key requirement is to write reports in a manner which provides needed technical information, yet is understandable to the general public. It is sometimes useful to have reports reviewed by an advisory committee who can point out confusing, biased, or unnecessary material in the report.

Brochures are usually brief (up to sixteen pages) and contain a description of the study, the issues involved in the study, and a summary of the opportunities for the public to participate in the study. Typically, brochures are used to reach new publics or inform known publics of the initiation of the study. The usefulness of a brochure is almost entirely dependent on its visual attractiveness and the skill with which it is written.

Information bulletins or newsletters are periodic reports to the public published as a means of maintaining a continuing interest in the study as well as documenting the progress in the study in a highly visible manner for the public. Information bulletins or newsletters are particularly important during portions of the study which are relatively technical in nature. During these periods the general public is less likely to be involved but should be kept informed of what is occurring through these media. The value of an information bulletin or newsletter rests almost entirely upon its ability to stir interest and encourage interaction. A drab, boring, bureaucratic sounding newsletter will usually not be worth the effort.

(a) Advantages of Publications: Publications are direct means of providing a substantial amount of information to a large number of people in a relatively economic manner. Publications are able to communicate a greater amount of information than almost any other form of communication. Publications serve as a permanent record of what has transpired in the public involvement program.

(b) Disadvantages of Publications: Preparation of attractive publications requires definite skills which are not available in all organizations, so may have to be purchased outside the organization. Because of cost factors publications still reach only a limited audience and cannot be considered the only means by which to inform and involve the general public.

(5) Conduct a Survey—Surveys are an effort to determine public attitudes, values, and perceptions on various issues employing a rigorous methodology to insure that the findings of the survey in fact represent the sentiment of the community being sampled. Surveys can be conducted by phone, by mail, by individual interviews, or in small group interviews. Firms that design surveys spend many hours and utilize complex procedures to insure that the survey does not contain bias and that the "sample" of people interviewed is in fact representative. As a result, surveys must be designed and conducted by persons experienced in survey design. Normally this means that someone outside the planning organization must be retained to design and conduct the survey. Federal regulations require OMB approval of all surveys or formal questionnaires conducted by federal agencies or with federal funds. These approvals are very difficult and time-consuming to obtain, virtually ruling this technique out for most federal agencies.

(a) Advantages of a Survey: Surveys can provide an expression of feeling from the total public, not just those publics which are most directly affected. Surveys can provide an indication of whether or not the active participants in a public involvement program are in fact representative of the broader public.

(b) Disadvantages of the Survey: Unless surveys are carefully designed, they do not produce reliable and meaningful data. The cost of developing statistically reliable surveys is high. Surveys cannot substitute for political negotiation between significant interests.

(6) Participatory Television — Because of the number of people reached by television, it holds considerable potential as a tool for both informing the public, and soliciting participation. Some experts see cable television as holding the answer to participation, since eventually cable television may be utilized in such a way that it allows for two-way communication. In the meantime, there have been several major uses of television programs. These include:

(a) Preparation of a half-hour or a one-hour television program describing alternative courses of action in a major study. Participants are asked to express their preference by mail or by a ballot that has been distributed in advance of the television program. In some cases discussion groups have been organized so that people watch the television program as a group, and discuss the program as a group, before marking the ballots.

(b) The agency could also obtain a block of time and conduct a call-in show on issues. One planning agency conducted a television program much like a telethon, with banks of telephone operators to receive calls from the public and have them answered by a panel of elected officials.

Although television reaches large numbers of people, it is unusual to be able to obtain sufficiently large blocks of time for a participatory television program on commercial television, although this has been accomplished in a few cases where the study was extremely controversial. The audience on educational, university or cable television is much smaller and something of an educational and social economic elite. This creates problems of representation. Any poll which is taken accompanying such a program would share these problems of representation.

(a) Advantages of Participatory Television: Participatory television reaches the largest audience of any community involvement technique. This technique is most convenient for the participants, because they do not have to leave their own home. Even if people do not participate by filling out a ballot or phoning in, there is a definite educational function to participatory television.

(b) Disadvantages of Participatory Television: The audience viewing the program may not be representative, and any votes or tallies taken as a result of the program may also be unrepresentative. Unless some participation occurs in designing the program, the public may not feel that the agency accurately or objectively described the issues. This kind of participation gives equal value to somebody who lives immediately near a problem as somebody who lives fifty miles away and is only peripherally affected.

(7) Cumulative Brochure — The cumulative brochure is a document which keeps a visible record of a series of repetitive public meetings, public brochures, workshops, and citizen committee meetings. At the beginning of the process, a brochure is prepared presenting various study alternatives along with the pros and cons for each of the alternatives. In a series of public meetings and workshops, individuals, agencies, and organizations are invited to submit their own alternatives which are then included in the brochure along with their descriptions of pros, cons, and a no-action alternative. The brochure is then republished with space provided in the brochure for individuals to react to the various alternatives

by writing their own pros and cons. These comments then become a part of the new brochure. With each round of meetings or other forums for public comment, the brochure grows by the addition of the public comment and technical response. The final document is quite thick, but does provide a visible record of the entire process.

(a) Advantages of a Cumulative Brochure: The process is very visible and allows the public to see how a decision was reached. The process encourages open communication between the various publics as well as between the public and the sponsoring agency. No special status is granted to any one individual or group over another.

(b) Disadvantages of the Cumulative Brochure: The final brochure is a large, cumbersome document and the many editions of the brochure can be expensive to produce. The effectiveness of the brochure depends on the ability of the sponsoring agency to address the issues in non-bureaucratic language. The format of the brochure forces public reaction into a pro or con response when there may be other positions as well. Since the sponsoring agency prepares the brochure, groups which are suspicious of that agency may question whether the brochure is biased.

(8) Mediation — Mediation is the application of principles of labor/management mediation to environmental or political issues. In mediation a group is established which represents all major interests which will be affected by a decision. Members of the mediation panel are all "official" representatives of the interests, and are appointed with the understanding that the organizations they represent will have the opportunity to approve or disapprove any agreements which result from the mediation. The basic ground rule which is established is that all agreements will be made by unanimity. A key element in mediation is the appointment of a third party mediator — someone skilled in mediation, who is not seen as an interested party to the negotiations. The mediator not only structures the deliberations, but often serves as a conduit for negotiations between the various parties. Mediation is only possible when the various interests in a conflict believe they can accomplish more by negotiation than by continuing to fight. Additional information on mediation will be provided later in this chapter.

(a) Advantages of Mediation: Mediation can result in an agreement which is supported by all parties to the conflict. Mediation may lead to quick resolution of issues which might otherwise be dragged out through litigation or other political processes.

(b) Disadvantages of Mediation: Mediation is an entirely voluntary process, so it will work only when all parties are willing to negotiate. Mediation requires a highly skilled third party mediator.

(9) Delphi — The Delphi process is a method for obtaining consensus on forecasts by a group of experts. It can be used as a technique for estimating possible environmental improvements of various aquifer restoration actions. The basic procedure is as follows. A questionnaire is submitted individually to each participant requesting them to indicate their forecasts concerning the topic. The responses to the questionnaire are consolidated and resubmitted to the participants with a request that they make an estimate of the probable occurrence of each event. The participants' responses are again collected and a statistical summary is prepared. The statistical summary is distributed to all participants and the participants are asked to give a new estimate now that they have seen the response

of the total group. Participants whose answers differ substantially from the rest of the group are asked to state the reasons behind their answers. The new responses are then summarized statistically and redistributed to the participants who are asked to prepare a final estimate. A final statistical summary is then prepared based on participants' comments.

    (a) Advantages of a Delphi Process: The Delphi process is an effective tool for achieving a consensus on forecasts among groups of experts. Delphi minimizes the disadvantages of group dynamics such as overdominance by a single personality or positions taken to obtain status or acceptance from the group.

    (b) Disadvantages of a Delphi Process: Delphi may have a tendency to homogenize points of view. The process of mailing questionnaires and redistributing summaries can be a time-consuming and cumbersome process. The public may be no more willing to accept the findings of an expert panel than it would of a single technical expert. The experts still may not be right.

(10) Simulation Games — There have been a number of simulation games which have been designed to allow people to simulate the effects of making particular policy choices and decisions, and in that process learn more about the impact of decisions and the interrelatedness of various features of an environmental or economic system. Simulation gaming provides an opportunity for people to try out their positions, and see what the consequences would be and how other groups react to them. Simulation games vary greatly in their complexity and length of time required to play them. Unfortunately, the closer the game resembles "reality," the more lengthy and complex it usually becomes. While simulation games can serve as an effective educational device — as a method for informing the public of the consequences of various choices — they typically do not provide opportunities for the public to provide comments specifically on study issues. As a result, simulation games could be used to educate an advisory group or leadership group of some sort, but must be used in conjunction with other public involvement techniques.

    (a) Advantages of a Simulation Game: Simulation games can provide the public with information about the consequences of various policy positions or decisions. Simulation games can also provide the public with an understanding of the dynamics of an economic or environmental system. Participation in a simulation game is usually fun, and participants develop a rapport and communication which can be maintained throughout the entire study.

    (b) Disadvantages of a Simulation Game: There are a number of simulation games on the market which are confusing, over-technical or misleading. Care must be exercised in selecting a simulation game appropriate for a particular study. No simulation games currently exist for aquifer restoration projects. While simulation games can be educational, they typically do not provide opportunities for direct public comment on a study. Since few games have a perfect fit with reality, citizens may apply the game rules inappropriately to the actual situation. People may become so engrossed in the game that they forget about the actual issues at hand.

Conflict management and resolution is becoming increasingly important in public participation programs. It could be important in the planning and implementation of an aquifer restoration project. As noted above, one technique for handling conflict is environmental mediation

(Creighton, 1981b). In order to develop conflict management and resolution strategies it is necessary to understand the different type of conflicts, and the fact that behavior which may contribute to resolution of one kind may exacerbate another. Creighton (1981a) delineated four types of conflicts:

(1) Cognitive Conflict: This is conflict which occurs when people have a different understanding or judgment as to the facts of a case. How extensive is the aquifer pollution? What are the costs for cleanup, etc.? When conflict is exclusively cognitive conflict, then it is possible to resolve the conflict if a process can be agreed upon to determine "the facts." Arguments over factors are often advanced, however, as part of values or interest conflict, so it is not always easy to distinguish the cognitive elements of a conflict from values or interest.

(2) Values Conflict: This is conflict over goals, whether or not a project (or outcome) is desirable/undesirable or should/should not occur. Obviously people with different values have a fundamentally different perspective from which they evaluate a proposed project. Values conflicts are also difficult to distinguish from cognitive or interest conflicts. People tend to accept those facts that support their values position; they also tend to adopt values consistent with their interest. This can quickly lead to a "chicken or the egg" argument—do facts cause people to tend to certain values positions? Do they then take on roles (which define their interest) based on their values? Or does their self-interest dictate their values, which in turn filter the facts to which they pay attention? At a minimum, they are intertwined.

(3) Interest Conflict: Since the costs and benefits resulting from an aquifer restoration project are rarely distributed equally, some people will have a greater interest in a project than others. Some may have an interest in assuring it does not occur. As a result it is possible to have agreement on facts, and on values, and still have conflict based on interest. An aquifer restoration project that will lead to major industrial development may be ardently supported by downtown businessmen and opposed by suburban interests. They may agree on the effects of the project, they may even all agree that industrial development is a desirable goal, but since the businessmen may be heavily favored by the project there may still be conflict between the developers and the suburbanites.

(4) Relationship Conflict: There are several psychologically oriented bases for conflict as well. Every time people communicate they communicate both content (information, facts) and relationship (how much someone is valued, accepted, etc.). Decision-making processes can also communicate relationship—decision-making processes may, for example, favor those groups which are well enough financed and organized to present scientific supporting data over those who primarily argue from a values base. The result is that there are a number of emotional motivations that lead to conflict on grounds other than disagreement on facts or values, or interest differences. One group may feel insulted or oppressed by another. A group or individuals may feel that the decision-making process gives an advantage to one group or another. Individuals or groups may react to others based on emotional symbols such as hairstyle, dress, or language. A group or individual may feel resentful that they were not consulted.

One approach for conflict management and resolution is to use environmental mediation via a third-party intervenor. Conflicts can arise in aquifer restoration projects over the significance of the pollution and the costs and cost allocations for cleanup. In some cases, the resolution of a conflict can be facilitated by a third party conciliator. In other cases, third party intervention may only serve to make the situation worse. The first decision, therefore, is whether third party intervention is appropriate in a particular dispute. Wehr (1979) identified five criteria which could be used in determining if third party intervention would be helpful; the criteria are:

(1) Accessibility: Can a third party gain entry into the conflict? Do the conflicting parties view it as a "private dispute?"
(2) Tractability: Does the conflict offer some hope of successful resolution?
(3) Divisibility: Can the conflict be subdivided into smaller, more manageable segments? Can the third party intervene in only one segment?
(4) Timing: Is it too early or too late to intervene?
(5) Alternatives: Is intervention risky to successful resolution? Would nonintervention be a less risky course of action?

As noted earlier, not all conflicts are amenable to third party intervention. Even with those that are, timing may be critical in the success or failure of the enterprise (Creighton, 1981a). As a result practical experience suggests that there are certain key conditions that must exist—or must be within the power of the third party to create—if the intervention is to be a success. These six key conditions are (Creighton, 1981a):

(1) Motivation toward resolution: The first requisite is that all critical parties must have motivations that make resolution desirable. If one major party can win more by no decision being made, the conditions for resolution do not exist. If one party can only lose if a decision is reached, the conditions do not exist. The motivation towards resolution must also be assessed at a relationship or emotional level. If people are actively nursing grudges, insults, or slights (real or imagined), then the timing may be wrong. If people are beginning to think they are wasting emotional energy with all the psychological games, then the timing may be right. It should also be remembered that different sides have different degrees of viability. If a "loss" on this issue threatens the continued existence of a group, then it is forced to play by different rules than the organization that will continue to exist, no matter the outcome. This can be a major factor in creating motivation towards resolution, or against it.
(2) Roughly equal power: Neither side is likely to compromise if they think they have the political or legal power to "win" outright. This can apply where one side believes it has a clear-cut legal basis for its point of view, or it can apply where a group or individual is confident of winning through the courts or through intervention of outside political power. In effect, people will only negotiate when they can win more (or endanger less) by doing so. In some cases the role of the third party is to insist that both sides be treated as equals. This assumes that the third party has some power in the situation.

(3) The risks of failure not too great: The old saying is that "it is better to have tried and lost than never to have tried at all." That depends on the consequences of failure. Sometimes the consequences of a failure at third party intervention may be that a controllable conflict may become totally dysfunctional. In other cases, when a visible effort at conciliation fails, this reinforces the negative perceptions on both sides. Even if the leadership is willing to continue negotiations they are often under intense pressure from their followers to take a hard line position. Since the other side is almost invariably seen as the cause of the failed negotiations, the anger and resentment towards the adversaries is further increased.

(4) Organizational authority: To be effective, a conciliator must usually speak for an organization that possesses authority and credibility. Even when an individual is hired as a "mediator," he is placed in that position with the authority and credibility of the organization doing the hiring. If the organization the conciliator represents is not credible to the antagonists, the conciliator will not be accepted.

(5) Negotiability of issues: One of the tactics engaged in by a conciliator is to attempt to enlarge the number of issues which are negotiable. The more issues which are negotiable, the more likelihood exists that a "positive-sum" solution can be found. This negotiability is a function of a number of things: (a) the strength of the leadership of each group within their own organization, (b) the consequences of a "loss" to the continued viability of a group or entity, (c) the external pressures on the groups to compromise, and (d) the skills of the conciliator.

(6) Control over the process: Experienced conciliators stress the importance of the conciliator's control over the communication process. This is particularly true in formal mediation, or at critical junctures in negotiations. Obviously the willingness of the antagonists to grant this kind of control is a function of the credibility and authority of the organization the conciliator represents, and the personal skills of the conciliator.

If it is decided that third party intervention is appropriate, the next step is selection of a conciliator (intervenor). There are two qualities which are essential in the intervenor: credibility and neutrality. The conflicting parties must believe that the intervenor is sincere in his job. He should be a person who has a good reputation, has successfully resolved disputes in the past, and has some authority. Secondly, he must have no vested interest in the dispute except the desire to see it settled. Baldwin (1978) identified some possible roles the intervenor can play in conflict management and resolution, including:

(1) Creating a climate of trust and a willingness to negotiate on the disputants' parts;
(2) Ensuring fair and adequate representation;
(3) Bringing the best available environmental information and expertise to the discussions;
(4) Breaking deadlocks by setting goals and deadlines;
(5) Suggesting solutions, or alternative solutions;
(6) Outlining implementation plans and helping create mechanisms for implementation and enforcement of the agreement.

Assuming that a third party intervenor is chosen, there are numerous things this individual can do to facilitate conflict resolution and management. Frost and Wilmot (1978) delineated the following seven suggestions for intervenors to facilitate their work:

(1) Be descriptive rather than judgmental. Describe behavior in terms of what your reaction is to it, rather than pinning a label on the other person.

(2) Encourage specificity. Feedback is more effective when it describes specific instances instead of general feelings.

(3) Deal with things that can be changed instead of "givens." Feedback is most effective when it concerns descriptions that are not inherent characteristics of persons or situations.

(4) Encourage parties to give feedback when it is requested. Or the intervention agent can ask parties to give feedback to each other when they need it, even if the need has not been articulated.

(5) Give feedback as close as possible to the behavior being discussed. Feedback sometimes serves as a kind of "instant replay" to a behavioral pattern that has been observed by the third party.

(6) Encourage feedback whose accuracy can be checked by others. In a group situation, sometimes people are unaware of their own word choices and behavior.

(7) Speak only for yourself. Encourage persons in conflicts to not speculate what other people think.

At some point during conflict resolution, the disputants must meet face-to-face to negotiate a mutually acceptable solution. This meeting must be carefully controlled, or it can degenerate into a shouting match. If negotiations are to be successful, all affected parties should be included. However, it is sometimes necessary to exclude all others. The presence of "observers" may retard progress because negotiators may assume extreme positions in the presence of witnesses. It is vital that the representatives of each party be persons with authority who can make commitments on behalf of their group. Creighton (1981a) has summarized the usual negotiating procedure into a four-step strategy. This strategy can be used whether the disputants are political factions, labor and management, countries, or project proponent and citizen. The four steps are as follows:

(1) Areas of Agreement: Parties enumerate all areas on which there is agreement. These are then eliminated from discussion. This step saves time later on and, most important, establishes common ground and fosters a feeling of mutual trust.

(2) Areas of Disagreement: Parties clearly define all areas of disagreement. Each party must state its position on each point of conflict, giving the underlying reasons for its position. This gives negotiators an idea of the magnitude of the problem. It also ranks points of conflict in a rough order of importance.

(3) Conflict Resolution Procedure: If possible, a procedure for resolving disagreements should be agreed upon. Doing this establishes a suitable climate in which agreements can be made.

(4) Negotiate Issue-by-Issue: It is not usually possible to resolve all outstanding disagreements at once. A more realistic approach is to try and solve points of conflict one at a time. It may be advisable to negotiate minor issues first and then progress to major ones. In this way, the negotiators will address the more difficult problems having a record of successful negotiations on less thorny issues.

An alternative approach for conflict management and resolution is to use the "Samoan Circle" (Aggens, 1981). The Samoan Circle meeting process is designed to facilitate the discussion of controversial issues when there is a large group of people interested in the topic. Its principal value is in the opportunity it affords for an exploration and exchange of knowledge and opinion where the large size of the group, or an environment of controversy, might disable other kinds of meetings. This meeting process is also useful when the possibility exists that no one person could be accepted as a fair moderator by all who might seek to be involved in the discussion.

In a Samoan Circle meeting, individuals can speak out on the issue without the need for oratorical skills, or the ability to put all their thoughts together into the one, short, cogent statement so often required by the dynamics of involvement in large group meetings. In this process, no one needs to be burdened with the responsibility of being the moderator of equitable participation, a judge of fairness, or the controller of other people's behavior. The advantages of small group discussion are afforded in the midst of a large group. The need and the opportunity for participants to use dramatics, "us/them" name-calling, and "cheerleading" in efforts to make their points are lessened.

The process does not resolve conflict. It is intended for the fullest possible exchange of information about an issue in anticipation of other group processes better designed for decision-making or conflict management. However, some users of the Samoan Circle have experienced the spontaneous resolution of conflicting views and agreement on actions required — apparently as a result of the contestants in a controversy having heard one another for the first time. It is not recommended that users of this process anticipate this result.

The most notable characteristic of the Samoan Circle is that there is no one who is the Chairman, or Moderator, or Facilitator. It is a leaderless meeting. Responsibility for discipline in this kind of meeting is vested in everyone, rather than in meeting leaders. Everyone has, and will quickly see that he or she has, a clear stake and part in maintaining an orderly environment for discussion. Five key issues related to the use of the Samoan Circle are room arrangements, starting the meeting, meeting dynamics, meeting records and evaluation, and ending the meeting (Aggens, 1981).

(1) Room Arrangements: As many chairs as seem needed for the meeting should be set up in concentric circles, with the inner circle big enough in diameter to allow for a round table with five chairs. There should be enough space around the central table and five chairs for people to walk around it without having to climb over the legs of those sitting in the first circle of seats. Four or more aisles should be left open to permit people to move easily from seats in the concentric circles to seats at the center table. For large groups, a microphone should be placed on the center table to insure that discussion across this table can be heard easily by everyone in the room—but it is destructive to the group dynamic intended if this microphone is handed around the table as each person takes a turn talking. People at the center table should be talking to one another, personally, at close range. They should not be coming to the center table only to gain access to the microphone in order to whip up enthusiasm among allies in the audience. An omni-directional microphone (taped down, if necessary) in the center of the table should be used—but only when this is absolutely necessary because of the size of the group.

(2) Starting the Meeting: After the group has been called to order by the person who will begin and end the meeting, it should be stated that the purpose of the meeting is to learn from one another as much as possible about the topic that is at issue in the community—including facts, problems, obstacles, needs, values, solution ingredients, suggestions for improvement, and new ideas. Representatives of the two or three sides that may be contesting over the issue could share in this introduction in an effort to strengthen the realization that the meeting process is not a contrivance or manipulation of one side by the other. The general rules for the meeting are, as stated by the organizing person:

(a) Anyone may participate by making a statement, asking a question, answering a question, taking exception to or confirming another person's opinion, making a rebuttal, and so on. But to do any of these things, the person who wants to say something must come to this center table and take a seat. Once there, he or she may interrupt, or wait for an opening in any discussion that is going on. The person taking a seat can join in the discussion or try to change its direction, or raise a new topic.

(b) The discussion at any one time is limited to the five people who can be seated at this table. If you come to the table, you may stay as long as you feel you have a contribution to make to the discussion. You may leave and return again as often as you wish. If there are no vacant seats at the table and you want to get in on the discussion, stand near the table until someone gives up a seat. The more people there are standing near the table waiting for seats, the more this should signal those sitting in the discussion to evaluate their own need to continue to participate. If you want to talk to one of the people at the table, stand directly behind that person's chair as a signal to the others at the table that you want one of their seats.

(c) If you want to cheer, or groan, or make any other noises to represent your opinion, please come to the table, take a seat here, and then do it. The discussion will go on until there is no one left at the table, or until the time for adjournment arrives.

(3) Meeting Dynamics: Once there are two or more people involved in the discussion, the talk takes on the "you-and-I" character of communication at short range. The oratory and belligerance that are common when "discussion" is taking place

across the width of a 30-foot room lessens when people of different persuasions close the physical distance between them. Discussion across the round table is usually (but not always) more relaxed, temperate, conversant, and instructive. If people feel the need to assault one another over their convictions, oceans of space will not prevent this. When all the seats at the table are filled and a person comes to the center to wait for a vacancy to develop, it is not uncommon for everyone at the table to stand up and leave. The sense of self-discipline invoked by this unchaired meeting process is very strong in most groups. On the other hand, when no one comes to the center table to wait for a vacancy, those sitting at the table feel free to expand their discussion and register their opinions and feelings several times. Sponsors of this meeting process usually have to suppress the inclination to rush into such situations and shut off a talkative person, or suggest, in the name of equity, that others might want to be heard. If and when such actions are needed, plenty of time should be given for the group to make its own interruption of a monologue, or to show its need for more participant involvement by individual actions to accomplish this. Any guidance needed from meeting sponsors should be given by someone who takes a seat at the table to express that need as a personal opinion.

(4) Meeting Records and Evaluations: A number of things can be added to the meeting process to make a record of transactions and to achieve some degree of "closure" that points at further action. Comments can be transcribed and the process used as a form of public hearing. (However, this meeting format seems inappropriate when formal, written statements are being read into the record.) Minutes can be kept. Decision-makers can be identified as auditors scattered throughout the audience. Comments can be written on newsprint on a wall. This can be done as a sequential list of opinions stated, or as a series of categorized lists—such as: "advantages" and "disadvantages"; or "strengths of the proposal" and "suggestions for improvement in the proposal."

(5) Ending the Meeting: Discussion can be allowed to run its course if there is no time required for adjournment. The meeting room will gradually empty until there may be no more people left except for an intensely interested group at the center table. If time limits, or the need to move on to another agenda topic require the ending of the discussion before everyone has left the center table, a number of things can be done. Someone who is responsible for the time limits can take a seat at the table and call attention to the disappearing time and remind the group of the agreed-upon time for ending the meeting. This often causes a flurry of activity by people who have been holding back, but who are still intent upon getting their point of view heard. An announcement of the need to close the discussion should be made early enough to accommodate this last-minute rush. When the time to end the meeting is about five to ten minutes away, the person who started the meeting can move to the table, wait for a seat to be vacated if none is already empty, and withdraw that chair. Continuing to stand near the table, the "meeting-ender" can withdraw each chair as it is vacated. This action is frequently acknowledged by the audience with understanding chuckles and, sometimes, by a last-minute rush. The message: "I need to end this meeting", is clear and nonthreatening; but the person ending the meeting should avoid cutting off last-minute participants from at least some chance at expression unless this is absolutely necessary and the need is obvious to all concerned. If any concluding comments are needed, these can be made when the person ending the meeting takes possession of the last chair.

The "Samoan Circle" has been used successfully by a variety of public agencies and private organizations (Aggens, 1981). Satisfaction with the meeting process seems to be related to a recognition by meeting sponsors and participants that it provided an environment for discussion of a controversial subject where other, more conventional, meeting processes had failed them in the past. In using this process, sponsors have modified it to fit peculiar circumstances, or make it "feel" better to the personalities involved.

PRACTICAL PLAN FOR IMPLEMENTING PUBLIC PARTICIPATION PROGRAM

Actual implementation of a public participation program involves a number of considerations. Delli Priscoli (1981) identified five major points to consider in creating an effective public involvement strategy:

(1) Implementation of citizen involvement programs must start by realizing that initial dissonance will arise. The roots of that dissonance and its likely effects must be understood and anticipated. Initial conflicts, such as between public affairs offices and public involvement staff, should be usefully managed. Overall management rewards should be commensurate with the way the staff actually allocates time.

(2) Decisions must be made about how much sharing of decision should be done and can be done. The "should" versus "can" distinction of these decisions is critical. Often staff analysis of the "can" in sharing comes packaged to executives at the "should" of decision sharing.

(3) Public involvement programs must be closely related to actual decision-making. Either managers get into citizen involvement programs, or line-staff are given more decision authority. Agencies will find some point in between these extremes. At any rate, consultants should be used only as resources to consult. When outside consultants are given the responsibility for citizen involvement, decision-makers become further isolated from the effects of their decisions. Consultants can provide critical staff support, training, evaluation, and critiques. But insofar as the success of citizen involvement depends on getting close to decisions, they should not replace decision-makers.

(4) Training is one of the best long-range techniques in implementing public involvement programs. Training should be coupled to strategies of recruiting new personnel. It must also be keyed to varying audiences within the agency. Effective training programs require enough flexibility to change as the agency and issues change in the process. Interactive training models offer even further public involvement opportunities. Joint training programs themselves can become public involvement techniques.

(5) Public involvement techniques must be appropriate—in time and money—to the type of decision being made. As such, funding can become a major consideration in the successful citizen involvement program. Public involvement techniques

must be clearly linked to the decision-making process. There is, of course, budgeting for line decision-making activities, such as interview, advertising, press releases, hearings, large and small meetings, workshops, surveys, and reports; but something called citizen involvement funding is difficult to conceptualize. It is more difficult to trace professional staff time in design, concern, and interaction for citizen involvement, because these attitudinal orientations should become part of the larger professional job definition. Table 8.7 provides general approximations of the costs of various techniques (Delli Priscoli, 1981).

Practical Suggestions for Implementing Program

The following list of items represents some very practical ideas and suggestions that can be useful in organizing a public participation program (Canter, 1977):

(1) Coordinate the various federal, state, and local agencies that have interests and responsibilities in the same geographic or technical areas of the study. Develop formal agreements or informal relationships.
(2) Develop lists of groups and citizens in the geographic area who have previously expressed interests or potential interests in the study.
(3) Assemble a newspaper clipping file on project needs and previous history of the project or study.

Table 8.7   Rough Cost Guide to Most Frequently Used Public Involvement Techniques (Delli Priscoli, 1981)

Technique	1978 Cost ($)
Interviews (per 20-min interview)	15 – 30
Newspaper advertising	250 – 750
Radio advertising	250 – 750
Press release	100 – 500
Public hearing	2,500 – 6,500*
Large public meeting	2,500 – 6,500*
Small meeting or workshop	2,000 – 4,000*
Publicity on radio or TV	250 – 500
50-page report	5,000 – 10,000
200-page report	10,000 – 50,000
Information bulletins (4 – 8 pages)	500 – 1,500
Conducting a survey:	
Per mailed questionnaire	3 – 5
Per telephone interview	10 – 15
Per personal interview	15 – 30

*May be reduced if a series of identical workshops or meetings is held.

(4) Try to convey the attitude—"What can we do in this study to assist you in your local planning problems? How can we coordinate with other local ground water efforts and projects?" This is in contrast to the attitude—"We are here to solve your problems and prepare plans and studies for you."

(5) Disseminate study information through the news media (newspaper, radio, and television) and through regular publication of a planning newsletter. The mailing list should encompass all state and federal interests as well as local groups and individuals who have participated in previous meetings or shown interest in the study.

(6) Every third or fourth issue of the newsletter should contain a mail-in coupon for persons wanting to continue to receive the newsletter. Each issue should contain a coupon for suggestions of other persons or groups that should receive the newsletter.

Some practical suggestions for actually conducting a public meeting are as follows (Canter, 1977):

(1) Keep data presentations simple. The purpose of data presentations is to inform, and not to confuse or disillusion.

(2) An outline of any project or alternative should include a discussion of location, features, benefits and costs, and beneficial and detrimental environmental impacts.

(3) Use simple visual aids. Slides of the actual area prove very beneficial.

(4) Discuss project timing (prior authorizations and resolutions) previous to the meeting and the anticipated timing of future required steps.

(5) Discuss general concept, features and requirements of benefit-cost ratio, or other relevant economic analysis.

(6) Only those persons who can speak on general matters, commingled with engineering expertise, should be considered for meeting with the public. It is well known that not everybody has the ability to speak, answer questions, and perhaps debate while holding a specific image in front of the public. The ability to speak well is not the only trait on which the selection should be based. A person who can answer questions quickly and confidently from an audience in which the sentiment is mixed or opposed can establish a profound positive relation with the audience.

(7) Avoid use of technical jargon or words that may be hard to understand, especially with local groups unaccustomed to engineering and hydrogeological terminology.

(8) Be familiar with the area.

(9) Be earnest, sincere, and willing to work on problems with individual groups.

Incorporation of Results in Decision-Making

The feedback loop from public participation must be used in the decision-making process, or the purpose of public participation will not have been fully satisfied (Canter, 1977). Public participation results can be useful in defining project need, describing unique features of the environmental setting, and identifying environmental impacts of potential alternatives. Results can also be utilized by the proposing agency in assigning significance (importance) values to both environmental items and values. Finally, selection of the most desirable alternative for meeting the identified aquifer restoration need can be aided by public participation.

Two levels are suggested for incorporation of public participation information in the decision-making process. First, all public meetings and the entirety of the planned and accomplished public participation program should be summarized. Any information relative to the objectives outlined earlier and obtained through questionnaire surveys or other public participation techniques should be summarized. Second, public preference can be used in selection of the alternative to become the proposed project for meeting the aquifer restoration need.

SELECTED REFERENCES

Aggens, L., "The Samoan Circle: A Small Group Process for Discussing Controversial Subjects", in "Public Involvement Techniques: A Reader of Ten Years Experience at the Institute of Water Resources", Creighton, J. L., and Delli Priscoli, J., editors, IWR Staff Report 81-1, 1981, U.S. Army Engineer Institute for Water Resources, Fort Belvoir, Virginia.

Arnstein, S. R., "A Ladder of Citizen Participation", *Journal of the American Institute of Planners*, Vol. 35, No. 4, July 1969.

Baldwin, P., "Environmental Mediation: An Effective Alternative?", *Proceedings of Conference Held in Reston, Virginia*, Jan. 11-13, 1978, RESOLVE, Center for Environmental Conflict Resolution, Palo Alto, California.

Bishop, A. B., "Structuring Communications Programs for Public Participation in Water Resources Planning", IWR Contract Report 75-2, May 1975, Institute for Water Resources, Fort Belvoir, Virginia.

Bishop, A. B., "Communication in the Planning Process", in "Public Involvement Techniques: A Reader of Ten Years Experience at the Institute of Water Resources", Creighton, J. L., and Delli Priscoli, J., editors, IWR Staff Report 81-1, 1981, U.S. Army Engineer Institute for Water Resources, Fort Belvoir, Virginia.

Canter, L. W., Miller, G. D. and Fairchild, D. M., "Sentry Missile Environmental Assessment Program", Report submitted to U.S. Army Ballistic Missile Division Systems Command, Huntsville, Alabama, Sept. 1982, 2424 pages.

Canter, L. W., *Environmental Impact Assessment*, 1977, McGraw Hill Book Company, Inc., New York, New York, pp. 220-232.

Council on Environmental Quality, "National Environmental Policy Act — Regulations", Federal Register, Vol. 43, No. 230, Nov. 29, 1978, pp. 55978-56007.

Creighton, J. L., "Acting as a Conflict Conciliator", in "Public Involvement Techniques: A Reader of Ten Years Experience at the Institute of Water Resources", Creighton, J. L., and Delli Priscoli, J., editors, IWR Staff Report 81-1, 1981a, U.S. Army Engineer Institute for Water Resources, Fort Belvoir, Virginia.

Creighton, J. L., "A Short Catalogue of Public Involvement Techniques", in "Public Involvement Techniques: A Reader of Ten Years Experience at the Institute of Water Resources", Creighton, J. L., and Delli Priscoli, J., editors, IWR Staff Report 81-1, 1981b, U.S. Army Engineer Institute for Water Resources, Fort Belvoir, Virginia.

Creighton, J. L., "Identifying Publics/Staff Identification Techniques", in "Public Involvement Techniques: A Reader of Ten Years Experience at the Institute of Water Resources", Creighton, J. L., and Delli Priscoli, J., editors, IWR Staff Report 81-1, 1981c, U.S. Army Engineer Institute for Water Resources, Fort Belvoir, Virginia.

Creighton, J. L., Chalmers, J. A. and Branch, K., "Integrating Planning and Assessment Through Public Involvement", in "Public Involvement Techniques: A Reader of Ten Years Experience at the Institute of Water Resources", Creighton, J. L., and Delli Priscoli, J., editors, IWR Staff Report 81-1, 1981, U.S. Army Engineer Institute for Water Resources, Fort Belvoir, Virginia.

Creighton, J. L., and Delli Priscoli, J., "Establishing Citizens' Committees", in "Public Involvement Techniques: A Reader of Ten Years Experience at the Institute of Water Resources", Creighton, J. L., and Delli Priscoli, J., editors, IWR Staff Report 81-1, 1981, U.S. Army Engineer Institute for Water Resources, Fort Belvoir, Virginia.

Delli Priscoli, J., "Implementing Public Involvement Programs in Federal Agencies", in "Public Involvement Techniques: A Reader of Ten Years Experience at the Institute of Water Resources", Creighton J. L., and Delli Priscoli, J., editors, IWR Staff Report 81-1, 1981, U.S. Army Engineer Institute for Water Resources, Fort Belvoir, Virginia.

Frost, J. H. and Wilmot, W. W., *Interpersonal Conflict*, 1978, Wm. C. Brown Company Publishers, Dubuque, Iowa.

Hanchey, J. R., "Objectives of Public Participation", in "Public Involvement Techniques: A Reader of Ten Years Experience at the Institute of Water Resources", Creighton, J. L., and Delli Priscoli, J., editors, IWR Staff Report 81-1, 1981, U.S. Army Engineer Institute for Water Resources, Fort Belvoir, Virginia.

Ragan, Jr., J. F., "An Evaluation of Public Participation in Corps of Engineers Field Offices", in "Public Involvement Techniques: A Reader of Ten Years Experience at the Institute of Water Resources", Creighton, J. L., and Delli Priscoli, J., editors, IWR Staff Report 81-1, 1981a, U.S. Army Engineer Institute for Water Resources, Fort Belvoir, Virginia.

Ragan, Jr., J. F. "Constraints on Effective Public Participation", in "Public Involvement Techniques: A Reader of Ten Years Experience at the Institute of Water Resources", Creighton, J. L., and Delli Priscoli, J., editors, IWR

Staff Report 81-1, 1981b, U.S. Army Engineer Institute for Water Resources, Fort Belvoir, Virginia.

Schwertz, Jr., E. L., "The Local Growth Management Guidebook", in 1979, Center for Local Government Technology, Oklahoma State University, Stillwater, Oklahoma.

U.S. Army Corps of Engineers, "Public Participation in Water Resources Planning", EC 1165-2-100, May 1971, Washington, D.C.

U.S. Department of Transportation, "Effective Citizen Participation in Transportation Planning: Community Involvement Processes", 2 volumes, 1976, Washington, D.C.

Wehr, P., *Conflict Regulation*, 1979, Westview Press, Boulder, Colorado.

CASE STUDIES AND APPLICATIONS OF GROUND WATER POLLUTION CONTROL

Case Studies of Ground Water
Pollution Control

Three case studies on the application of aquifer restoration measures are presented in this appendix. The three cases are discussed according to a description of the ground water pollution problem, sources of contaminants and remedial actions. Information is also included on the aquifer restoration program of the New Jersey Department of Environmental Protection (NJDEP). Finally, summary information on 15 cases dealing with restoration using biological processes is presented. Case A on Tar Creek in Oklahoma summarizes an acid mine drainage problem and its adverse effect on the ground water. The Boone formation in the Tar Creek area is already affected and the contaminants are likely to spread to the Roubidoux formation unless effectively contained. No remedial actions have actually been implemented, although plans have been developed to restore the aquifer and to prevent further contamination. Case B presents information on the Gilson Road site in New Hampshire. Treatability studies have been conducted for the Gilson Road site, and some remedial actions have been implemented. Case C deals with the Rocky Mountain arsenal and is drawn from work conducted by the U.S. Army. Remedial actions have been implemented to clean up the alluvial aquifer, and further studies are in progress on source control measures.

Detailed information on aquifer restoration measures is not easily accessible. No assessment can be made of the long-term effectiveness of treatment methods. An economic assessment of various aquifer restoration strategies cannot be made due to a lack of detailed data on initial (capital) costs and operation and maintenance costs.

CASE A: TAR CREEK IN OKLAHOMA

Tar Creek is a small tributary of the Neosho River in northeastern Oklahoma. The creek runs through the Picher field, which is now an abandoned mine area, but was at one time one of the most productive lead and zinc mining districts in the United States. Actual production of lead and zinc from this area began around 1904 and continued until the late 1950s when all major operations were terminated. Mining began again in the early 1960s, but lasted only a few years. When the mining operations ceased and the dewatering pumps within the mines were removed the mines started to flood due to natural recharge and surface water inflow. By 1979 the majority of the workings were completely flooded and surface discharge of poor quality water began.

The hydrogeology of the Tar Creek area is depicted in Figure A.1 (Hittman Associates, Inc., 1981). The major fresh water aquifer is the Roubidoux formation, which is comprised of 160 ft of cherty dolomite interspersed with several sandy sequences. The depth to this aquifer from the land surface is between 900 and 1000 ft. Overlying the Roubidoux are two dolomite formations called the Jefferson City and the Cotter. Overlying these dolomite deposits is the Boone formation, a 370-ft-thick cherty limestone with spent lead and zinc deposits found from the old Picher mining district. The Boone formation was probably the major source of fresh water in the area before the era of large-scale mine watering. Overlying the Boone formation through most of the mining district is the Krebs group which consists principally of black fissile shale. The Krebs group serves as a confining layer, thus allowing artesian conditions to develop in the Boone formation.

The Tar Creek area is faced with two major problems as a result of the flooding of the mines and the resultant generation of acid mine water.

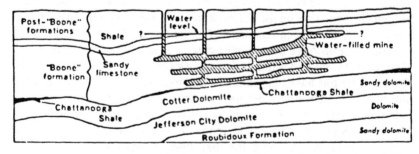

Figure A.1. Hydrogeology of the Tar Creek area (Hittman Associates, Inc., 1981).

First, the acid mine water follows the subsurface hydraulic gradient and moves laterally outward in the Boone formation and possibly downward into the Roubidoux formation. Second, after periods of intense precipitation, low-quality mine waters discharge into the surface waters of the Tar Creek area. If the low-quality water moves into the underlying Roubidoux formation it could start moving laterally and become a threat to this major fresh water aquifer. Dilution and mixing of the mine water with the more alkaline water of the Roubidoux formation will increase the pH; this could lead to precipitation of some of the heavy metal constituents in the mine water. However, it is anticipated that the effects of dispersion, adsorption and precipitation will not be enough to reduce the heavy metal concentrations below drinking water standards.

Although still in the planning stages, the Tar Creek situation represents an interesting case study of aquifer restoration and raises the following issues (Knox, Stover and Kincannon, 1982):

(1) Tri-State Involvement—The geographical location of both the affected aquifer (Boone) and the threatened aquifer (Roubidoux) encompasses parts of three states: Oklahoma, Kansas, and Missouri. One of the key steps yet to be undertaken is the quantification of the amount of leakage, if any, from the polluted Boone formation down to the fresh water Roubidoux formation. Any reasonable estimate of leakage will require integrating well data from all three states.

(2) Surface Water Impacts—Tar Creek represents a good example of polluted ground water adversely affecting the quality of surface water. Discharge of degraded quality water has been documented (Hittman Associates, Inc., 1981). This adds another facet to the problem since cleanup of the polluted ground water will not be enough. Measures will need to be implemented to prevent surface water pollution. This could become a large task in that it will require identifying and sealing all avenues of escape for the ground water. These avenues can range from mine shaft openings to abandoned wells or boreholes from previous oil and gas field activities in the area.

(3) Financial Responsibility—Another issue concerning Tar Creek, and one that is becoming all too common, is the question of financial responsibility, or who pays. The pollution problem cannot be attributed to a single company or individual. Hence any attempt to assign financial responsibility would be useless. Current preliminary funding has come from the "Superfund" program. However, preliminary estimates of the costs of cleanup range from $17 to 22 million. The "Superfund" program will not be able to pay for many cleanup programs of this magnitude.

(4) Abandonment—One alternative that should be addressed in any aquifer restoration study is that of abandoning the polluted aquifer and development of a new source of fresh water. This has not been given much consideration in the Tar Creek area. Current planning involves a series of pump tests and new well networks to assess the potential for contaminant transport, to be followed by the development of specific alternatives for removing and treating the polluted water. These alternatives will then be compared to the alternative of abandoning the Roubidoux aquifer. The order should be reversed. Large sums of money could

probably be saved by first examining the possibility of abandonment. It would be a waste if the aquifer characterization study showed no leakage from the Boone to the Roubidoux, with the polluted wells being due to faultily cased wells through the Boone formation.

Sources of Contaminants

Lead and zinc mining activities expose sulfide-bearing minerals to moist, oxygen-rich air which oxidizes the iron sulfide minerals. This has occurred in the Picher mining district. Pyrite-rich waste rock was discarded in mined out portions of drifts, with other sources of waste rock derived from sections of the shaley roof rock present in several mines. These rocks provided additional supplies of pyrite for oxidation. The oxidation of pyrite to form acid occurs in accordance with the following stoichiometric reaction (Hittman Associates, Inc., 1981):

$$FeS_2(s) + {}^7\!/_2\,O_2 + H_2O \rightleftarrows Fe^{+2} + 2SO_4^{2-} + 2H^+$$

For the above reaction to continue, both oxygen and water must be present at the mineral surface. As the Picher mine area flooded, the production of acid slowed because the water tended to restrict the diffusion of O_2 to the mineral surfaces. Additionally, the limestone-dolomite rocks within the formation allowed for natural neutralization to take place. As the limestone removed the hydrogen ions to form water and carbon dioxide, calcium ions went into solution, thus increasing both the total dissolved solids (TDS) and hardness. Eventually, the calcium ions precipitate as $CaSO_4$. This precipitate forms a thin film on the mineral surfaces, thus precluding further neutralization but also further inhibiting the production of acid mine water. Therefore, acid mine waters have been produced and are still being produced, albeit at a lower rate due to local natural phenomena.

Discharge of the acid mine water can disrupt the quality of surface waters. The surface waters will present a source of oxygen for the oxidation of the dissolved ferrous iron to ferric iron. The ferric iron then forms an insoluble ferric hydroxide with the release of more acidity in accordance with the following reactions (Hittman Associates, Inc., 1981):

$$Fe^{+2} + {}^1\!/_4\,O_2 + H^+ \rightleftarrows Fe^{+3} + {}^1\!/_2\,H_2O$$
$$Fe^{+3} + 3H_2O \rightleftarrows Fe(OH)_3(s) + 3H^+$$

The net result of these reactions is to decrease the pH of Tar Creek and to decrease its dissolved oxygen concentration. Also, the exposure of the low-pH waters to the native creek bed minerals has resulted in increased aqueous concentrations of iron, lead, zinc, and cadmium.

Cessation of mining in the mid-1960s ended more than 50 years of water removal from the Boone formation. The many years of pumping resulted in a large cone of depression under water table conditions within the Boone formation. The cone of depression began filling via natural recharge through abandoned mine shafts, collapsed geological features, and abandoned oil and gas wells. These shafts and abandoned wells penetrate the overlying Krebs group confining layer that crops out at the surface. The water recharging the mine workings came in contact with the oxidation products that had formed earlier, and dissolution occurred rapidly. The majority of the mineralized water was confined to the approximate area where it was formed until the water level in the Boone formation reached equilibrium (Hittman Associates, Inc., 1981).

The potentiometric surface of the Boone formation continued to rise until it reached the base of the Krebs group. The shaley Krebs group became a confining layer, and the Boone formation began to behave as a confined aquifer as the potentiometric surface rose above the base of the Krebs group. By 1979, the potentiometric surface of the Boone formation exceeded the ground surface elevation in several locations. Surface discharge of acid mine water began to occur via abandoned or partially plugged oil or gas well exploration holes and mine air shafts. Confined conditions within the Boone formation are known to exist west of a north-south line connecting Commerce and Miami, Oklahoma. Surface discharge will continue as long as the potentiometric surface of the Boone formation exceeds the ground surface and pathways through the Krebs group exist (Hittman Associates, Inc., 1981).

Discharge of the mine water to Tar Creek has already increased the TDS and heavy metal concentrations and, as noted earlier, decreased the dissolved oxygen concentration and pH. The pH has, however, remained moderate because there is not sufficient oxygen present to rapidly oxidize the ferrous ions, and because sewage treatment plant effluent, with a higher pH, is also discharged to Tar Creek or its tributaries (Hittman Associates, Inc., 1981).

Studies to Determine Remedial Actions

As a result of a preliminary study, the following three alternative remedial measures have been identified (Hittman Associates, Inc., 1981): (1) collection of surface water discharges with limited ground water pumpage and treatment; (2) ground water pumpage from flooded mine workings and collection of surface discharges and treatment; and (3) ground water pumpage from flooded mine workings and treatment with a surface sealing and diversion program to minimize direct recharge to the ground water system. To arrive at these alternatives, assumptions were made regarding the ground water recharge rate; necessary treatment plant capacity, operating life, heavy metals recovery, and siting criteria; and siting criteria for collection wells.

A state-of-the-art technology for treatment of acid mine drainage has been identified to consist of the following steps (Hittman Associates, Inc., 1981):

(1) Neutralization—an ion exchange reaction that, in the case of acid mine drainage, combines hydroxyl ions (OH⁻) with acidic hydronium ions (H⁺).

(2) Aeration—a gas transfer process to increase the oxygen transfer rate and convert ferrous ions to ferric ions.

(3) Precipitation—a chemical reaction based on decreasing the solubility of toxic and other metal ions.

(4) Sedimentation—a physical process involving settling to remove suspended solids.

Carbon adsorption, ion exchange, reverse osmosis, electrodialysis, and ozonation were candidate technologies investigated before arriving at lime neutralization as the desired treatment method. The goals for the treatment systems included the production of an acceptable effluent quality and minimum sludge volume with a maximum system reliability and acceptable capital and operating costs. Also, each treatment alternative has been planned to minimize the amount of land disturbed for the unit processes (Hittman Associates, Inc., 191).

The Alternative I treatment system consists of aeration, flocculation and clarification (sedimentation). The conceptual design of this alternative is presented in Figure A.2 (Hittman Associates, Inc., 1981). Surface discharges from several locations would be collected and transported to the treatment plant via pipelines. The treatment plant processes are shown in Figure A.3 (Hittman Associates, Inc., 1981). Alternative I is designed principally to abate the surface discharge problem. Heavy metal concentrations in the treatment plant effluent are expected to be low; however, the effluent would be likely to have high hardness, high sulfate concentrations, and a low buffering capacity. The treated water will not be potable, but it would be suitable for agricultural and industrial applications. To clean up the acidic system in the Picher mining area, an estimated 36 years would be required.

Figure A.4 contains the conceptual design for Alternative II (Hittman Associates, Inc., 1981). As was the case for Alternative I, the treatment

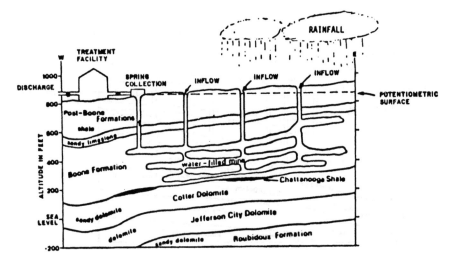

Figure A.2. Conceptual design of Alternative I (Hittman Associates, Inc., 1981).

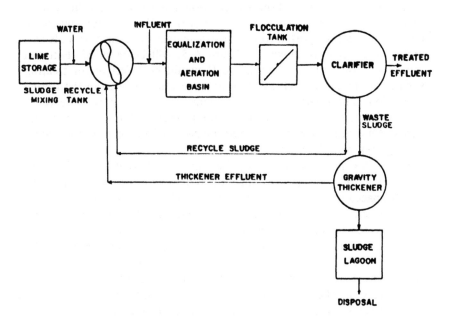

Figure A.3. Acid mine drainage treatment plant—unit processes (Hittman Associates, Inc., 1981).

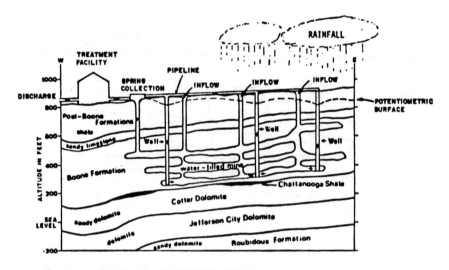

Figure A.4.   Conceptual design of Alternative II (Hittman Associates, Inc., 1981).

plant also consists of aeration, flocculation and clarification. Both natural discharges and pumped acid mine water from flooded shafts would be combined and transported to the treatment facility. Collector wells and monitoring wells, as shown in Figure A.5, will be required (Hittman Associates, Inc., 1981). Alternative II would be effective for treating surface discharges as well as the acid mine water located in the Boone formation. The treatment plant effluent would not be suitable for a potable water supply, but it could be used for agricultural and industrial needs. An estimated 23 years would be required to clean up the Picher mining area with Alternative II.

Figure A.6 shows the conceptual design for Alternative III (Hittman Associates, Inc., 1981). The collection system would consist of 16 wells widely spaced to intercept and collect as much of the acid mine water as possible. Surface discharge holes would be sealed off using a packer and bentonite cement slurry and/or acid-resistant grout. The treatment plant processes would be the same as proposed for Alternatives I and II. Alternative III has been designed to seal surface discharges, divert inflow away from the mine area, and pump and treat ground water through a well collection system and treatment plant. As was the case for Alternatives I and II, the treatment plant effluent water quality would permit its usage only for agricultural and industrial purposes. Alternative III is also expected to take 23 years to clean up the Picher mining area.

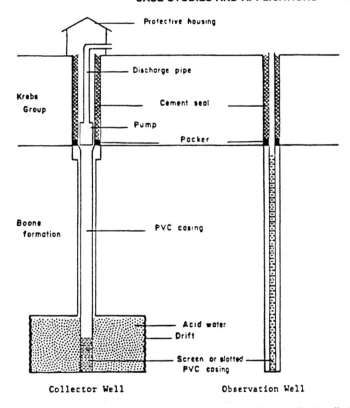

Figure A.5.   Conceptual design for collector and monitoring (observation) wells (Hittman Associates, Inc., 1981).

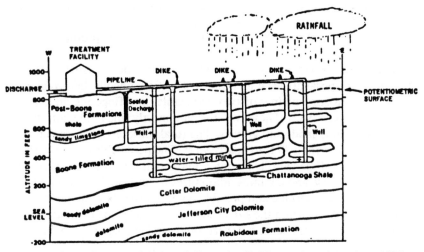

Figure A.6.   Conceptual design of Alternative III (Hittman Associates, Inc., 1981).

Proposed Remedial Action

After evaluation of the three alternatives, Alternative II has been proposed as the optimum course of action for abatement of the Tar Creek acid mine drainage problem. Summaries of the capital and annualized costs for the three alternatives are in Tables A.1 and A.2 (Hittman Associates, Inc., 1981). Alternative II is designed to collect and treat the surface water discharges with the existing state-of-the-art technology. In addition, a system of 12 wells will withdraw 1200 gpm of acid mine water. The collector system facilitates acid mine water removal, thus preventing further contamination of the Boone formation and minimizing the potential for contamination of the Roubidoux formation. Final design will require the further conduction of problem characterization studies. Table A.3 lists some of the required studies (Knox, Stover and Kincannon, 1982).

CASE B: GILSON ROAD IN NEW HAMPSHIRE

A 6-acre site located in Nashua, New Hampshire was used as a borrow pit where, during the late 1960s, after much of the sand had been removed, an unapproved waste disposal operation was initiated, apparently to fill the excavation. Household refuse, demolition materials, chemical sludge, and hazardous liquid chemicals were all dumped at the site. This activity was first discovered in late 1970, and in the late 1970s all further disposal of hazardous wastes on the site was prohibited (Roy F. Weston, Inc., 1982).

The Gilson Road disposal site and property downgradient of the immediate disposal area is underlain by stratified, unconsolidated glacial desposits consisting mainly of two permeable, interfingering units (Roy F. Weston, Inc., 1982). These two units are mostly fine to medium sands or fine to coarse sands and gravels. The total thickness of these deposits ranges from less than 20 ft to nearly 90 ft. The permeable sands overlie a thin sequence of glacial tills having a maximum thickness of around 12 ft. It is very dense and contains admixtures of unstratified silt, and sand and gravel. Metamorphic and possibly igneous bedrock underlie the site.

Ground water beneath the site occurs under unconfined or water table conditions in the permeable stratified sands and gravels. It probably also

Table A.1   Summary of Capital Coss for Tar Creek Alternatives I, II, and III (Hittman Asssociates, Inc., 1981)

Capital Costs, 1981 Dollars	Alternative 1 1 MGD Surface Water Collection & Treatment	Alternative 2 2 MGD Surface & Ground Water Collection & Treatment	Alternative 3 2 MGD Goundwater Collection & Treatment with Diversions
Collection System			
Collection Wells	40,000	104,000	128,000
Observation Wells	12,000	39,000	39,000
Pumps	54,000	116,000	108,000
Piping	167,000	410,000	469,000
Pump Housing	10,000	28,000	32,000
Trenching	178,000	408,000	457,000
Spring Collection	40,000	40,000	
Diversion System			
Sealing			50,000
Diversion			400,000
Consulting and Design	45,000	104,000	124,000
Subtotal	546,000	1,249,000	1,807,000
Treatment Facility			
Unit Processes			
Influent Pumping	211,000	430,000	430,000
Flow Equalization	156,000	219,000	219,000
Preliminary Treatment	83,000	187,000	187,000
Chemical Addition	34,000	108,000	108,000
Clarification	156,000	314,000	314,000
Outfall Pumping	48,000	114,000	114,000
Laboratory/Maintenance Building	236,000	480,000	480,000
Gravity Thickening	47,000	101,000	101,000
Sludge Handling	61,000	156,000	156,000
Non-Construction Costs			
Consulting and Design	62,000	120,000	120,000
Land	50,000	50,000	50,000
Other non-construction costs	68,000	138,000	138,000
Subtotal	1,212,000	2,417,000	2,417,000
Total Capital Costs	1,758,000	3,666,000	4,224,000

Table A.2 Summary of Annualized Costs for Tar Creek Alternatives I, II, and III (Hittman Associates, Inc., 1981)

Capital Costs, 1981 Dollars	Alternative 1 1 MGD Surface Water Collection & Treatment	Alternative 2 2 MGD Surface & Ground Water Collection & Treatment	Alternative 3 2 MGD Goundwater Collection & Treatment with Diversions
Collection System O&M			
Manpower	10,000/yr	20,000/yr	20,000/yr
Electrical	12,000/yr	18,000/yr	21,000/yr
Replacement Costs	9,000/yr	40,000/yr	37,000/yr
Pipeline Replacement	3,000/yr	10,000/yr	10,000/yr
Subtotal	34,000/yr	88,000/yr	88,000/yr
Treatment Facility O&M			
Manpower	120,000/yr	120,000/yr	120,000/yr
Electrical	35,100/yr	70,200/yr	70,200/yr
Chemicals	73,000/yr	142,200/yr	142,200/yr
Maintenance & Repair	72,000/yr	138,000/yr	138,000/yr
Subtotal	300,000/yr	470,400/yr	470,400/yr
Total Annual Operation & Maintenance Costs	334,100/yr	558,400/yr	558,400/yr
Plant Life	36 years	23 years	23 years
Total O&M Costs Over Life of Alternative	12 million	12.8 million	12.8 million
Construction Grant Interest Rate	7-3/8%	7-3/8%	7-3/8%
Annualized Capital Costs	141,000/yr	336,000/yr	387,000/yr
Total Equivalent Uniform Annual Costs (EUAC)	475,100/yr	894,400/yr	945,400/yr
Total Cost Over Life of Alternative	17.1 million	20.6 million	21.7 million

occurs under semiconfined conditions in secondary fractures in the bedrock. It occurs under a gradient which is generally much less than 1%. Ground water testing and monitoring by atomic absorption and gas chromatograph/mass spectrometry showed contaminants in several wells near the dump site, including heavy metals, volatile organics, and

Table A.3   Additional Studies Needed for Tar Creek (Knox, Stover and Kincannon, 1982)

Phase	Step
A. Field Investigation	I. Stream system and spring water analysis.
	II. Ground water investigation.
	III. Mine water investigation
	IV. Investigation of pollution causation.
	V. Drainage and diversion analysis.
	VI. Public health impacts verification.
	VII. Evaluation of the extent of the pollution problem.
B. Feasibility Study	I. Evaluate feasibility of treatment systems.
	II. Evaluate alternative water supplies.
	III. Comparison of abatement activities and alternative water supplies.
	IV. Treatment plant feasibiity (if needed).
	V. Treatment plant design specifications.
	VI. Alternative water supply feasibility (if needed).
	VII. Tar Creek Task Force review of contractor recommendations.
	VIII. Remedial actions.

extractable organics. The concentrations of heavy metals, volatile organics and extractable organics found in the ground water create concern for both the ambient air and water qualities. The primary compounds found in the contaminated ground water are shown in Table A.4 (Stover and Kincannon, 1982).

Sources of Contaminants

The major source of contaminants in ground water and the nearby surface waters, is the Gilson Road waste disposal site. The disposal site is located less than 1000 ft from Lyle Reed Brook, a small stream tributary

Table A.4 **Primary Contaminants of Concern in Ground Water at the Gilson Road Site (Stover and Kincannon, 1982)**

Heavy Metals	Volatile Organics	Extractable Organics
Arsenic	Tetrahydrofuran	1,4-Dichlorobenzene
Barium	1,1,1-Trichloroethane	1,2-Dichlorobenzene
Cadmium	Benzene	Naphthalene
Chromium	Trichloroethylene	2-Chlorophenol
Lead	Methyl Isobutyl Ketone	2-Nitrophenol
Mercury	Xylenes	Phenol
Selenium	Toluene	2,4-Dimethylphenol
Silver	Ethylbenzene	O-Cresol
Copper		M-Cresol
Iron		Benzoic Acid
Manganese		Pentachlorophenol
Nickel		
Zinc		

to the Nashua River. Any surface discharge from the site would likely flow into this brook, which has a total drainage area of about 1.5 mi.² The flow from Lyle Reed Brook enters the Nashua River about seven miles upstream of the confluence with the Merrimack River. The ground water from underneath the disposal site migrates to Lyle Reed Brook and thus contaminates it. Volatile organic compounds are the main concern due to their presence, in extremely high concentrations, in the ground water. Volatilization of the compounds from surface waters into the ambient air may have adverse effects on the surrounding residents, flora and fauna (Roy F. Weston, Inc., 1982).

Studies to Determine Remedial Actions

A grant from the U.S. Environmental Protection Agency funded under the Comprehensive Environmental Response, Compensation, and Liability Act of 1980 (CERCLA or Superfund) for cleanup of the site required that an interim report be prepared for dealing with the feasibility of ground water treatment. On the basis of the environmental effects, monetary costs, technical feasibility, and long-term operational considerations, the most cost-effective solution for alleviating the condition of contaminated ground water at the hazardous waste dump site was to be determined.

Ambient air and water quality criteria were evaluated, and appropriate ambient standards were defined. A thorough literature search was conducted to identify all applicable treatment technologies (Stover and Kincannon, 1982). The best treatment process was to be selected based on process feasibility, demonstrated reliability and costs. Selection of the most cost-effective alternative was based on a comparison of five options. The lowest-cost, technically feasible option was considered to be the most cost-effective alternative (Knox, Stover and Kincannon, 1982).

Conventional wastewater treatment technologies directly applicable for cleanup of this contaminated ground water were identified to be chemical precipitation, air stripping, steam stripping, activated carbon adsorption, and biological treatment. These processes could be aggregated in a variety of combinations to achieve various treatment levels. The evaluation of these unit processes by laboratory bench-scale treatability studies presented a challenge in selecting the best unit processes and their combinations. The treatability studies indicated that the heavy metals, volatiie organics and extractable organics of concern could be successfully treated (removed) by two alternative processes (Stover and Kincannon, 1982). One process consisted of metals removal by chemical precipitation followed by steam stripping and activated carbon adsorption. The alternate process consisted of the combined physical/chemical and biological treatment scheme of chemical precipitation, steam stripping and biological treatment.

Heavy metals can be removed from water by chemical precipitation, at an optimum pH. Three types of chemical systems used have been identified as carbonate systems, hydroxide systems, and sulfide systems. Jar test studies on the Gilson Road site ground water indicated that lime (hydroxide) addition was the best choice. Variation of pH and chemical dosages indicated the best operating pH to be 10.0. The pH titration characteristics of raw water and treated water for metals removal are shown in Figure A.7 (Stover and Kincannon, 1982). The levels of heavy metals in the raw water and in the treated water at the optimum pH of 10 are shown in Table A.5 (Stover and Kincannon, 1982). The sludge produced during the metals removal step will have to be removed from the liquid stream and treated. Activated carbon was also effective in removal of heavy metals, with Table A.5 showing the results (Stover and Kincannon, 1982). The effectiveness of activated carbon in removal of heavy metals was found to be comparable to lime treatment. Some of the important design factors identified in this treatability study were:

(1) Best chemical addition system
(2) Optimum chemical dose
(3) pH considerations
(4) Rapid mix requirements
(5) Flocculation requirements

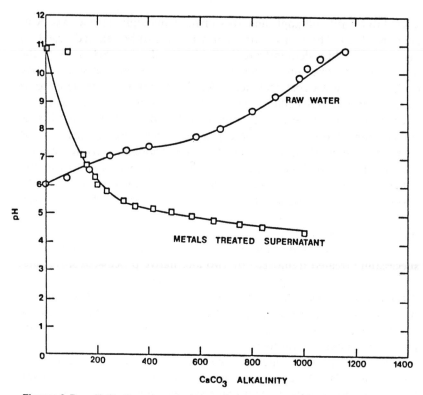

Figure A.7. pH titration characteristics of raw and treated ground water from the Gilson Road site (Stover and Kincannon, 1982).

(6) Sludge production

(7) Sludge flocculation, settling and dewatering characteristics

Volatile organic compounds are the contaminants of major concern in ground water from the Gilson Road site. The treatment methods considered included air stripping, steam stripping, activated carbon adsorption, ultraviolet light/ozone, biological treatment, wet oxidation, flash evaporation, distillation, and membrane separation processes. Due to the excessive contamination, the alternatives were narrowed down to air stripping and steam stripping as potentially the most effective. The results of the stripping experiments in terms of total organic carbon removals (TOC) are shown in Figure A.8 (Stover and Kincannon, 1982). As shown in Table A.6, steam stripping was found to be the most successful in removal of all the volatile organics (Stover and Kincannon, 1982). Addition of activated carbon increased the overall TOC removal and removal of extractable organics.

Table A.5   Test Results from Metals Removal Studies of Ground Water from the Gilson Road Site (Stover and Kincannon, 1982)

Parameter, mg/l	Raw Water	Lime Metals Treatment	Activated Carbon, mg/l		
			3,600	7,200	14,400
Arsenic	0.10	0.03	0.03	0.03	0.03
Barium	0.25	0.16	0.05	0.05	0.04
Cadmium	0.06	0.001	0.001	0.001	0.001
Chromium	0.07	0.01	0.01	0.01	0.01
Lead	0.01	0.006	0.006	0.005	0.004
Mercury	<0.001	<0.001	<0.001	<0.001	<0.001
Selenium	<0.001	<0.001	<0.001	<0.001	<0.001
Silver	<0.001	<0.001	<0.001	<0.001	<0.001
Copper	0.06	0.03	0.03	0.03	0.03
Iron	390	0.40	0.40	0.27	0.27
Manganese	75	0	0	0	0
Nickel	1.23	1.30	1.30	1.30	1.30
Zinc	0.69	0.31	0.30	0.30	0.30

Table A.6   Volatile Organics Removal During Stripping Studies of Ground Water from the Gilson Road Site[a] (Stover and Kincannon, 1982)

Volatile Organic Compound, µg/l	Raw Water	Air Stripping	Ozone and Air Stripping	Steam Stripping
Tetrahydrofuran	22,000	ND[b]	ND	ND
1,1,1-Trichloroethane	150,000	7,600	ND	ND
Benzene	68,750	36,000	375	ND
Trichloroethylene	338,000	189,000	5,195	ND
Methyl Isobutyl Ketone	76,400	18,500	1,000	ND
Xylenes	517	ND	ND	ND
Toluene	92,000	6,600	5,300	ND
Ethylbenzene	23,500	ND	ND	ND

[a]Results are from best runs at higher air-to-water ratios.
[b]None detected (no peaks on chromatograms).

Extractable organics present in ground water from the Gilson Road site include phenols, cresols, naphthalenes, phthalates, organic acids, and alcohols. Steam stripping is effective in their removal, with Table A.7 showing the test results of the steam stripping runs (Stover and Kincannon, 1982). Activated carbon adsorption isotherm studies were conducted with raw water, metals treated water and steam stripped water. Activated carbon was not effective in removing high levels of TOC, but was useful for certain extractable organics as shown in Table

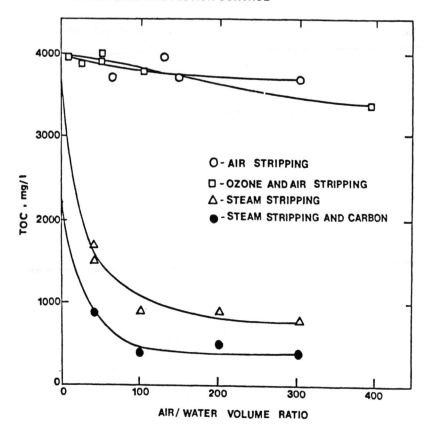

Figure A.8.   TOC removal by air and steam stripping of ground water from the Gilson Road site (Stover and Kincannon, 1982).

A.8 (Stover and Kincannon, 1982). Activated carbon does not readily adsorb organic acids and alcohols. Biological treatment was also considered as an alternative for removal of extractable organics. An activated sludge system was used with seed sludge from several systems already in operation. The system was operated by feeding metals treated water on a food-to-microorganism (F/M) ratio of about 0.3–0.5 (BOD_5/VSS), and wasting sludge on a daily basis by the fill-and-draw operating procedures. The test results are presented in Table A.9 (Stover and Kincannon, 1982). Activated sludge was found to have an excellent potential for removal of TOC and specific extractable organic compounds.

More detailed information is required prior to the final design of a treatment system to ensure an effective, reliable design. Pilot-scale studies are required to refine the final design of the steam stripper, and since

Table A.7   Removal of Extractable Organics by Steam Stripping of Ground Water from the Gilson Road Site (Stover and Kincannon, 1982)

Extractable Organic Compound, μg/l	Raw Water	Treated Water
1,4-Dichlorobenzene	35	ND[a]
1,2-Dichlorobenzene	5	ND
Naphthalene	≤1	≤1
2-Chlorophenol	540	ND
2-Nitrophenol	15	6
Phenol	370	20
2,4-Dimethylphenol	20	ND
O-Cresol	80	25
M-Cresol	220	ND
Benzoic Acid	1,230	ND
Pentachlorophenol	40	2

[a]None detected (no peaks on chromatograms).

Table A.8   Test Results from Various Activated Carbon Studies of Ground Water from the Gilson Road Site (Stover and Kincannon, 1982)

Extractable Organic Compound, μg/l	Raw Water	Lime Treated[a]	PAC Treated[b]	Lime and PAC Treated[c]	Steam Stripped and PAC Treated
1,4-Dichlorobenzene	35	35	ND[d]	≤1	ND
1,2-Dichlorobenzene	5	5	ND	2	ND
Naphthalene	≤1	≤1	ND	≤1	ND
2-Chlorophenol	540	540	8	6	≤1
2-Nitrophenol	15	15	≤1	8	ND
Phenol	370	370	≤1	ND	ND
2,4-Dimethylphenol	20	20	2	≤1	ND
O-Cresol	80	80	10	ND	ND
M-Cresol	220	220	≤1	ND	ND
Benzoic Acid	1,230	1,230	≤1	ND	ND
Pentachlorophenol	40	40	≤1	ND	ND

[a]Lime treated water to pH 10.0.
[b]PAC—Nuchar Aqua (Westvāco) and Carborundum PAC 20.
[c]Lime and PAC samples adjusted to pH 10.0, then the supernatant was adjusted to pH 6.5 with sulfuric acid for PAC treatment.
[d]None detected (no peaks on chromatograms).

metals removal is required prior to steam stripping, both of these processes must be piloted at the Gilson Road site. These two processes will accomplish removal of specific organics and inorganics of primary concern; however, the treated water will remain high in TOC, COD, and

Table A.9   Test Results from Batch Activated Sludge Studies of Ground Water from the Gilson Road Site (Stover and Kincannon, 1982)

Volatile Organic Compound, μg/l	Batch Run Number 1		Batch Run Number 2	
	Feed (Time 0)	Time (12 hr)	Feed (Time 0)	Time (8 hr)
Tetrahydrofuran	ND[a]	ND	ND	ND
1,1,1-Trichlorethane	3,776	78	2,513	ND
Benzene	13,800[b]	330	1,058	1,897[b]
Trichloroethylene		3,920	3,316	
Methyl Isobutyl Ketone	866	143	3,385	1,098
Xylenes	ND	ND	ND	ND
Toluene	156	11	2,115	25
Ethylbenzene	ND	ND	1,123	ND
Extractable Organic Compound, μg/l				
1,4-Dichlorobenzene	28	ND	$\leq$10	ND
1,2-Dichlorobenzene	$\leq$10	ND	$\leq$10	ND
Naphthalene	$\leq$10	ND	$\leq$10	ND
2-Chlorophenol	50	ND	30	ND
2-Nitrophenol	ND	ND	ND	ND
Phenol	20	ND	12	ND
2,4-Dimethylphenol	ND	ND	ND	ND
O-Cresol	60	ND	25	ND
M-Cresol	ND	ND	ND	ND
Benzoic Acid	ND	ND	3	ND
Pentachlorophenol	15	ND	ND	ND

[a]None detected (no peaks on chromatograms).
[b]Benzene and trichloroethylene peaks combined into one peak.

BOD. Biological treatment will be required following steam stripping to reduce these parameters to acceptable levels. Additional biological treatability studies will be required to define the biokinetic descriptive constants and the treatment performance reasonably achievable by biological treatment. This information will be required for full-scale design and can be developed from laboratory bench-scale, continuous-flow systems.

Remedial Actions Implemented

The remedial actions considered for implementation consisted of installation of a clean water exclusion system consisting of a slurry wall and surface cap containing a twenty-acre site determined to be highly contaminated. A ground water collection system would then be installed to deliver the contaminated ground water to a treatment plant for removal of the heavy metals, volatile organics, and extractable organics. The treated water would then be returned to the contained area for the purpose of flushing the remaining contaminated ground water. The specific selected plan consisted of the following (Roy F. Weston, Inc., 1982):

(1) A 3-ft-thick slurry wall, fully enclosing the 20-acre site.
(2) An impermeable surface cap consisting of a 40-mil polyethylene liner.
(3) Three to four ground water collection wells.
(4) A ground water treatment facility.
(5) A ground water reinjection system using surface trenches.

The treatability studies for designing the ground water treatment facility are summarized in Table A.10 (Stover and Kincannon, 1982). From an analysis of this treatability program, it was determined that the combined unit processes of metals removal, steam stripping, and biological treatment was the most feasible treatment alternative. The second alternative of metals removal, steam stripping, and activated carbon adsorption would remove the specific inorganic and organic compounds of concern; however, the residual organics measured as TOC would still be too high.

One essential aspect of the evaluation of the lifetime cost of the treatment facility and the cleanup or restoration potential of the aquifer was the length of time required to achieve a desired level of treatment. Therefore, a laboratory-scale experiment was designed to simulate flushing of the aquifer to estimate the number of times the ground water would have to be turned over, or flushed through the contaminated soil, to achieve the desired level of water quality. In this experiment a sample of uncontaminated soil from the site was soaked in contaminated ground water under controlled conditions, and then flushed with clean water under controlled conditions (Stover and Kincannon, 1982). From this experiment, a relationship between pumping flow rate and required aquifer flushing time was determined for achieving the desired ground water quality.

The construction of the slurry wall enclosing the twenty-acre site and the impermeable surface cap have been completed. Additional second phase pilot-scale studies have been initiated to finalize the design for metals removal and steam stripping. The effluent from this pilot plant will then be subjected to the bench-scale biological studies (Knox, Stover and Kincannon, 1982).

Table A.10   Summary of Treatability Studies of Ground Water from the Gilson Road Site (Stover and Kincannon, 1982)

Parameter, μg/l	Raw Water	Metals Treatment	Steam Stripping	Activated Carbon (6,000 mg/l)	Activated Carbon (24,000 mg/l)	Biological Treatment
Tetrahydrofuran	22,000		ND[a]			ND
1,1,1-Trichloroethane	150,000		ND			ND
Benzene	68,750		ND			1,897[b]
Trichloroethylene	338,000		ND			1,098
Methyl Isobutyl Ketone	76,400		ND			ND
Xylenes	ND		ND			25
Toluene	92,000		ND			ND
Ethylbenzene	23,500		ND			ND
1,4-Dichlorobenzene	35	35	17	ND	ND	ND
1,2-Dichlorobenzene	5	5	ND	ND	ND	ND
Naphthalene	≤1	≤1	≤1	≤1	≤1	ND
2-Chlorophenol	540	540	40	≤1	≤1	ND
2-Nitrophenol	15	ND	6	ND	ND	ND
Phenol	370	ND	20	ND	ND	ND
2,4-Dimethylphenol	20	ND	ND	ND	ND	ND
O-Cresol	80	80	25	ND	ND	ND
M-Cresol	220	220	ND	≤1	ND	ND
Benzoic Acid	1,230	1,230	ND	ND	ND	ND
Pentachlorophenol	40	40	40	8	2	ND

[a]None detected (no peaks on chromatograms).
[b]Benzene and trichloroethylene peaks combined into one peak.

CASE C: ROCKY MOUNTAIN ARSENAL

The Rocky Mountain Arsenal was established in 1942 by the U.S. Army, and it is located 27 miles northeast of Denver, Colorado (Warner, 1979). Figures A.9 and A.10 show the location of the arsenal and the surrounding area (Warner, 1979; and Shukle, 1982). During World War II incendiary munitions and various toxic chemicals were produced. As a result of waste leakage from open pits and lagoons, the ground water in the vicinity of the arsenal became contaminated over the years. Since 1947, private industry has leased many of the arsenal facilities for production of pesticides. Liquid wastes generated were discharged to a series of unlined waste storage basins. Crop damage detected at the northwest

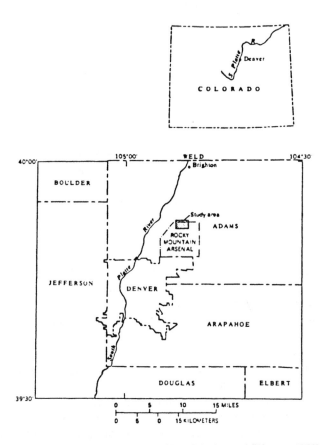

Figure A.9. Location of Rocky Mountain Arsenal (Warner, 1979).

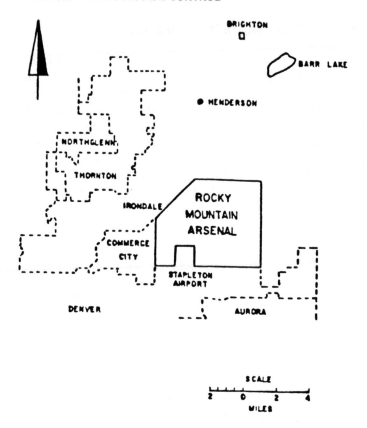

Figure A.10.   Vicinity map of Rocky Mountain Arsenal (Shukle, 1982).

boundary of the arsenal in 1954 was suspected to be due to seepage from the basins. The crops were being irrigated with water from the alluvial aquifer underlying the basins (Shukle, 1982).

The alluvium consists of silts and clays overlying sands and gravels, with thickness ranging from 20 to 30 ft. An illustration of ground water contours and bedrock highs in the arsenal area is presented in Figure A.11 (Warner, 1979). The Denver formation is composed of clay shale and clay stone with some sand units. Alluvium is saturated within 5–15 ft of the ground surface and an estimated 64,000 gpd of the ground water flows at the north boundary of the arsenal. Ground water flows into the South Platte river, which appears to act as a boundary for further underground migration (Shukle, 1982).

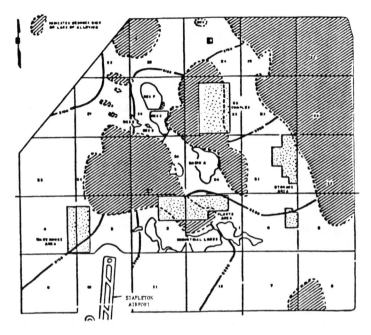

Figure A.11.   Ground water contours and bedrock highs at the Rocky Mountain Arsenal (Warner, 1979).

In April 1975, the Colorado Department of Health issued orders to the U.S. Army and Shell Chemical Company relative to the identification of diisomethylphosphonate (DIMP) and dichloropentadiene (DCPD) in seepage from the north boundary of the arsenal. The orders required: (1) cease the discharge of these compounds to surface and subsurface waters; (2) develop a plant to accomplish the cessation of the discharge; and (3) implement a monitoring program to demonstrate compliance with the orders (Shukle, 1982). The stringent water rights within the state of Colorado also required that treated ground waters be reinjected so as not to disrupt subsurface flow patterns (Shukle, 1982).

Sources of Contamination

As mentioned earlier, the main source of ground water contamination at the arsenal is the seepage from the waste storage basins. A monitoring program which consists of approximately 92 ground water wells and 19 surface water locations has provided a good understanding of the conditions, and has identified new organic contaminants and new sources and

Table A.11   Ground Water Contaminants at North Boundary of Rocky Mountain Arsenal (Hager and Loven, 1982)

Compound	Concentration (μg/l)	
	PW2[a]	PW3[b]
Aldrin	<2.0	<2.0
Dieldrin	4.5	<2.0
Dichloropentadiene	1000	82
Diisomethyl phosphonate	530	2800
Endrin	8.6	<2.0
Dibromochloropropane (DBCP)	7.6	<1.0
p-Chlorophenylmethyl		
sulfide	68.3	<10.0
sulfoxide	53.3	<10.0
sulfone	40.5	<10.0

[a]Monitoring well PW2.
[b]Monitoring well PW3.

areas of contamination. In addition to the disposal basins, the railroad yard, the manufacturing and chemical storage area, and miscellaneous burial sites have also impacted ground water under the arsenal. Various chemical compounds have been found in the ground and surface waters sampled near the arsenal. A list of the chemicals found is provided in Table A.11 (Hager and Loven, 1982). Diisomethylphosphonate (DIMP) and dichloropentadiene (DCPD) were identified as the major contaminants present (Hager and Loven, 1982). The extent of ground water contamination in the arsenal area has been identified as shown in Figure A.12 (Shukle, 1982). The contaminants in the alluvial aquifer are not uniformly distributed and some wells encounter higher concentrations than others. There are also variations in the types of contaminants in different wells (Shukle, 1982). Other than the reservoirs (basins) present, the sewage treatment plant could also be a source of contamination. A plan view of the north boundary area of the arsenal is provided in Figure A.13 (Hager and Loven, 1982).

Studies to Determine Remedial Actions

A number of studies have been conducted on containment and treatment options for the ground water underlying the Rocky Mountain Arsenal. Of the many treatment techniques evaluated, activated carbon adsorption was found to be the most cost effective (Hager and Loven,

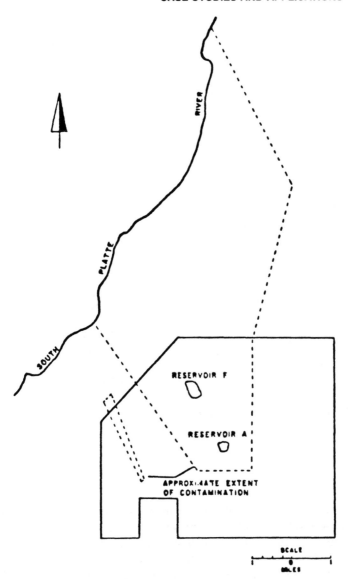

Figure A.12. Extent of ground water contamination at Rocky Mountain Arsenal (Shukle, 1982).

1982). Testing on dynamic columns of granular activated carbon indicated that diisomethylphosphonate (DIMP) was the first to appear in the effluent of the column. An interim containment and treatment system was installed in 1978, with the location shown in Figure A.13 (Hager and

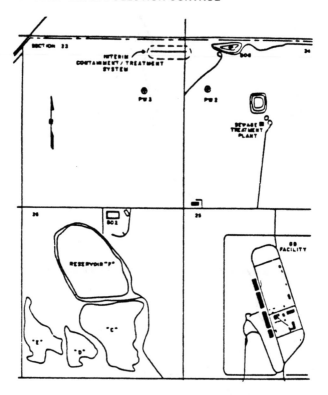

Figure A.13.   North boundary area of the Rocky Mountain Arsenal (Hager and Loven, 1982).

Loven, 1982). A 1,500-foot-long containment system consisted of six 12-in. withdrawal wells on 200-ft centers, and twelve 6-in. recharge wells on 100-ft centers. A full-scale granular activated carbon system was also installed. Following treatment the recharge is by gravity feed. Between the dewatering and recharge wells a 2-ft-thick bentonite slurry wall, 1500 ft long, was constructed to prevent recirculation of treated waters.

After withdrawal of contaminated ground water from the north boundary area, the water is passed through a dual-media filter to remove suspended matter, then to activated carbon columns, and finally to post filtration prior to recharge. Calgon adsorption equipment was used with monitoring over a three-year period from 1978 to 1981. The principal operating parameter used was a maximum effluent DIMP concentration

Table A.12   Information on Interim Granular Activated CarbonTreatment System
(1/7/78 through 6/30/81) Used at North Boundary of Rocky Mountain
Arsenal (Hager and Loven, 1982)

Parameter	Information
Adsorption Mode	Single stage fixed bed
Carbon Bed Volume, ft³	666
Flow Rate, gpm	70 – 100
Contact Time, min	50 – 300
Type of Carbon	Calgon Service Carbon
DIMP Concentration	
Influent	1200 μg/l average
Effluent	nondetectable
Carbon Use Rate	
Projected	0.33 lb/1000 gal
Actual	1.00 lb/1000 gal

of 500 μg/1. Carbon was replaced periodically. Some of the salient features of the activated carbon system are listed in Table A.12 (Hager and Loven, 1982). Over the operating period of three years, it was observed that the granular activated carbon adsorption process was effective in removing most organic compounds. The system operated unattended except for periodic monitoring, backwashing and carbon replacement.

While the interim containment and treatment system was being tested, analytical work sponsored by the U.S. Army was in progress. A digital transport model for contamination of ground water at Rocky Mountain Arsenal was developed (Warner, 1979). This model was used to determine the effects on ground water movement, and on DIMP concentrations in ground water, of the bentonite barrier in the aquifer near the north boundary of the arsenal. The ground water flow equation was coupled with a method-of-characteristics solution of the solute-transport equation, and solved by an iterative-alternating-direction-implicit procedure. This transport model assumes steady-state ground water flow and conservative (nonreactive) transient transport of DIMP. In simulations, it was assumed that the bentonite barrier was impermeable and penetrated the entire thickness of the aquifer. It was estimated that the 1,500-ft barrier would intercept about 50% of the DIMP that would otherwise cross the north boundary. Of the ground waters with DIMP concentrations greater than 500 μg/l, 72% would be intercepted by the barrier. It was also found that the amount of DIMP underflow intercepted may be increased to 65% by doubling the pumpage, or to 73% by doubling the length of the barrier. The report concludes that the containment of

DIMP-contaminated ground water could also be further accomplished by a hydraulic barrier consisting of both withdrawal and recharge wells. The hydraulic barrier could be just as effective as the bentonite barrier, with the only difference being the additional pumping required to account for recirculation between the line of withdrawal wells and recharge wells. The simulation results of this modeling study are summarized in Table A.13 (Warner, 1979).

Remedial Actions Implemented

Following the encouraging results from the interim containment and treatment system, a permanent facility was installed in 1981 to contain the ground water contamination at the north boundary. This facility included a 6700-ft-long clay slurry wall which was installed 200 ft inside the arsenal perimeter. Dewatering (withdrawal) and reinjection (recharge) wells which are present along the barrier remove water for treatment and deliver purified water into the aquifer. The 6700-ft-long boundary includes the 1500-ft pilot system as well as 29 new alluvial dewatering wells and 38 new discharge wells (Shukle, 1982). There are three separate water collection systems, with different organic compositions in each. Therefore, for treatment flexibility three different adsorption trains were used, rather than treating a mixed influent. This is illustrated in the process diagram in Figure A.14 (Hager and Loven, 1982). The three adsorbers in parallel are connected to cartridge filters ahead of and following them. There are storage vessels for both fresh and exhausted carbon with two pressure transfer vessels for moving carbon between the equipment in a measured slurry volume. The carbon adsorbers in the permanent facility are up-flow pulsed bed units, whereas the interim plant had down-flow carbon columns which required replacement of all carbon when organic breakthrough occurred. The treatment system characteristics are listed in Table A.14 (Hager and Loven, 1982).

In addition to the north boundary containment and treatment system, the need for a smaller system on the northwest boundary of the property was identified in early 1980 as a result of contamination being found in private wells (Shukle, 1982). The area is a rural residential and agricultural mixture. Subsequent investigation identified a plume containing an organic contaminant at the 1.0 ppb level originating from a railroad yard on arsenal property. The geology was identified as silty sand and clay overlying sand and sandy gravels. The saturated thickness in the area of the system is on the order of 6–10 ft, with the alluvium ranging from 50 to 60 ft in thickness. Ground water movement in the area was projected

Table A.13 Simulation Results of a Digital-Transport Modelling Study at Rocky Mountain Arsenal (Warner, 1979)

Model Simulations	Underflow Intercepted by Barrier	DIMP Intercepted by Barrier	Water Level Effects South of the Barrier and on the Arsenal
Simulation 2: 1,500-ft bentonite barrier near D Street.	Barrier would intercept about 0.09 ft³/s or about 25% of the total ground water underflow.	Barrier would intercept about 50% of the total DIMP and about 72% of ground water with DIMP concentrations greater than 500 μg/l.	No significant changes in water table altitude or in direction and rate of ground water movement.
Simulation 3: 1,500-ft bentonite barrier near D Street with twice the pumpage and recharge.	Barrier would intercept about 0.18 ft³/s or about 50% of the total ground water underflow.	Barrier would intercept about 65% of the total DIMP and about 90% of the ground water with DIMP concentrations greater than 500 μg/l.	A 1- to 2-ft drawdown in the water table altitude would occur on the arsenal over most of the model area. The minimum drawdown would be about 5 ft near the barrier and decrease to about 3 ft approximately 1,000 ft south of the barrier.
Simulation 4: Bentonite barrier near D Street extended 1,500 ft to west	Barrier would intercept about 0.10 ft³/s or about 28% of the total ground water underflow.	Barrier would intercept about 57% of the total DIMP.	South of the barrier only minor changes in the water table altitude would result and a maximum rise of slightly more than 3 ft would occur.
Simulation 5: Bentonite barrier near D Street extended 1,500 ft to east.	Barrier would intercept about 0.18 ft³/s or about 50% of the total ground water underflow.	Barrier would intercept about 73% of the total DIMP.	No significant changes.

Table A.13, continued

Model Simulations	Water Level Effects North of the Barrier and Off the Arsenal	DIMP Concentrations of Underflow Across the North Boundary of the Arsenal	DIMP Concentrations of Ground Water Off the Arsenal
Simulation 2	No significant changes.	The maximum DIMP concentration of ground water underflow across the boundary would not be reduced by this barrier configuration. Ground water with DIMP concentrations greater than 500 µg/l would cross the boundary both east and west of the barrier.	A plume of ground water with DIMP concentrations less than 200 µg/l would develop within 4 years after initial operation of the barrier system and would extend 1,500 to 2,000 ft north of the arsenal.
Simulation 3	A maximum rise in the water table altitude of about 8 ft would occur just north of the barrier caused by doubling the recharge.	The barrier would intercept almost all of the ground water with DIMP concentrations greater than 500 µg/l. Doubling the pumpage would result in a faster removal of the DIMP-contaminated ground water from the aquifer and may possibly result in operation of the barrier being discontinued from 2 to 6 years earlier than in simulation 2.	A plume of ground water with DIMP concentrations less than 200 µg/l would develop in approximately 2 years—one-half the time required in simulation 2.
Simulation 4	North of the westward extension of the barrier, a part of the aquifer would become dewatered and a maximum decline of more than 10 ft would occur.	Extension of the barrier to the west would result in an area of ground water with large DIMP concentrations becoming partly trapped in a zone south of the barrier near the bedrock high. This would likely result in a prolonged operating life of the barrier.	Extension of the barrier would prevent any ground water underflow west of the barrier. Although no new DIMP-contaminated ground water would cross the arsenal boundary, the ground water with already large DIMP concentrations north of the arsenal would remain in place.

| Simulation 5 | No significant changes. | This barrier configuration would prevent any ground water with DIMP concentrations greater than 500 μg/l from crossing the arsenal boundary east of the barrier. | The spread of the plume of ground water with DIMP concentrations less than 200 μg/l that would develop north of the barrier would be greater than for other barrier configurations considered and would be much more effective in reducing DIMP concentrations of ground water off the arsenal. |

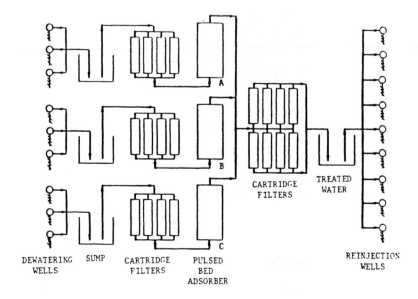

Figure A.14. Permanent granular activated carbon treatment facility at north boundary of Rocky Mountain Arsenal (Hager and Loven, 1982).

Table A.14   Information on Permanent Granular Activated Carbon Treatment System Used at North Boundary of Rocky Mountain Arsenal (Hager and Loven, 1982)

Parameter	Information
Adsorption Mode	Three pulsed beds operating in parallel
Carbon Bed Volume (each vessel)	1000 ft³
Volume of Carbon Pulse	70 ft³
Type of Carbon	WV-G 12 × 40
Design Flow Rate, gpm	Average 250/Maximum 350
Contact Time, min	Average 84/Minimum 30

Influent Organic Contaminants, μg/l	Process Stream A	Process Stream B	Process Stream C	Effluent Objectives
DIMP	700 – 1200	100 – 500	10 – 100	500
DCPD	10 – 2000	500 – 2400	24	24
DBCP	1 – 5	1 – 5	1 – 5	0.2

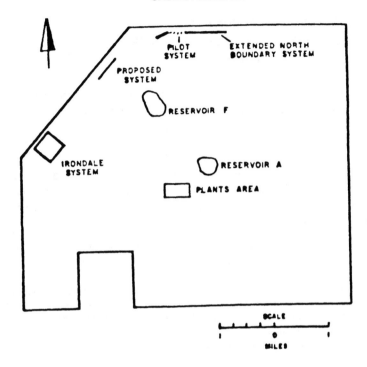

Figure A.15. Containment system locations at Rocky Mountain Arsenal (Shukle, 1982).

at about 5 ft/day in a northwesterly direction. The control system, called the Irondale system, became operable in December, 1981, and consists of three wells each in two lines for dewatering, and one line of 14 recharge wells. All wells are spaced on 100-ft centers with about 800 ft between rows of dewatering wells and 600 ft between dewatering and recharge wells. The completed system initially encountered flows of 1,600 gpm with subsequent rates approaching 900 gpm. Piezometric levels existing around the system indicate a very effective hydrological barrier to contaminant migration. Improvement in the quality of downgradient waters has also been observed.

As shown in Figure A.15, the U.S. Army is evaluating a third system to be located between the two existing systems (Shukle, 1982). Subsequent efforts will then be focused on source control which will likely include the disposal basins, the railroad yard, the manufacturing and chemical storage area plus burial sites which over years of use have impacted ground water on the arsenal (Shukle, 1982).

NEW JERSEY DEPARTMENT OF
ENVIRONMENTAL PROTECTION AQUIFER
RESTORATION PROGRAM

The Ground Water Quality Protection Division in the New Jersey Department of Environmental Protection (NJDEP) was started 6 years ago to address the complaints from the public concerning leaking gasoline storage tanks (Kasabach and Althoff, 1983). Over 300 different contamination problems in the state have been, or are being, investigated by the Division. The success and efficiency of NJDEP in avoiding long and involved legal actions concerning ground water contamination problems can be identified as a combination of the following three factors (Althoff, 1983):

(1) NJDEP has legislated authority to protect the waters of the state. This authority at least requires that pollutors listen to NJDEP and not dismiss them as unable to force action.

(2) A competent staff, which is respected by the companies approached concerning their ground water contamination problems. The companies work to alleviate the problems rather than argue about the validity of claims made against them.

(3) To avoid a negative public image, companies have moved fast to alleviate their ground water contamination problems.

The sources of ground water contamination in New Jersey have been classified as industrial, hydrocarbon loss, landfills, illegal dumpsites, and miscellaneous. The number of ground water investigations and their categorization is included in Table A.15 (Kasabach and Althoff, 1983). The investigations have also been classified in terms of contaminant types as shown in Table A.16 (Kasabach and Althoff, 1983). A summary of aquifer restoration programs in New Jersey, as of January, 1983, is listed in Table A.17 (Kasabach and Althoff, 1983). Discharge from contaminated wells to sewer systems, for treatment on site or at a local wastewater treatment plant, is an important decontamination option in New Jersey. Hydrocarbon recovery systems have not been included in Table A.17, though they are in use in industry-related cleanup mostly for organic solvents, metals and inorganic contaminants. Hydrocarbon losses to aquifers have been identified to be due to a variety of reasons such as service station contamination of domestic or municipal wells, home heating oil leaks, fuel spills at military installations, and poor housekeeping at railroad yards.

To serve as an illustration of some aquifer restoration efforts in New Jersey, summary information on six industrial case studies is listed as follows by company, pollutant, remedial action, cost, and key issues and comments:

Table A.15 Ground Water Contamination Investigations in New Jersey—By Source (Kasabach and Althoff, 1983)

Source	Number of Investigations	Percent of Total (308)
Industrial[a]	195	63
Hydrocarbon Loss[b]	52	17
Landfill[c]	26	8
Illegal Dump Site	21	7
Miscellaneous	14	5

[a]Chemical, petrochemical and plating industries predominate.
[b]Includes all types, from major recovery programs at some refineries to service station and home heating oil spills. Some cases involve petroleum contamination on "industrial" sites.
[c]New Jersey has 300 landfills. The Division of Water Resources investigates selected sites referred to the Division by other agencies.

Table A.16 Ground Water Contamination Investigations in New Jersey—By Contaminant Type (Kasabach and Althoff, 1983)

Waste Type	Number of Occurrences[a]	Percent Occurrence In Ground Water Cases
Organics[b]	209	56
Hydrocarbons[c]	61	17
Metals	59	16
Inorganics	36	2
Radioactive	4	1

[a]A single ground water investigation often involves more than one contaminant type, e.g., plating wastes and cutting oils. Thus, the combined totals exceed caseload of 308 investigations in March 1983.
[b]Volatile organic solvents account for most organics reaching New Jersey aquifers, and of these six to ten compounds dominate the list.
[c]Hydrocarbon cleanups include removal of floating product and, increasingly, decontamination for dissolved fractions such as benzene, toluene and xylene.

(1) *Company*: Rockaway Township
   Pollutant: trichloroethylene; town's only well field (3 wells) affected.
   Remedial Action: Two 20,000-lb GAC units installed at well field.
   Cost: Estimated as of Sept., 1981 – $0.30 million.
   Issues and Comments:
      Employed aeration prior to GAC—Were expecting big slug of ethers.
      Going to discard GAC because water from air stripping is so clean it picks up organics from GAC.

Table A.17   Aquifer Restoration Programs in New Jersey—January, 1983 (Kasabach and Althoff, 1983)

Treatment Methodology	Number of Uses	Percent of Total Cleanup Programs (36[a])
Discharge to Sewer For Treatment[b]		
On-Site	10	28
Off-Site	11	31
Aeration[c]	7	19
GAC[d]	3	8
Aeration Plus GAC	1	3
Biological	1	3
Miscellaneous	3	8

[a]Includes systems both in operation and under construction. Hydrocarbon recovery programs are not included.
[b]Contaminants are intercepted either by wells or by interceptor trenches. Contaminated production wells are used in some programs for plume control. One decontamination program utilizes both aeration and sewer discharge from separate wells.
[c]Includes both natural schemes as well as systems engineered to enhance aeration via towers, nozzles and other equipment.
[d]Granular activated carbon.

(2) *Company*: Mideast Aluminum
Pollutant: 1,1,1-trichloroethane, and others
Remedial Action: Air stripping with recharge via infiltration ponds. Pumping about 104 gpm. Organic levels decreasing.
Cost: Estimated as of Mar., 1984 — $0.4 million.
Issues and Comments:

Feed varies — 620 ppb (April 1980) to about 100–250 ppb currently.
Lot of work done to determine ground water flow — rate and direction — concern about pollutant moving toward municipal wells.

Legal issue — Use of chlorinated hydrocarbons violated no statute at the time. Felt a community responsibility to participate in resolution. Wanted to share monitoring costs.

Company also required to line existing lagoons and remove settled solids.
Air stripping is cheaper than GAC by (2–4):1. Also had to get air pollution permit.

(3) *Company*: Allied Chemical
Pollutant: Chloroform, and carbon tetrachloride
Remedial Action: Program consists of two 200 gpm interceptor wells.
Cost: Estimated as of Sept., 1981 — $0.25 million
Issues and Comments:

Suspected old buried drums of spill — magnetometer study showed no drums. Pump to waste treatment plant, very little reaches surface drainage because both compounds evaporated rapidly upon exposure to air.

(4) *Company*: American Cyanamid

Pollutant: Host of synthetic organic chemicals resulting from decades of manufacturing activity. Shallow monitor wells near waste lagoons confirmed contaminated ground water. Also have reached bedrock aquifer.

Remedial Action: Several dozen wells in overburden and 9 bedrock monitor wells, large demand of plant prevents off-site migration. Water is withdrawn, used for cooling and treated with other process effluent via large-scale activated carbon before surface water discharge.

Cost: Estimated as of Sept., 1981 – $0.20 million (probably underestimated; program to continue for several years).

Issues and Comments:

Continue pumping to confine movement and locate and reduce sources within plant.

(5) *Company*: Saytech (Hexcell)

Pollutant: 1,2 dichloroethane

Remedial Action: Interceptor trench and discharge to sewer.

Cost: Estimated as of Sept., 1981 – $0.1 million

Issues and Comments:

Trench – Excavate down to clay (8 ft), gravel placed in bottom, 3-in. perforated pipe set above bottom, gravel again to 1 ft above seepage line. Trench slumps badly, requires shoring during construction. Impervious cover installed on top of the stone or gravel to prevent clogging of gravel pore spaces.

Trench will be removed at end of decontamination period and filled with bentonite.

(6) *Company*: Biocraft

Pollutants: Acetone, methylene chloride, and butyl alcohol.

Remedial Action: Injection-recovery trench system with biostimulation treatment. Start up in June, 1981.

Cost: Estimated as of Sept., 1981 – $0.15 million.

Issues and Comments:

Chemical reactions in aquifer. Example – chlorinated hydrocarbons reacting with naturally occurring organic matter has been proven to be the source of dangerous trihalomethane.

Add air and nutrients to help biodegradation.

CASE STUDIES OF AQUIFER RESTORATION VIA BIOLOGICAL PROCESSES

Based upon a review of published literature and contacts with professionals in the field, 15 case studies involving restoration using biological processes were identified in sufficient detail to allow their summarization. This summary information is presented as follows in accordance with location, company affiliation, background, action taken, and effectiveness. The first six case studies involve the enhancement of natural microbial populations in the subsurface environment; cases (7) through

(12) encompass the addition of adapted microbes to the subsurface environment. Cases (13) through (15) involve the use of biological treatment following withdrawal of polluted ground water.

(1) *Location*: Ambler, Pennsylvania (Raymond et al., 1975; Raymond et al., 1976; Jamison et al., 1975; and Jamison et al., 1976).
Company Affiliation: Sun Ventures, Inc., Marcus Hook, Pennsylvania.
Background: A pipeline break spilled 3,186 barrels of high octane gasoline to the ground water. The aquifer was composed of a highly fractured dolomite that was a very good source of ground water although silt and clay filled many of the fractures. The gasoline was contained by pumping nearby water wells. Physical recovery of the gasoline overlying the water table was begun and continued until no longer productive; and estimated 1000 barrels remained which might take 100 years to recover.
Action Taken: The activity of the gasoline-utilizing microbial population was stimulated by the addition of nitrogen and phosphorus in the form of $(HN_4)_2SO_4$, Na_2HPO_4, NaH_2PO_4, and dissolved oxygen. Dissolved oxygen was supplied by sparging air into wells with diffusers connected to paint sprayer-type compressors. Ground water flow was controlled by a series of injection and withdrawal wells.
Effectiveness: Estimates based on the amount of nitrogen and phosphate retained in the aquifer suggested that from 744 to 944 barrels of gasoline were degraded. The amount of free gasoline in the wells declined as the nutrient addition program continued. The levels of gasoline in the produced water were not reduced during the period of nutrient addition, but no gasoline was found in the produced water ten months later. No direct evidence from the magnitude of conversion of gasoline to bacterial cells in the aquifer was found, although bacterial levels increased greatly. The biostimulation program met its objective of removing the hydrocarbons from the ground water to the satisfaction of the governmental agency overseeing it.

(2) *Location*: Millville, New Jersey (Raymond et al., 1978).
Company Affiliation: Suntech, Inc., Marcus Hook, Pennsylvania.
Background: A gasoline spill from a storage tank was found when gasoline fumes were noticed in basements adjacent to the gasoline station. An undetermined amount of gasoline was lost to the shallow, unconsolidated, sandy aquifer of high porosity and permeability. An extensive clay bed lay beneath the aquifer. Institution of a physical recovery program resulted in the removal of 7000 gallons, but the gasoline recovery rates decreased dramatically after two months to the point where only a few gallons were being recovered daily. The state agency in charge of the cleanup directed the operation to continued recovery operations until no trace of liquid gasoline remained on the ground water surface in order that Millville well field be protected.
Action Taken: A biostimulation program was begun. Nutrients including ammonium sulfate, disodium phosphate, monosodium phosphate, sodium carbonate, calcium chloride-dihydrate, magnesium sulfate-heptahydrate, manganese sulfate-monohydrate, and ferrous sulfate-heptahydrate were added to the wells continuously or in batches. Air was introduced into the wells by a carborundum diffuser with an output of 10 scfm and by diffusers constructed from DuPont Viaflow® tubing that generated 1 scfm. Producing wells were used to control ground water flow.

Effectiveness: The biostimulation program was not completely successful; residual gasoline was found at the last sampling period, but no free hydrocarbon was observed in any of the wells after the biostimulation program ended. The gasoline concentrations in cores taken from the aquifer did not seem to change substantially during the biostimulation program. Phenol concentrations decreased to acceptable levels as more aerobic conditions were established in the aquifer. Gasoline levels in the produced water were generally quite low; although following an interruption in the biostimulation program, gasoline was found in the produced water which indicated that pockets of gasoline remained in the aquifer. One well which did not receive nutrients and air was still contaminated with gasoline. This demonstrates that the biostimulation program did have an effect on the degradation of the gasoline. After the area near this well was treated, no free gasoline was found in any of the wells, and operations ceased with state approval. Microbial growth and utilization of nutrients and dissolved oxygen backed up the conclusion that enhanced microbial degradation was responsible for the removal of the gasoline.

(3) *Location*: Parkside, Pennsylvania (Suntech, 1978).

Company Affiliation: Environmental Group of Suntech, Inc., Marcus Hook, Pennsylvania.

Background: Gasoline fumes from a gas station spill reached the Parkside Elementary School and forced its closure. Physical recovery was able to remove much of the gasoline and to lower the water table so that fumes did not reach the school.

Action Taken: Nutrients and oxygen were supplied to the ground water microorganisms for six months.

Effectiveness: Tests showed that no excess hydrocarbon remained, but no data were presented to judge this claim. The fume problem was alleviated.

(4) *Location*: LaGrange, Oregon (Minugh et al., 1983).

Company Affiliations: Environmental Emergency Services Company, Portland, Oregon; Chevron U.S.A., Concord, California; and consultants; Suntech Group, Marcus Hook, Pennsylvania.

Background: Gasoline fumes were detected in the basements of two restaurants and led to their closure. The source of contamination was traced back to a bulk plant storage tank which until recently had been owned by Chevron. A field investigation determined the boundaries of the gasoline plume and the hydrogeologic features of the site. The shallow ground water (about 13 ft) roughly paralleled the land surface and was overlain by a top layer of topsoil, gravel, and silt at the surface; a five-foot section of assorted gravel, medium to coarse sand with some clay and silt; and a third layer about 4 ft in thickness which was composed of medium to large cobbles, gravel, and coarse sand. A number of stream channel deposits crossed the site and provided highly permeable pathways for ground water flow.

Action Taken: Laboratory biostimulation studies showed that the following nutrients were needed to stimulate the microbial population to degrade the gasoline: ammonium chloride, sodium phosphates, iron, manganese, magnesium, and calcium. A system was designed to draw down the static water levels 6 to 10 ft, to recycle the produced water (thereby recycling nutrients and eliminating the problem of disposing of the water), and to supply nutrients and dissolved oxygen to the microbes. Ground water was cycled through the site by installing three recovery wells in highly contaminated areas and injecting produced water into trenches upgradient of the area of known contamination. Air was supplied through a 2-in. line at the bottom of the injector trench with porous stone diffusers every 40 ft.

Effectiveness: The physical recovery system accounted for recovery of 3,266 gal of free hydrocarbon. Nutrients broke through the production wells after only a few days, but high dissolved oxygen levels did not reach the recovery wells for five months due to its removal by microorganisms. Bacterial levels increased up to six million times the initial levels. The system was operated for a year to encompass the complete range of seasonal water fluctuations. After nine months of treatment, soils in the highly contaminated tank storage area still showed signs of gasoline contamination at levels of 500 to 100 ppm and the average concentrations of dissolved carbon (DOC) was 20 ppm. After the bio-stimulation program ended the DOC in the water had fallen to the point where 71% of the measurements fell below 5 ppm and 50% were under 2 ppm. Since the DOC levels initially were not given, it is difficult to evaluate the efficiency of the cleanup. Free product was no longer present in the recovery or monitoring wells, and the vapor problems at the restaurants were mitigated. The cleanup effort was ended with the establishment of an appropriate ground water monitoring program.

(5) *Location*: Waldwick, New Jersey (Jhaveria and Mazzacca, 1982).

Company Affiliation: Ground Water Decontamination Systems, Waldwick, NJ; Consultants; Suntech, Inc., Marcus Hook, PA; Geraghty and Miller, Syosset, NY; and Princeton Aqua Science, New Brunswick, NY.

Background: An estimated 33,000 gal of a mixture of methylene chloride, acetone, n-butyl alcohol, and dimethylanaline was spilled at the Biocraft Laboratories, Inc., pharmaceutical manufacturing plant in Waldwick, NJ. The contamination was first detected in a small stream into which the storm sewers from Biocraft discharged. Following the installation of several monitoring wells at the Biocraft site, a monitoring program revealed that the ground water was contaminated and was the source of the pollution in the storm sewers. The contamination problem was confined to an 8- to 15-ft layer of glacial till composed of a poorly sorted mixture of boulders, cobbles, pebbles, sand, silt, and some clay. A 40-ft-thick semiconsolidated layer of silt and fine sand lay below the glacial deposit and acted as an aquitard to separate the shallow aquifer from a shale layer which served as the drinking water supply for much of the region. Ground water movement was fairly rapid at 0.4 ft/day in the shallow water table. The permeability of the glacial till varied from 1 to 35 gpd/ft².

Action Taken: An average of 10,000 gal of contaminated ground water was collected per month and sent to an approved disposal facility. Other options were investigated; a series of biodegradation tests indicated that the native microbial population could be stimulated to degrade the contaminants. A hydrogeological survey was performed so that a system to confine the contaminated ground water on site and decontaminate it could be designed. The biostimulation-decontamination system which was designed consisted of a trench and series of wells to withdraw the ground water, four tank wagons used in the biostimulation process, two reinjection trenches, and nine wells to inject air from the blower into the aquifer. The contaminated water was pumped from the dewatering system into two of the tank wagons. In these stirred activating tanks, air was supplied at 50 scfm, nutrients added as needed, and the temperature maintained at 18–22°C. Following a residence time of 16–18 hr in the activating tank, the effluent was pumped to the two settling tanks. About 200 gal of the activated sludge slurry was recycled to the activating tanks daily. The effluent from the settling tanks which was enriched with organisms, nutrients, and some oxygen was supplied with additional oxygen and recirculated via the reinjection trenches. Nine wells provided air to the ground water as it flowed from the reinjection trench to the dewatering system. The following compounds were used in the nutrient solution: 500 mg/1 NH_3Cl_2, 270 mg/1 KH_2PO_4, 410 mg/1 K_2HPO_4, 18 mg/1 $MgSO_4$, 9 mg/1 Na_2CO_3, 1.8 mg/1 $MnSO_4$, 0.45 mg/1 $FeSO_4$, and 0.9 mg/1 $CaCO_3$. The system treated 10,000 gpd with a 9.5-gpm feed stock flow and retention time in the activating tank of 17.5 hr.

Effectiveness: The above ground system was able to remove about 95% of the organics, but some of this may have been due to air stripping the volatile organics instead of microbial activity. The contaminant levels in one of the pumping wells declined from an average of 91 ppm methylene chloride and 54 ppm acetone to less than 1 ppm within a year of treatment. The COD of the ground water from the pumping well was significantly reduced also. The area near the pumping wells was the last to be treated and still showed fairly high levels of contamination (up to 40 mg/1 of methylene chloride), but the monitoring wells closer to the reinjection system had very low levels of methylene chloride; a reduction from about 70 ppm to less than 0.02 ppm was noted for two for the monitoring wells. A separate investigation showed that the acclimated organisms from the site were able to biodegrade methylene chloride.

(6) *Location*: Karlsruhe, West Germany (Nagel et al., 1982).

Company Affiliation: Federal Ministry for Research and Technology.

Background: A number of spills and losses during transfers had contaminated the train yard of the German Railroad System in Karlsruhe with petroleum products. The hydrocarbons reached the ground water table and caused a nearby water well to be closed. Cyanide was also found in the ground water. The aquifer was composed of coarse sands and gravel mixed with pebbles and had a permeability factor of 2.1×10^{-3} m/s.

Action Taken: A system was constructed in which contaminated ground water was pumped out, treated with ozone, and then reinfiltrated via five injection wells. This treatment system controlled ground water flow, improved the biodegradability of the contaminants, and provided oxygen to increase the activity of the microorganisms in the aquifer. Approximately 1 g of O_3/g of DOC was added to the ground water and allowed a 4-min contact time. The oxygen content increased to 9 mg/1 with a residual 0.1–0.2 g ozone/m^3 in the treated water.

Effectiveness: This treatment was effective in restoring the aquifer. Dissolved oxygen levels increased in the wells with the most rapid changes noted in the monitoring wells closest to the infiltration wells. Oxygen consumption in the heavily contaminated zone reached 40 kg/day, but began to decline as the DOC levels dropped. The DOC concentrations fell to about 1.5 g/m^3 and virtually no mineral oil hydrocarbons or cyanide was detected. Total bacterial counts increased about tenfold, but potentially harmful bacteria did not increase. The aquifer was restored to the point where drinking quality water was produced. The destruction of the hydrocarbons by ozone may have been more important in the cleanup than microbial activity. Microbial activity within the aquifer may have been limited since ozone is a toxicant (it is used as a disinfectant in some water distribution systems) and levels of inorganic nutrients were not supplemented.

(7) *Location*: Indiana (Polybac Corporation, 1983).
Company Affiliation: Polybac Corporation, Allentown, PA.
Background: An acrylonitrile spill contaminated the soil and ground water of a site in Indiana.
Action Taken: Treatment was by pumping ground water from several wells to a biotreator seeded with mutant bacteria from polybac and then injection into the ground water table.
Effectiveness: The concentrations of acrylonitrile fell from 100 to 1 ppm within three months. The report did not contain sufficient details to judge the importance of microbial activity in the removal of the acrylonitrile.

(8) *Location*: Midwest Metropolitan Area (Walton and Dobbs, 1980).
Company Affiliation: Polybac Corporation, Allentown, PA.
Background: The ground water beneath a rail yard was contaminated by a leaking rail car which released about 7000 gal of acrylonitrile. The aquifer contained significant amount of silt and clay and consequently had a low permeability. The ground water table was quite shallow—only about 5 ft.
Action Taken: The ground water was treated by air stripping the recovered ground water and, after concentrations had been reduced enough to allow microbial growth, by adding mutant bacteria.
Effectiveness: The degradation of acrylonitrile occurred rapidly, as the levels fell from 1000 ppm to the limits of detection within a month. No data were presented that conclusively linked the drop in the concentration of acrylonitrile to microbial activity by the mutant bacteria.

(9) *Location*: Central Michigan (Walton and Dobbs, 1980).
Company Affiliation: Polybac Corporation, Allentown, PA.
Background: A mixture of phenol and its chlorinated derivatives was spilled and reached the ground water.
Action Taken: Activated carbon filters were used to reduce the high concentrations of pollutants to levels tolerable by microbes. Mutant bacteria were injected into the soil and a surface runoff containment pond to degrade the pollutants.
Effectiveness: Degradation of phenol was rapid, but slower for o-chlorophenol. Once again the absence of carefully controlled experiments prevented any firm conclusions on the role of the mutant bacteria, although the data suggested that there was a correlation between the removal of the contaminants and the introduction of microbes.

(10) *Location*: Midwest (Quince and Gardner, 1982a and 1982b).
Company Affiliation: O. H. Materials, Findlay, OH.
Background: A transportation accident spilled more than 100,000 gal of various organic compounds including ethylene glycol and propyl acetate over a 250,000-ft² area. Three of the compounds penetrated the soil rapidly and were not removed when the contaminated surface soils were excavated. The soil consisted of a thick silty clay that extended to a depth of 50 ft. The water table was quite shallow and most of the contaminants were confined within it at a depth of 6–10 ft.
Action Taken: A series of 200 recovery wells were installed and the ground water pumped to a treatment system employing clarification, aeration, and granular activated carbon adsorption. Once levels of the contaminants fell below 200 ppm, a biostimulation program was begun using special bacteria, nutrients, and air which were injected into the surface.

Effectiveness: The system was able to reduce the concentrations of ethylene glycol from 1200 mg/1 and propyl acetate from 500 mg/1 to less than 50 mg/1 and the total concentrations of spilled compounds from 36,000 to less than 100 mg/1. After the treatment had reduced the concentration of the contaminants to levels acceptable to the regulatory agencies overseeing the treatment, a further monitoring program revealed that the contaminants were below detectable limits. The importance of the microbes in the removal of the contaminants cannot be judged, although the authors noted that the inoculation and biostimulation program reduced the time necessary to reach the contaminant levels that met the regulatory agencies' approval.

(11) *Location*: Unknown (Quince and Gardner, 1982a and 1982b).

Company Affiliation: O. H. Materials, Findlay, OH.

Background: Ground water contamination which resulted from leaks, overfills, and spillage at an industrial facility included dichlorobenzene, methylene chloride, and trichloroethane at levels up to 2,500 mg/1. The site geology consisted of a fractured limestone bedrock overlain by weathered glacial deposits.

Action Taken: The primary treatment was air stripping. The treated water was inoculated with a commercial hydrocarbon degrading bacteria and nutrients injected into the subsurface.

Effectiveness: The levels of bacteria increased until optimum conditions were established in the reactor and then were injected into the soil. In $2^{1}/_{2}$ months, the levels of methylene chloride fell from 2,500 mg/1 to less than 100 mg/1 and dichlorobenzene fell from 800 mg/1 to less than 50 mg/1 in a monitoring well. The inoculated bacteria were expected to continue to degrade the contaminants beyond the 95% reduction reached before the treatment was terminated. The importance of microbial activity could not be determined from the data presented.

(12) *Location*: Unknown (Ohneck and Gardner, 1982).

Company Affiliation: O. H. Materials, Findlay, OH.

Background: An accidental spill of 130,000 gal of organic chemicals entered a 15-ft-thick shallow unconfined aquifer and produced contaminant levels as high as 10,000 ppm. A 50- to 60-ft-thick gray silty clay separated the permeable aquifer from the main aquifer which was a major source of drinking water.

Action Taken: A treatment system employing clarification, granular activated carbon adsorption, air stripping, and reinjection was assembled. An investigation into stimulating microbial activity showed that the contaminants were biodegradable at concentrations below 1000 ppm. Both the indigenous microflora and a specific facultative hydrocarbon degrader were able to biodegrade the materials in a soil/water matrix rapidly when supplied with nutrients. A biodegradation program was initiated that inoculated the treated water with the specific hydrocarbon degraders, nutrients, and oxygen.

Effectiveness: Microbial growth responded positively to the presence of additional nutrients. The biological treatment process accelerated the removal of the compounds as shown by a series of soil borings during the treatment process and reduced the levels of contaminants in the ground water to less than 1 ppm. The specific hydrocarbon degraders did not increase degradation in laboratory tests beyond that of the native microbes and may not have significantly increased in situ biodegradation. No data were presented that showed that the number of the specific hydrocarbon degraders increased in the subsurface or that they were able to outcompete the indigenous microflora in degrading contaminants.

(13) *Location*: Mapleton, Illinois (Lindorff and Cartwright, 1977).
Company Affiliation: Illinois State Geological Survey, Urbana, IL.
Background: A train derailment spilled 20,000 gal of acrylonitrile which contaminated nearby soil and ground water and affected several drinking water wells.
Action Taken: Sixteen monitoring wells were installed at the site and five showed contamination. The contaminated ground water was withdrawn and sent to a sewage treatment plant for treatment. Five to eight feet of contaminated soil was also excavated and taken to a landfill for disposal.
Effectiveness: Ground water quality improved to the point where the case was closed. The effectiveness of the sewage treatment process was not discussed.

(14) *Location*: Leon, Kentucky (Wentsel et al., 1981).
Company Affiliation: Atlantic Research Corporation, Alexandria, VA; Consultant, O. H. Materials, Findlay, OH.
Background: Four railroad tank cars derailed near Leon, Kentucky, spilling 390,000 l of acrylonitrile. About one-third of this was recovered, and much of the remainder was burned or lost to the soil or a nearby river. The area had soils of a Muskingreen-Montivalle-Ramsey association.
Action Taken: O. H. Materials was hired as a consultant for the cleanup and installed 16 aeration ponds to treat the contaminated water. The treated water was then injected into the soil. When the concentrations of acrylonitrile fell below 3 ppb, the water was released to the nearby river.
Effectiveness: The treatment process reduced the levels of acrylonitrile to less than 2 ppb, but the importance of biological activity was not estimated.

(15) *Location*: Ott/Storey site in Muskegon, Michigan (Shuckrow and Pajak, 1982).
Company Affiliation: Michael Baker, Jr., Inc., Beaver, PA.
Background: The ground water at Ott/Story site is contaminated with a variety of volatile organics, phenols, aromatics, and other compounds at levels up to hundreds of mg/1 and has a TOC that ranges from 600 to 1500 mg/1.
Effectiveness: Activated sludge cultures could not be acclimated to the ground water even with the addition of nutrients and trace elements and adjustment of the pH or the addition of 10,000 mg/1 powdered activated carbon. A commercial microbial culture was also ineffective. Coupling aerobic activated sludge with granular activated carbon adsorption (GAC) was effective in removing 95% of the TOC when the GAC worked effectively, but as GAC removal efficiency declined, so did the overall treatment effectiveness. The levels of most of the organic priority pollutants were also removed to below the limit of detection (0.01 mg/1). Some of the removal in the activated sludge process was due to air stripping, but microbial activity accounted for a large portion. The combination of GAC and activated sludge did not work well for ground water from a less contaminated well having a TOC of only 200–300 mg/1.

SELECTED REFERENCES

Hager, D. G. and Loven, C. G., "Operating Experiences in the Containment and Purification of Groundwater at the Rocky Mountain Arsenal", *Proceedings of the National Conference on Management of Uncontrolled Hazardous*

Waste Sites, 1982, Hazardous Materials Control Research Institute, Silver Spring, Maryland, pp. 259-261.

Hittman Associates, Inc., "Surface and Ground Water Contamination from Abandoned Lead-Zinc Mines, Picher Mining District, Ottawa County, Oklahoma", Oct. 1981, Oklahoma Water Resources Board, Oklahoma City, Oklahoma.

Jamison, V. W., Raymond, R. L. and Hudson, J. O., Jr., "Biodegradation of High-Octane Gasoline in Groundwater", *Developments in Industrial Microbiology*, 1975, Vol. 16, pp. 305-311.

Jamison, V. W., Raymond, R. L. and Hudson, J. O., Jr., "Biodegradation of High Octane Gasoline", *In* J. M. Sharpley and A. M. Kaplan, ed., *Proceedings of Third International Biodegradation Symposium*, Applied Science Publishers, 1976, pp. 187-196.

Jhaveria, V. and Mazzacca, A. J., "Bioreclamation of Ground and Groundwater by the GDS Process", Groundwater Decontamination Systems, Inc., Waldwick, New Jersey, 1982.

Kasabach, N. and Althoff, W. F., personal communication, April 1983.

Knox, R. C., Stover, E. L. and Kincannon, D. F., "Examples of Aquifer Restoration", *Proceedings of the Ninth Annual Conference of the Groundwater Management Districts Association*, 1982, Scottsdale, Arizona, pp. 145-163.

Lindorff, D. E. and Cartwright, K., "Ground Water Contamination: Problems and Remedial Action", *Environmental Geology Notes*, May 1977, No. 81, Illinois State Geological Survey, Urbana, Illinois.

Minugh, E. M., et al., "A Case History: Cleanup of a Subsurface Leak of Refined Product", *Proceedings of 1983 Oil Spill Conference: Prevention, Behavior, Control and Cleanup*, San Antonio, Texas, Feb. 28-Mar. 3, 1983, pp. 397-403.

Nagel, G., et al., "Sanitation of Ground Water by Infiltration of Ozone Treated Water", *GWF-Wasser/Abwasser*, 1982, Vol. 123, pp. 399-407.

Ohneck, R. J. and Gardner, G. L., "Restoration of an Aquifer Contaminated by an Accidental Spill of Organic Chemicals", *Ground Water Monitoring Review*, Fall 1982, Reprint.

Polybac Corporation, Product Information Packet, Allentown, Pennsylvania, 1983.

Quince J. R. and Gardner, G. L., "Recovery and Treatment of Contaminated Ground Water: Part 1", *Ground Water Monitoring Review*, Fall 1982a, Reprint.

Quince, J. R. and Gardner, G. L., "Recovery and Treatment of Contaminated Ground Water: Part 2", *Ground Water Monitoring Review*, Fall 1982b, Reprint.

Raymond, R. L., Jamison, V. W. and Hudson, J. O., "Final Report on Beneficial Stimulation of Bacterial Activity in Ground Waters Containing Petroleum Products", Committee on Environmental Affairs, American Petroleum Institute, Washington, D.C., 1975a.

Raymond, R. L., Jamison, V. W., and Hudson, J. O., "Beneficial Stimulation of Bacterial Activity in Ground Waters Containing Petroleum Products", *AIChE Symposium Series*, 1976b, Vol. 73, pp. 390–404.

Raymond, R. L., et al., "Field Application of Subsurface Biodegradation of Gasoline in a Sand Formation — Final Report", American Petroleum Project No. 307-77, 1978.

Roy F. Weston, Inc., "Sylvester Hazardous Waste Dump Site Containment and Cleanup Assessment", Final Report Prepared for New Hampshire Water Supply and Pollution Control Commission, 1982.

Shuckrow, A. J. and Pajak, A. P., "Bench Scale Assessment of Concentration Technologies for Hazardous Aqueous Waste Treatment", In S. W. Shultz, ed., *Proceedings, Land Disposal of Hazardous Waste*, Philadelphia, Pennsylvania, Mar. 16–18, 1981, EPA-600/9-81-002b, U.S. Environmental Protection Agency, Cincinnati, Ohio, pp. 341–351.

Shuckrow, A. J. and Pajak, A. P., "Studies on Leachate and Groundwater Treatment at Three Problem Sites", In D. W. Shultz and D. Black, ed., *Proceedings, Land Disposal of Hazardous Waste*, Ft. Mitchell, Kentucky Mar. 8–10, 1982, EPA-600/9-82-002, U.S. Environmental Protection Agency, Cincinnati, Ohio, pp. 346–359.

Shukle, R. J., "Rocky Mountain Arsenal Ground Water Reclamation Program", *Proceedings of the Second National Symposium on Aquifer Restoration and Ground Water Monitoring*, May 26–28, 1982, pp. 366–368.

Stover, E. L. and Kincannon, D. F., "Treatability Studies for Aquifer Restoration", Presented at the 1982 Joint Annual Conference, Southwest and Texas Sections of America Water Works Association, Oct. 1982, Oklahoma City, Oklahoma.

Suntech Group, "Vapors Invade School; Bioreclamation Process Cleans Up Groundwater, Microorganisms Devour Spilled Hydrocarbons", *Petroleum Marketer*, July-Aug. 1978, Reprint.

Walton, G. C. and Dobbs, G., "Biodegradation of Hazardous Materials in Spill Situations", *Proceedings, 1980 National Conference on Control of Hazardous Material Spills*, 1980, Louisville, Kentucky, pp. 23–29.

Warner, J. W., "Digital-transport Model Study of Diisopropyl-Methylphosphonate (DIMP) Groundwater Contamination at the Rocky Mountain Arsenal, Colorado", U.S. Geological Survey, Denver, Colorado, 1979.

Wentsel, R. S., et al., "Restoring Hazardous Spill-Damaged Areas. Technique Identification/Assessment. Final Report", EPA-600/2-7-81-1-208, U.S. Environmental Protection Agency, Cincinnati, Ohio, 1981.

Annotated Bibliography for Ground Water Pollution Control

This appendix contains the abstracts of 225 references on both general and specific aspects of ground water quality protection and treatment. The contained references were identified via computer-based literature searches, planning and attendance at technical conferences, and personal contacts. The references are presented in alphabetical order by author. Due to the fact that some references address multiple aquifer restoration technologies, no classification of the contained references by technology has been developed.

Absalon, J.R. and Hockenbury, M.R., "Treatment Alternatives Evaluation for Aquifer Restoration", *Proceedings of the Third National Symposium on Aquifer Restoration and Ground Water Monitoring*, 1983, National Water Well Association, Worthington, Ohio, pp. 98-104.

This paper describes the essential steps necessary to adequately define a ground water contamination problem and then evaluate the technically feasible treatment alternatives from a performance, cost and O&M point of view. The approach described is based on actual field studies conducted in the last two years and the recommended solutions for a number of similar problems. The first critical step in solving a ground water contamination problem is the quantitative identification of the extent and nature of the problem. The primary objectives of the identification phase include:

- Define site environmental setting
  - Surface
  - Subsurface
- Locate potential contaminant sources

- Define potential contaminant migration pathways
- Determine contaminant extent and concentration

Given the ground water contamination information collected in the first phase of work, the study should proceed to an identification of treatment alternatives which may be useful for rehabilitating the aquifer. These would typically include physical, chemical, biological or other treatment processes depending on the chemical characteristics of the contaminated water and the level of treatment desired. Candidate treatment processes in each of the four categories which would be considered include:

- Physical treatment
  - Phase separation
  - Filtration
  - Gravity sedimentation

- Chemical or combined physical/chemical treatment
  - Chemical coagulation
  - pH adjustment
  - Carbon adsorption (liquid and vapor phase)
  - Air stripping
  - Steam stripping
  - Resin adsorption
  - Chemical oxidation

- Biological treatment
  - Aerobic fixed film
  - Aerobic suspended growth
  - Anaerobic

- Residuals treatment
  - Incineration
  - Sludge thickening/dewatering
  - Chemical recovery/reuse

Once the candidate treatment alternatives are identified, then bench- or pilot-scale treatability studies would be conducted to establish design criteria for the various treatment processes. These data will also demonstrate the performance capabilities of the various candidate treatment processes. In this paper, previously observed effectiveness on removal of organics (e.g., chlorinated solvents, hydrocarbons, etc.) will be presented for each treatment process. The results of these treatability studies are then used to develop a conceptual treatment process for the specific problem.

Adams, W.R., Jr. and Atwell, J.S., "A Dual Purpose Cleanup at a Superfund Site", *Proceedings of the National Conference on Management of Uncontrolled Hazardous Waste Sites*, 1983, Hazardous Materials Control Research Institute, Silver Spring, Maryland, pp. 352-353.

In this paper, the authors discuss the successful results generated by cooperative efforts between two state agencies, one faced with the need to construct a highway over a Superfund site and another responsible for the cleanup of that site. Also addressed are the engineering consultant's assessment of remedial alternatives and subsequent selection of the most cost-effective and environmentally sound solution to the problem posed by the construction. The hazardous waste site is the Pine Street Canal site in Burlington, Vermont, which was designated as Vermont's priority Superfund site because of its proximity to both business and residential areas and because of its location adjacent to Lake Champlain—the city's source of drinking water.

Alexander, W.J., Miller, D.G., Jr. and Seymour, R.A., "Mitigation of Subsurface Contamination by Hydrocarbons", *Proceedings of the National Conference on Management of Uncontrolled Hazardous Waste Sites*, 1982, Hazardous Materials Control Research Institute, Silver Spring, Maryland, pp. 107-110.

Subsurface contamination by hydrocarbons is a recurring problem. Sources range from home-fuel tanks to terminals and transcontinental pipelines. Subsurface releases may go undetected for months or years. An understanding of the various transport phenomena is beneficial in developing initial working hypotheses for a given site. Site-specific definition of the extent of contamination is ultimately necessary, however, for implementing cost-effective mitigative measures. In this paper, the authors describe two cases involving hydrocarbon leaks from terminals which have substantially different hydrogeological settings and consequently required different approaches for evaluation and mitigation.

Allen, R.D. and Parmele, C.S., "Treatment Technology for Removal of Dissolved Gasoline Components from Ground Water", *Proceedings of the Third National Symposium on Aquifer Restoration and Ground Water Monitoring*, 1983, National Water Well Association, Worthington, Ohio, pp. 51-59.

IT Enviroscience has conducted a technical and economic evaluation of technologies for removal of dissolved gasoline components from ground water. The components considered were benzene, toluene, xylene, ethylbenzene, t-butyl alcohol and methyl t-butyl ether. In an earlier literature search and evaluation, activated carbon adsorption and air stripping (or a combination of these technologies) were identified as the most likely treatment alternatives. IT Enviroscience conducted laboratory evaluations of these technologies in order to identify optimum design conditions. Activated carbon adsorption was evaluated over a range of feed concentrations, hydraulic loadings and regeneration conditions. The technology of steam regeneration was evaluated for regeneration of activated carbon used in this application.

Air stripping was evaluated over a range of feed concentrations, temperatures and gas/liquid ratios. This paper describes the results of the laboratory evaluations and identifies the optimum design and operating conditions, It shows the relative treatment and cost effectiveness of the carbon adsorption and air stripping technologies and combinations thereof for removing dissolved gasoline components from ground water.

Althoff, W.F., Cleary, R.W. and Roux, P.H., "Aquifer Decontamination for Volatile Organics: A Case History", *Ground Water*, Vol. 19, No. 5, Sept.-Oct. 1981, pp. 445-504.

Organic solvents, including 1,1,1-trichloroethane at concentrations up to 40,000 ppb, were detected in the Old Bridge aquifer under an industrial plant in South Brunswick Township, New Jersey. A hydrogeologic investigation defined the ground water flow system and a plume of contamination which extended, at a concentration of at least 100 ppb, for a distance of about 1000 feet downgradient of the plant. A contamination abatement system was designed and installed to prevent this plume from reaching a municipal well located about 2500 feet from the site. The system includes seven extraction wells and a water treatment facility. The locations of the extraction wells and their combined pumping rate were determined in part by a computer simulation of the aquifer, which was subsequently checked by a 20-day pumping test of the system. The on-site treatment facility uses two cooling towers in series to air-strip the volatile organics, and two infiltration ponds to return the treated water to the aquifer. The abatement system has been operating for about 10 months. Groundwater monitoring results show that the plume is now significantly smaller and less concentrated than before the abatement system was installed. It is projected that in several years the aquifer will be largely decontaminated.

Althoff, W.F. and Reuter, G., "Hydrocarbon Spills Into the Ground Waters of New Jersey: Two Case Histories", *Proceedings of the Conference on Oil and Hazardous Material Spills: Prevention, Control, Cleanup, Recovery, Disposal*, 1979, Hazardous Materials Control Research Institute, Silver Spring, Maryland, pp. 151-159.

Ground water contamination from hydrocarbons has only recently been given the attention these incidents warrant. Dramatic events involving explosions, evacuations or injuries receive the greatest publicity. But smaller spills, inventory losses, accidents and other causes occur with almost routine frequency and pass unnoticed by the general public. Typically, an organized response is lacking, large quantities of hydrocarbons are released into the environment and damage to life, property and water supplies all too often result.

The number and importance of these cases have forced national recognition of the problem. In New Jersey, for example, a coordinated spill response program has existed for several years. The need is clear. In fiscal 1978, more than 1400 separate incidents were reported within the state. Although many of these proved to be unimportant, the totals are startling. The amount of material spilled in this period is estimated at 1.5 million gallons. More than 75% of this total, or more than 1.1 million gallons, involved petroleum compounds. Problems related to hydrocarbon spills into ground water are presented through two case histories. The first incident involved gasoline contamination of a major bedrock aquifer and is, perhaps, an extreme example of the frustrations associated with such spills. In contrast, a successful recovery in unconsolidated material provides an illustration of the clean-up process and the effectiveness of the geologist in spill response programs.

American Petroleum Institute, "Underground Spill Cleanup Manual", API Publication 1628, June 1980, Washington, D.C.

When gasoline, kerosene, or similar petroleum product is lost into the ground, its behavior and ultimate fate will depend on the hydrogeology, or underground water conditions. Although spills involving hazardous materials or large volumes of fluids should be handled by an experienced professional familiar with this kind of problem, an understanding of certain elements of geology and ground water hydrology can be useful to anyone dealing with an underground spill. This manual discusses these elements and certain spill cleanup methods.

American Petroleum Institute, "The Migration of Petroleum Products in Soil and Groundwater—Principles and Countermeasures", Publication No. 4149, Dec. 1982, Washington, D.C.

This publication summarizes the methods being used by industry to control and prevent the migration of petroleum products in soil and ground water. The circumstances surrounding spills of petroleum products vary widely, and it is thus impossible to outline procedures that would be universally applicable. This publication does include, however, six case histories that describe procedures used in actual spills. These should serve as a useful guide to others who may encounter similar situations.

Anderson, D.C., Brown, K.W. and Green, J., "Organic Leachate Effects on the Permeability of Clay Liners", *Proceedings of the National Conference on Management of Uncontrolled Hazardous Waste Sites*, 1981, Hazardous Materials Control Research Institute, Silver Spring, Maryland, pp. 223-229.

Saturated conductivity or permeability is the primary laboratory measurement made on compacted clay soils to assess their suitability for use in constructing compacted clay liners for hazardous waste landfills or surface impoundments. The value obtained for the permeability is used to judge whether a compacted clay soil liner will prevent the movement of leachates into water bodies below or adjacent to the disposal facility. There is, however, little information available on the impact of organic fluids, likely to be placed in such confinements, on the permeability of the liner. Also, there has been no simple permeability test method developed that could be suitable for use with the range of possible waste fluids. Furthermore, a study evaluating the permeability of a range of typical clay soils with a spectrum of potential waste fluids would generate a valuable data base, useful for hazardous waste disposal permit applicants, writers and reviewers. Consequently this study was undertaken with the following objectives: (1) to construct from available information a delineation of the physical classes of fluid-bearing hazardous waste, the leachates they generate, and the predominant fluids in these leachates; (2) to develop a simple, inexpensive and rapid method for the comparative permeability testing of compacted clay soils, that would be suitable for use with a wide range of possible waste fluids; and (3) to evaluate the permeability of a range of typical clay soils to a spectrum of potential waste fluids.

Anderson, D.C. and Jones, S.G., "Clay Barrier-Leachate Interaction", *Proceedings of the National Conference on Management of Uncontrolled Hazardous Waste Sites*, 1983, Hazardous Materials Control Research Institute, Silver Spring, Maryland, pp. 154-160.

Interactions between clay barriers and leachate components may cause deterioration of the barrier's ability to contain the leachate. Two important dynamic clay barrier properties that may be affected by interactions with leachate are: effective pore volume, the pore space that transmits leachate percolating through a clay barrier; and permeability, the rate at which leachate percolates through a clay barrier at a given hydraulic gradient. Either a decrease in the effective pore volume at a constant permeability or an increase in the permeability of a clay barrier will result in an early appearance of leachate in the subsoil. Increased permeability of a barrier would result in an increased rate of leachate migration. A decrease in the volume of pores transmitting the leachate at a constant permeability would result in a decrease in the time it would take for the leachate to move through the barrier. A decrease in effective pore volume would occur if a clay barrier experienced shrinkage cracking. The leachate would preferentially move through the crack instead of through the whole clay barrier matrix. This would result in leachate breaching the clay barrier earlier than would be expected with a given permeability where the leachate moved uniformly through the clay barrier. Two of the most important potential failure mechanisms that involve barrier-leachate interaction are: dissolution and piping:

dissolution of clay barrier constituents that bind particles together and the subsequent piping (movement of clay particles with the percolating leachate); and development of soil structure: pulling together of soil particles into aggregations of many particles resulting in the formation of large interaggregate pores. By understanding the mechanisms by which clay barriers fail, it may be possible to alter leachate to lessen its impact on the dynamic properties of the clay barriers. Knowing leachate components likely to affect barrier properties makes it possible to identify wastes that require pretreatment. Discussions of these two clay barrier failure mechanisms and the leachate components that may cause these failures are in this paper.

Arlotta, S.V., Jr., "The 'Envirowall Cut-Off' Vertical Barrier", *Proceedings of the National Conference on Management of Uncontrolled Hazardous Waste Sites*, 1983, Hazardous Materials Control Research Institute, Silver Spring, Maryland, pp. 191-194.

The migration of polluted ground water or leachate from contaminated sites or waste disposal areas, especially abandoned sites, is an environmental problem of national proportions. A new and unique concept called the "Envirowall Cut-Off" has been developed to construct a vertical, impermeable barrier to prevent this migration. "Envirowall Cut-Off" is a hybrid cut-off wall. It is constructed with high density polyethylene (HDPE) and sand backfill and is installed by using the slurry trench construction method. Once installed, a very low permeability, composite, vertical barrier is established with several unique engineering properties. The overall concept and construction methodology was demonstrated during a full scale construction test at a sanitary landfill in New Brunswick, New Jersey during the fall of 1982. The test was planned to demonstrate the feasibility of the techniques and materials used to construct the Envirowall Cut-Off under generally difficult (by design) conditions. The test also demonstrated the major advantages of the Envirowall Cut-Off over other types of cut-off wall construction.

Ayres, J.A., Barverik, M.J. and Lager, D.C., "The First EPA Superfund Cutoff Wall: Design and Specifications", *Proceedings of the Third National Symposium on Aquifer Restoration and Ground Water Monitoring*, 1983, National Water Well Association, Worthington, Ohio, pp. 13-22.

Rapidly moving organic solvents with concentrations in the tenth of a percent range forced fast-track installation of the first cutoff wall to be funded under EPA's Superfund Program. The wall, located in Nashua, New Hampshire, totaled nearly 20,000 m^2 in cross section, reached a maximum depth of 33 meters and surrounded a plume more than 80,000 m^3 in area. Excavation was accomplished using a modified Koehring backhoe capable of trenching in excess of 20 meters deep in combination with a 12-ton clam shell at depths more than 20 meters. The 1-meter wide slurry trench excavation was keyed into bedrock and backfilled with a soil/bentonite mixture. The entire excavation and backfilling process was completed in less than eight weeks. The project necessitated slurry wall specifications based on performance criteria due to warrantees on materials, workmanship and long-term performance required of the contractor. Performance criteria were established for hydraulic conductivity of the wall ($\leq 1 \times 10^{-7}$ cm/sec); homogeneity of the backfill (no "windows"); long-term physical characteristics (cracking, shrinking, etc.); and the required depth of wall (top of bedrock). These performance criteria were further supplemented with minimum design standards to allow field control during construction. These included limits on the gradation of the backfill mix (≥ 5 percent bentonite and ≥ 30 percent fines); the type of bentonite used (prequalified suppliers); quality control testing, and final verification testing. Field observation and quality control testing results obtained during construction indicate satisfactory installation of the cutoff wall, with respect to both excavation to top of bedrock and backfill characteristics. An average backfill hydraulic conductivity of 5×10^{-8} cm/sec with a maximum value of 4×10^{-7} cm/sec was attained during construction.

Bailey, L.R., "Control and Recovery of Acrylonitrile from the Ground Water Near a Residential Area: A Case History", *Proceedings of the Conference on Oil and Hazardous Material Spills: Prevention, Control, Cleanup, Recovery, Disposal*, 1979, Hazardous Materials Control Research Institute, Silver Spring, Maryland, pp. 175-177.

The threat to life, property, and the environment when hazardous materials are spilled on the land surface or into water dictate that corrective action be immediate and effective. The technology for handling such occurrences is advancing rapidly as a result of the experience gained and documented in recent years. However, when hazardous fluids enter the subsurface and/or ground water as results of spills, the problem takes on broader ramifications. The time and cost of remedial measures are necessarily increased. Each situation is unique and depends upon the geologic and hydrologic conditions at the spill site, but basic principles governing the movement of ground water are the dominant controls on movement of the hazardous fluid. This paper presents a case history of a recovery operation involving a spill of 34,000 gallons of acrylonitrile onto the land surface, and subsequently a large portion of the product which entered the ground water system beneath the residential community of Guilford, Indiana. The community of Guilford, with a population of 97 persons, is served by a public water utility; however, a few of these residents have wells.

Bareis, D.L., Cook, L.R. and Parks, G.A., "Safety Plan for Construction of Remedial Actions", *Proceedings of the National Conference on Management of Uncontrolled Hazardous Waste Sites*, 1983, Hazardous Materials Control Research Institute, Silver Spring, Maryland, pp. 280-284.

   The USEPA and the Army Corps of Engineers have an interagency agreement for the Corps of Engineers to manage Superfund design and construction contracts and provide other technical assistance as requested in support of remedial action at hazardous waste sites. The Corps of Engineers primary responsibilities under this agreement take place when the USEPA has determined that Federal cleanup is appropriate (as opposed to remedial measures managed by states or private entities), and the USEPA has selected the remedial alternative to be implemented. The Corps then contracts for actual design and construction work and serves as the contract manager, reviewing designs and monitoring construction. A crucial part of successful remedial actions is insuring adequate safety and health precautions to protect both on-site employees and off-site populations. Construction of remedial measures often involves extensive physical disturbance of hazardous material; this creates hazards in addition to those normally encountered in site investigations. Another factor increasing the likelihood of a serious accident during construction activities involving hazardous waste is that construction workers often lack the specialized training given to site investigation teams. For these and other reasons, it became evident that existing safety manuals written to cover hazardous waste site investigations would not be directly applicable to construction activities. Therefore, the Corps of Engineers has developed guidelines for design contractors to prepare site-specific construction safety plans.

Barto, R.L. and Cleath, T.S., "Stringfellow: A Case History of Closure and Monitoring at a Hazardous Waste Disposal Site", *Proceedings of the Second National Symposium on Aquifer Restoration and Ground Water Monitoring*, 1982, National Water Well Association, Worthington, Ohio, pp. 323-339.

   Leaking of hazardous wastes from the Stringfellow Class I Disposal Site located in a small canyon near Riverside, California, has had an adverse effect on the adjacent ground water basin. The site received approximately 32 million gallons of plating wastes, industrial solvents and DDT manufacturing residue between 1956 and 1972. Operations at the site were discontinued after observations of nearby surface-water contamination in 1969 and ground water contamination in 1972. The California State Water Resources Control Board retained the firm of James M. Montgomery Consulting Engineers, Inc. in 1976 to study site conditions and evaluate alternatives for closure of the site. Hydrogeologic conditions were investigated using exploratory drilling, trenching, seismic refraction, soil analyses and water quality analyses. Results of the exploration confirmed that pollutants had contaminated the local ground water basin for a distance of more than 4,000 feet downstream from the site. Excessive concentrations of inorganic and organic constituents within the plume were indicated by total dissolved solids concentrations in excess of 50,000 mg/l, pH levels frequently less than 3.5 and total organic halogen concentrations in excess of 700 mg/l. Corrective measures were designed to inhibit subsurface flow from the site and extract contaminated ground water downstream. Remedial measures included a clay barrier, subsurface grout curtain and interceptor wells. The effectiveness of these measures will be evaluated by systematic analyses of ground water samples from a network of monitoring wells. Disposal alternatives of contaminated ground water include full treatment, hauling, deep well injection, evaporation and partial treatment with ocean outfall.

Bilello, L.J. and Dybevick, M.H., "A Computer Model for Optimization of Groundwater Decontamination", *Proceedings of the National Conference on Management of Uncontrolled Hazardous Waste Sites*, 1983, Hazardous Materials Control Research Institute, Silver Spring, Maryland, pp. 248-252.

The design and cost estimating methods used in this paper allow two processes to be compared by specifying the waste stream flow rate and three parameters for each compound: partition coefficient, mass transfer coefficient and carbon adsorption capacity. For single-component waste streams, the most economical choice is usually clear; for waste streams with several contaminants for which there are varying response objectives, several alternatives, including combined technologies, may be evaluated quickly and on an equivalent basis to determine the most effective system or the most likely candidates for pilot study. In none of the cases evaluated for this paper were the economics of a combined system more favorable than a system using only one of the technologies.

Bituminous Coal Research, Inc., "Studies on Limestone Treatment of Acid Mine Drainage", Jan. 1970, Monroeville, Pennsylvania.

Four actual coal mine waters have been neutralized with limestone both on a batch scale and by utilizing a continuous flow apparatus. Variations in treatment procedure were necessary depending on the characteristics of the individual waters. A standardized test was established to evaluate the reactivity of the limestones. The following variables are of importance in evaluating limestones for coal mine water neutralization: (a) particle size, (b) Ca and Mg content, and (c) surface area. Ferrous iron oxidation has been accomplished with both synthetic and actual coal mine water at low pH in the presence of coal-derived activated carbon. Electrophoretic mobility studies on precipitates obtained by both lime and limestone neutralization of coal mine water yielded information which can be applied for more effective sludge removal. Magnetic sludges were prepared using two different iron-bearing waters. The conversion of precipitates to a magnetic form results in significant reductions in settled sludge volumes as well as increases in solids content. Data obtained in these studies indicate that the limestone process offers considerable promise for an improved lower cost method for treating several types of coal mine waters.

Bituminous Coal Research, Inc., "Studies of Limestone Treatment of Acid Mine Drainage: Part II", Dec. 1971, Monroeville, Pennsylvania.

Laboratory investigations have demonstrated the feasibility of using limestone in place of pure lime for treating acid mine drainage for the neutralization of acidity and removal of iron to acceptable limits. The process consists of: equalization; addition of pulverized limestone and mixing; aeration; slurry recirculation to the mixing area; sludge settling; and sludge dewatering and disposal. Total costs are given. Advantages and disadvantages of limestone over lime are also discussed.

Bixler, B., Hanson, B. and Langner, G., "Planning Superfund Remedial Actions", *Proceedings of the National Conference on Management of Uncontrolled Hazardous Waste Sites*, 1982, Hazardous Materials Control Research Institute, Silver Spring, Maryland, pp. 141-145.

The remedial action program is a key part of the Superfund mandate to clean up uncontrolled hazardous waste sites across the nation. The USEPA has developed a remedial planning process embodied in the National Contingency Plan (NCP) that ensures rapid, consistent, and rational decision-making on the appropriate extent of remedy at priority hazardous waste sites. In this paper, the authors review five elements of the remedial planning process: (1) the remedial action master plan, (2) remedial investigations, (3) feasibility studies, (4) selection of a remedial alternative, and (5) remedial design. For each of these activities, the authors discuss both the technical considerations and the procedural requirements for conducting and submitting a complete product. The emphasis in this paper is on remedial investigations and feasibility studies, which form the basis for determining the appropriate extent of remedy and the cost-effective remedial alternative at an uncontrolled hazardous waste site.

Blake, S.B. and Lewis, R.W., "Underground Oil Recovery", *Proceedings of the Second National Symposium on Aquifer Restoration and Ground Water Monitoring*, 1982, National Water Well Association, Worthington, Ohio, pp. 79-76.

When gasoline, kerosene or similar petroleum products are lost through leaks or spills to the ground, its behavior and ultimate fate become dependent on the local hydrogeologic conditions and physical properties of the petroleum products. If the soil in the spill areas has a high clay content and low permeability, the product will penetrate very slowly or not at all. However, a highly permeable sandy soil will allow rapid percolation of the product. In situations where oils are present and the local ground water is relatively shallow, the downward percolating products will tend to "accumulate" on the water table. The weight of the petroleum product will depress the ground water table surface and form an "oil pool" having a convex lens shape. With time, this mound will disperse laterally on the surface of the ground water. The rate of dispersion of any given product will depend on viscosity and gravity of the product; hydraulic properties of the aquifer; slope or gradient of the ground water table; and quantity of product that reaches the water table aquifer. For retrieval of fairly shallow accumulation of oil, techniques such as interception trenches, pumped trenches and French drain systems can be very effective. For recovery of products at depths greater than 15 feet, wells are generally the most cost-effective. In this instance techniques such as well points, single pumps with separators, and two-pump systems can be employed. In each oil recovery program the oil types and the hydrogeologic environment must be evaluated to determine which recovery method will be cost-effective and achieve the best results.

Blasland, W.V., Jr., et al., "The Fort Miller Site: Remedial Program for Securement of an Inactive Disposal Site Containing PCB's", *Proceedings of the National Conference on Management of Uncontrolled Hazardous Waste Sites*, 1981, Hazardous Materials Control Research Institute, Silver Spring, Maryland, pp. 215-222.

In September 1980, the General Electric Company and the New York State Department of Environmental Conservation (NYSDEC) signed the first major agreement in the nation in which a corporation agreed to pay for remedial action to clean up abandoned hazardous waste dump sites where, in the past, it had disposed of its industrial waste. The agreement involves remedial actions at seven land disposal sites in Saratoga, Rensselaer, and Washington Counties. General Electric has agreed to carry out all engineering studies and the necessary remedial action at four locations and has agreed to pay a percentage of the cost of the engineering studies and necessary remedial action at the three other sites. One of the sites included in the agreement is known as the Fort Miller Site. It is located in the Town of Fort Edward, Washington County, approximately 7.5 miles south of Hudson Falls, New York. This paper summarizes the remedial program for the Fort Miller Site.

Blasland, W.V., Jr., et al., "Evaluation of Groundwater Hydraulics with Respect to Remedial Design", *Proceedings of the National Conference on Management of Uncontrolled Hazardous Waste Sites*, 1983, Hazardous Materials Control Research Institute, Silver Spring, Maryland, pp. 123-129.

In a hydrogeologic investigation of the Kingsbury landfill, ground water contamination attributed to discharges from the refuse was documented. In order to abate significant current and future releases or migration of hazardous wastes from the site, a remedial design involving in-place containment was developed. The remedial design consisted of a ground water cutoff wall keyed into a clay layer beneath the site and a low permeable cap installed over the refuse. The purpose of this in-place containment design is to effectively isolate the site from the hydrogeologic environment and thereby mitigate off-site migration of contaminated ground water. An analysis of the projected post-construction ground water hydraulics of the site was an integral part of the remediation design process and provided necessary documentation of the effectiveness of the proposed remediation to state regulatory officials. This analysis of the projected post-construction ground water hydraulics provided: an evaluation of the remedial design effectiveness in isolating the site from the hydrogeologic environment; an evaluation of the potential off-site migration of contaminated ground water and its potential impacts on the surrounding hydrogeologic environment; and information critical to the design and the remediation measures.

Bond, F.W., Cole, C.R. and Sanning, D., "Evaluation of Remedial Action Alternatives – Demonstration/Application of Groundwater Modeling Technology", *Proceedings of the National Conference on Management of Uncontrolled Hazardous Waste Sites*, 1982, Hazardous Materials Control Research Institute, Silver Spring, Maryland, pp. 118-123.

The LaBounty landfill was an active chemical waste disposal site from 1953 to 1977. During this period, it is estimated that over 181,000 m³ of chemical wastes were disposed in this 4.86 ha site located in the flood plain of the Cedar River in Charles City, Iowa. Compounds that are known to have been dumped in significant quantities include arsenic, orthonitroaniline (ONA), 1,1,2-trichloroethane (TCE), phenols, and nitrobenzene. Disposal ceased in December 1977 following the discovery that wastes from the site were entering the river. Since that discovery, the LaBounty landfill has been under intensive investigation. One aspect of these investigations was the ground water modeling effort described in this paper. The overall objective of this modeling effort was to demonstrate the use of modeling technology for evaluating the effectiveness of existing and proposed remedial action alternatives at the site. The primary emphasis of the project was on technology demonstration which included: modeling of the ground water system and prediction of the movement of contaminants; development of criteria to determine which water sources impacted by the site would require remedial action and the level of action necessary to insure risks are at acceptable risk levels; and the evaluation of the costs and effectiveness of various remedial action alternatives.

Boutwell, S.H., et al., "A Model Based Methodology for Remedial Action Assessment at Hazardous Waste Sites", *Proceedings of the National Conference on Management of Uncontrolled Hazardous Waste Sites*, 1983, Hazardous Materials Control Research Institute, Silver Spring, Maryland, pp. 135-139.

In the past, selection and design of remedial actions for uncontrolled hazardous waste sites has largely been accomplished through field data collection, simple analyses and engineering judgment. Such approaches are generally satisfactory for most sites. There is a subset of sites, however, where conditions are so complex that simple analyses may not provide enough guidance to allow for the proper selection and design of remedial actions. Given the high economic costs to government and industry associated with site cleanup combined with the future societal cost of inadequate actions, it is important that effective, economical actions be chosen. Thus, there is a need for a remedial action assessment methodology which is flexible enough to provide the appropriate level of analysis and accurate enough to provide confidence in the decisions made. It is clear from experience to date that two levels of sophistication are needed in hazardous waste site and remedial action assessments: (1) simplified, "desktop" methods (e.g., nomographs, analytical equations and hand-held calculator programs) for use where resources such as data, time, money and user expertise are limited and, (2) comprehensive numerical models for use on complex sites that will require costly remedial actions. Consequently, a methodology is being developed that provides: (1) a comprehensive modeling capability by interfacing sophisticated models for the three hydrologic zones, (2) flexibility by incorporating simplified, "desk top" methods which require limited resources in their application and (3) guidance on when to apply each method. In this paper the authors focus on the first issue: the selection and interfacing of sophisticated models.

Bradley, E., "Trichloroethylene in the Ground-Water Supply of Pease Air Force Base, Portsmouth, New Hampshire", Water Resources Investigations Open-File Report 80-557, 1980, U.S. Geological Survey, Boston, Massachusetts.

TCE (trichloroethylene) in concentrations that may be hazardous to health occurs in the water of an ice-contact largely sand and gravel aquifer (called the main aquifer) underlying much of the PAFB (Pease Air Force Base). In 1977 and 1978, the highest TCE concentration was found surrounding the most productive well, Haven well. Large quantities of TCE were used for degreasing between 1955 and 1965 and, so far as is known, TCE was used extensively up until 1973. Data on how, where, and when TCE got into the aquifer are lacking, but geochemical evidence and TCE analyses show a strong relation between the high TCE concentrations near Haven well and the extensive storm-drain system that underlies the parking apron and runway. The drains help recycle some of the TCE-contaminated ground water from the Haven well area. This study locates TCE contamination in the main PAFB aquifer and discusses three possible water-use alternatives: (1) abandonment of Haven well, (2) treatment of Haven well water to remove TCE, and (3) treatment of a portion of Haven well water for domestic consumption and use of the remainder for nondomestic uses. Continued use of the Haven well, with treatment of all or part of its yield for TCE removal, is advantageous, both because the Haven well is a high-yield source of water and because cessation of pumping from it might allow TCE-contaminated ground water to move downgradient (southward), endangering other wells.

Brown, K.W. and Anderson, D., "Effect of Organic Chemicals on Clay Liner Permeability: A Review of the Literature", *Proceedings of the Sixth Annual Research Symposium on Disposal of Hazardous Waste*, 1980, U.S. Environmental Protection Agency, Cincinnati, Ohio, pp. 123-134.

Permeability is the primary criteria for evaluating the suitability of clay soils for lining of waste impounds. A review of available information has revealed the near-complete lack of knowledge about the possible impact typical waste impoundment contents have on the permeability of clay liners. The aims of this review have been to develop a list of organic chemicals found in waste impoundments; develop a list of clay minerals used to line impoundments, and review possible mechanisms for the failure of these liners.

Brunotts, V.A., et al., "Cost Effective Treatment of Priority Pollutant Compounds with Granular Activated Carbon", *Proceedings of the National Conference on Management of Uncontrolled Hazardous Waste Sites*, 1983, Hazardous Materials Control Research Institute, Silver Spring, Maryland, pp. 209-216.

Carbon adsorption has played a key role in many wastewater treatment installations in helping them meet their treatment objectives. Based on a decade of experience, the authors discuss the effectiveness of activated carbon for the removal of a variety of pollutants when utilized for treatment in a representative sample of 17 industrial projects. Data from 29 ground water and spill control applications is presented to demonstrate the efficiency of carbon adsorption in this area of treatment. The cost effectiveness of carbon adsorption, while accomplishing high removal efficiencies, is also discussed. In light of new treatment challenges (ground water contamination, for example) and future comprehensive priority pollutant regulations, Calgon has developed a new technology for analyzing contaminated water quickly. The newly developed Accelerated Column Test (ACT) has several advantages over conventional test methods in addition to the speed with which it can be run. The advantages of the ACT are discussed along with an example of how the ACT correlates to current testing methods.

Brunsing, T.P. and Cleary, M., "Isolation of Contaminated Ground Water by Slurry-Induced Ground Displacement", *Proceedings of the Third National Symposium on Aquifer Restoration and Ground Water Monitoring*, 1983, National Water Well Association, Worthington, Ohio, pp. 28-38.

Ground water supplies throughout the U.S. are threatened by leaching and migration of contaminants from numerous sources including solid, chemical and hazardous waste sites. Complete isolation of these sources and their corresponding high concentration migrating plumes is a major step in mitigating the danger of ground water contamination and is often a cost effective step in the remedial cleanup process. A technique referred to as "block displacement" has been developed for complete isolation of large tracts of contaminated ground. A full-scale demonstration was recently conducted on a 50,000 cubic foot block of unconsolidated soil. The block displacement technique places an impermeable barrier around and beneath the contaminated zone. In the process, the ground is physically displaced upward by multiple well injection of a barrier material in a slurry state. The block displacement process is composed of separate bottom barrier and perimeter barrier construction processes, each of which is described in detail. The interaction between perimeter and bottom construction processes is explained. Application guidelines including a "mini-displacement" test are outlined. A cost comparison with other cleanup methodologies and additional cost considerations of this system are also discussed.

Brunsing, T.P. and Grube, W.E., Jr., "A Block Displacement Technique to Isolate Uncontrolled Hazardous Waste Sites", *Proceedings of the National Conference on Management of Uncontrolled Hazardous Waste Sites*, 1982, Hazardous Materials Control Research Institute, Silver Spring, Maryland, pp. 249-253.

A technique for complete in situ isolation of uncontrolled hazardous waste sites has been developed. This technique is intended to emplace a seal at low cost around the sides as well as underneath contaminated ground. Demonstration of the Block Displacement Method (BDM) was conducted over a four month period adjacent to the Whitehouse Oil Pits site in White-house, Florida. The purpose of the demonstration project was to add this technique to the list of available construction technologies applicable to chemical waste remedial action by: demonstrating the BDM in the geologic conditions of an existing chemical waste site; developing BDM specifications for "A User's Guide for Evaluating Remedial Action Technologies"; evaluating the applicability of the BDM to existing site geologic/hydrologic conditions; and establishing BDM implementation procedures for remedial action designers and contractors.

Burt, E.M., "The Use, Abuse and Recovery of a Glacial Aquifer", *Ground Water*, Vol. 10, No. 1, Jan.-Feb. 1972, pp. 65-72.

The inter-relationships between an industrial plant and a shallow sand aquifer of glacial origin are described. These relationships include industrial and potable water supplies, industrial and human waste water disposal systems, the hydraulics of the pollution of the ground water aquifer and the types of corrective actions taken to re-establish the wise use of the groundwater resource. Also reviewed are the ground water movement-quality relationship, well designs, method of drilling, well redevelopment and ground water recharge.

Cavalli, N.J., Arlotta, S.V. and Druback, G., "Subsurface Pollution Containment with the 'Envirowall' Vertical Cutoff Barrier", *Proceedings of the Third National Symposium on Aquifer Restoration and Ground Water Monitoring*, 1983, National Water Well Association, Worthington, Ohio, pp. 23-27.

This paper presents a new and unique concept to construct a vertical impermeable barrier to prevent the migration of polluted ground water or leachate from a contaminated site or waste disposal area. Developed by the ICOS Corporation of America, "Envirowall" is a hybrid cutoff wall. It is constructed with high density polyethylene (HDPE) and sand backfill, and it is installed using the slurry trench construction method. Once installed, a very low permeability, composite, vertical barrier is established with several unique engineering properties. These properties and engineering analyses are described in the paper. The "Envirowall" essentially functions as a cutoff wall with redundant features for controlling horizontal seepage. The HDPE is used to form an envelope which lines the walls of an excavated trench. The bentonite slurry used during construction keeps the trench stable for placement of the HDPE and sand backfill and also forms a filter cake on the walls of the trench. This filter cake further decreases the permeability of the composite in-place barrier. The sand backfill, besides serving as ballast, provides an internal, porous medium for monitoring seepage and, if necessary, withdrawal of intruding pollutants.

Depending on site stratigraphy, cutoff walls are often keyed into an underlying soil stratum of low permeability to effectively seal off the horizontal migration of contaminated ground water. This is often done at landfill sites or hazardous waste disposal sites as a remedial measure. The "Envirowall" has several advantageous features when used in this capacity. Recent research indicates that changes in permeability of clay soil often used for constructing liners and vertical barriers can be a major concern upon exposure to certain chemicals and subsequent permeation. The HDPE envelope provides a stable, low permeable (10^{-12} cm/sec) membrane and, when installed in the trench, it provides improved resistance to permeation and low overall permeability for the "Envirowall" system. The slurry trench construction method allows installation of the "Envirowall" in areas with difficult site conditions and to relatively deep elevations. An important component of the "Envirowall" is the sand backfill. As the HDPE envelope is placed in its trench, it is backfilled with sand and water to submerge it and to help the HDPE conform to the shape of the trench. The sand provides a medium in which monitoring wells may be placed to monitor water quality within the wall on a long-term basis. If leakage is ever noted by detection of contamination in a monitoring well, the porous sand medium can be pumped to evacuate and control the contaminated water and prevent further migration.

A full-scale construction test project of the "Envirowall" was performed at an existing sanitary landfill in New Jersey in the fall of 1982, to demonstrate the overall construction procedure for fabricating the HDPE liner and placing it into a trench filled with slurry by backfilling with sand. The project was conducted in three phases: 1) the excavation of a triangular shaped trench filled with the bentonite slurry mixture through the landfill waste, 2) the cutting and seaming of HDPE material to form an envelope with the prescribed geometry, and 3) the placement of this HDPE envelope in the trench and subsequent backfilling with sand and water. The success of the operation under difficult conditions proved that the theory and construction concept of the "Envirowall" can be utilized to construct a composite cutoff wall. Much information was gained for use in future applications.

Chaffee, W.T. and Weimar, R.A., "Remedial Programs for Ground-Water Supplies Contaminated by Gasoline", *Proceedings of the Third National Symposium on Aquifer Restoration and Ground Water Monitoring*, 1983, National Water Well Association, Worthington, Ohio, pp. 39-46.

This paper describes programs for investigating and cleaning ground water supplies contaminated by gasoline. Two case histories where leaks from buried gasoline storage tanks at service stations impacted public and private drinking water supplies are discussed. The procedures taken to investigate the gasoline contamination, including installation of monitoring wells, methods of gasoline lens detection and contaminant plume detection are described. In describing techniques for detecting spills, a discussion of contaminant transport and the geophysical phenomena involved with gasoline movement is provided. Emphasis is placed on gasoline movement above the ground water surface in the capillary fringe. Various methods used in detecting contaminant plumes are discussed and relate to a specific case history. Final cleanup plans are described for both sites. The criteria used for assessment of remedial alternatives are similar for both sites. However, the final cleanup techniques and schedules are different. Methods for on-site gasoline removal and treatment are discussed in detail. In one case, the final cleanup method has been approved by local and state regulatory authorities and is in the final design stage. This system is composed primarily of a recirculation cell treatment technique with air stripping, vapor phase granular activated carbon (GAC) treatment and liquid phase GAC treatment for the hydrocarbon contamination in the recirculation cell water. A computer model designed for use on an interactive graphics computer system was developed to aid in the recirculation cell analysis.

Clark, T.P., "Survey of Ground Water Protection Methods for Illinois Landfills", *Ground Water*, Vol. 13, No. 4, July-Aug. 1975, pp. 321-331.

This paper summarizes the experience in Illinois of development of a rational program for protection of the State's ground and surface water from the indiscriminant disposal of solid wastes. Classes of solid waste sites recognized in the State and their position in terms of the hydrogeologic environment are presented. Means of controlling landfill leachate, either by natural renovation in subsurface materials or by engineered collection and treatment are discussed. Examples are presented which illustrate the in-field implementation of protective systems and monitoring devices for ground water in Illinois, which should be generally applicable to other areas of the country of similar geology and climate.

Cochran, S.R., et al., "Survey and Case Study Investigation of Remedial Actions at Uncontrolled Hazardous Waste Sites", *Proceedings of the National Conference on Management of Uncontrolled Hazardous Waste Sites*, 1982, Hazardous Materials Control Research Institute, Silver Spring, Maryland, pp. 131-135.

With the passage of the Comprehensive Environmental Response, Compensation, and Liability Act (Superfund) and the release of the National Contingency Plan, the USEPA has developed a systematic methodology for assigning liability and conducting remedial actions at uncontrolled hazardous waste management facilities. It is clear from past experiences such as Love Canal that short term and long term environmental and health related hazards exist when inadequate technologies are used during the handling and disposal of hazardous materials. Because new remediation techniques are continually evolving and known technologies are constantly being retrofitted and adapted for remedial actions use, the alternatives available for cleanup are continuing to expand. In order to assess the effectiveness and limitations of these remedial actions techniques, USEPA is conducting case study examinations of these technologies which have been employed in field situations.

Cohen, R.M. and Miller, W.J., "Use of Analytical Models for Evaluating Corrective Actions at Hazardous Waste Disposal Facilities", *Proceedings of the Third National Symposium on Aquifer Restoration and Ground Water Monitoring*, 1983, National Water Well Association, Worthington, Ohio, pp. 85-97.

Many land-based hazardous waste disposal facilities have been cited by the U.S. Environmental Protection Agency (EPA) as endangering human health and the environment because they are leaking hazardous contaminants into ground water. The operators of these facilities are being required to prevent contaminant migration and to institute measures to clean up contaminated ground water beyond a specified distance from the facility (i.e., the compliance point which is dictated by the EPA in the facility operating permit). As such, engineering solutions will be implemented as corrective actions to control contaminant migration and to remove contaminants in ground water beyond the compliance point. These measures will often be large scale and costly. Therefore, it is economically advantageous to evaluate the effectiveness of corrective actions prior to their implementation. The most common corrective actions include the installation of slurry cutoff walls, French drains, clay caps and interceptor wells. During the design phase of corrective actions, analytical models allow a first-cut evaluation of the effectiveness of a particular measure or a combination of measures. To exemplify this approach, a variety of analytical solutions applied to specific corrective action scenarios are presented with emphasis on their critical assumptions and limitations.

Cole, C.R. and McKown, G.L., "The Use of Mathematical Models to Assess and Design Remedial Action for Chemical Waste Sites", *Proceedings of the National Conference on Management of Uncontrolled Hazardous Waste Sites*, 1981, Hazardous Materials Control Research Institute, Silver Spring, Maryland, pp. 306-312.

It is estimated that there may be 30,000-50,000 old disposal sites and sites contaminated from spills in the United States, of which 1,000-2,000 are sufficiently contaminated to pose a hazard to the public. Restoration costs are likely to exceed $1,000,000 per site. Faced with the complexity and high costs of addressing such a large number of sites, it is clear that formalized methods are required to expedite the restoration process and identify the most cost-effective means of site remedial response. This paper describes a methodology which utilizes mathematical models describing fluid flow and contaminant transport. These constructs have been developed over a period of nearly two decades and employed to solve a variety of surface and ground water contamination problems. They offer the ability to organize, interpret and better understand what is presently occurring, and in addition, provide the ability to predict what will occur in the foreseeable future. These capabilities are of particular value in designing remedial action since, through the use of these tools, one can: (1) determine when a specific site is sufficiently characterized and understood such that when implemented the selected remediation program will be effective; (2) allow one to design a cost effective site characterization and data gathering effort; (3) rapidly and cost effectively assess a wide variety of remedial options without moving any dirt; and (4) design a cost effective site surveillance and monitoring plan which will maximize the ability to detect any deviation from the remediation plan.

Continental Oil Company, "Microbiological Treatment of Acid Mine Drainage Waters", Sept. 1971, Ponca City, Oklahoma.

The purpose of the study was to determine if the abilities of certain bacteria to oxidize ferrous iron or to convert sulfate to hydrogen sulfide could be applied to the neutralization and subsequent removal of iron from difficult-to-treat mine drainage waters. If one or both of these concepts could be successfully utilized, the expense of adequately treating these types of problem waters might be significantly reduced. Laboratory studies demonstrated that both pure cultures and fresh field cultures of acidophilic iron bacteria could readily oxidize ferrous iron in both synthetic and natural acid mine drainage waters. Approximate requirements of oxygen, carbon dioxide, nitrogen and phosphorus by the iron bacteria were established. Limestone neutralizations of partially oxidized acid mine waters were conducted. Although sulfate-reducing bacteria were isolated, attempts to grow the cultures or produce hydrogen sulfide at pH values below 5.5 were unsuccessful.

Cooper, J.W. and Shultz, D.W., "Development and Demonstration of Systems to Retrofit Existing Liquid Surface Impoundment Facilities with Synthetic Membrane", *Proceedings of the National Conference on Management of Uncontrolled Hazardous Waste Sites*, 1982, Hazardous Materials Control Research Institute, Silver Spring, Maryland, pp. 244-248.

Surface impoundment facilities such as pits, ponds, and lagoons are used extensively throughout the United States to store, treat and dispose of hazardous wastes. These impoundments usually are designed to contain fluids, utilizing native materials or liners and in the past, many were constructed with little concern for preventing seepage from the facility. In fact, many such facilities were designed to seep water into the ground as a means of disposal. Unfortunately, many of these facilities are presently causing ground water contamination problems. With the establishment of regulations passed under authority of the Resource Conservation and Recovery Act (RCRA) legislation, retrofitting in-service disposal and treatment impounds may be required to prevent continued seepage. Retrofitting with impermeable liners may accomplish this objective. Liners may also be required to protect against wind and water erosion, another objective of

RCRA regulation requirements. Retrofitting an in-service impoundment with a flexible membrane liner generally implies the emplacement of liner material on the bottom and sides of a facility to mitigate leakage without removing any in situ fluid (hazardous material) in the impoundment. Such action may be desirable as a temporary measure, while a new impoundment is constructed, or as a permanent step to upgrade an untrustworthy facility. In this paper, the authors report on an ongoing project that Southwest Research Institute is conducting for the USEPA, for the development and demonstration of systems to retrofit existing liquid surface impoundment facilities with synthetic membrane liners.

Crystal, R.M. and Heeley, R.W., "Comprehensive Aquifer Evaluation and Land-Use Planning for Aquifer Restoration and Protection in Acton, Massachusetts", *Proceedings of the Third National Symposium on Aquifer Restoration and Ground Water Monitoring*, 1983, National Water Well Association, Worthington, Ohio, pp. 175-185.

A crucial, but often neglected, part of aquifer restoration programs should be to ensure that contamination incidents are not repeated. Acton, Massachusetts, moved to protect its ground water after losing 40 percent of its water supply in 1978 due to organic chemical pollution by a W.R. Grace chemical plant. Grace signed consent decrees with EPA and the town to pay for restoration of the Sinking Pond aquifer, to be coordinated with townwide, town-financed planning and geohydrologic studies. Acton engaged Lycott Environmental Research to conduct a two-part study: a geohydrologic analysis to evaluate ground water quality and quantity and identify alternative supplies, and a planning assessment to develop regulations and actions to protect water supply. The geohydrologic analysis provided an extensive data base. Eight geohydrologic maps were compiled: data locations, bedrock topography, hydrologic materials, surficial geology, saturated thickness, transmissivity, water table elevation and water quality hazards. Twenty-five monitoring wells were installed and sampled, and a long-term monitoring program was developed. Potential well sites were identified. The planning analysis had three key steps: mapping the most important areas for protection, conducting a modeling analysis to determine if current and saturation land use would significantly elevate ground water nitrate levels, and revising zoning development regulations and management actions. An aquifer protection map was prepared for use in a zoning overlay district defining restricted uses. Engineering requirements and performance standards covering hazardous materials, fuel storage and solid and industrial waste disposal were proposed. Nitrate modeling indicated that the federal 10 mg/l standard would be exceeded in several aquifers at saturation development, but recommended zoning changes could meet the standard.

Acton is currently implementing the plan, giving priority to rezoning to restrict industry in the Sinking Pond aquifer so it can be used for water supply after restoration.

D'Appolonia, D.J., "Slurry Trench Cutoff Walls for Hazardous Waste Isolation", undated, Engineered Construction International, Inc., Pittsburgh, Pennsylvania.

Slurry trench cut-off walls have been used extensively in the past decade as foundation and embankment cut-offs for water retaining structures and to control seepage into excavations. More recently, slurry trench cut-offs have been successfully employed to prevent underground seepage of liquid wastes and to isolate leachate from solid wastes from contaminating ground water. Advances in technology have enabled construction of exceptionally low permeability cut-offs which remain stable in a contaminated environment and resistant to chemical attack by the waste being contained. The Resource Conservation and Recovery Act has resulted in stringent requirements for new disposal areas and imposes new requirements for upgrading isolation of existing and abandoned landfills. In many cases, slurry trench construction is the only practical means of isolating existing waste disposal areas short of excavating and transporting in-place waste materials to new disposal sites. This paper reviews the various systems for isolating hazardous waste materials and points out the geological conditions where slurry trench construction is most applicable. The important elements of slurry trench design and investigation necessary to insure an effective waste isolation barrier are reviewed. These include mainly the connection between the slurry trench cut-off and the underlying aquaclude and the design of the slurry trench backfill material itself. Important considerations include not only the permeability of the isolation barrier, but also the long-term effects on the engineering properties of the barrier resulting from contact with or leaching by the contained hazardous waste material.

Dobbs, R.A. and Cohen, J.M., "Carbon Adsorption Isotherms for Toxic Organics", EPA-600/8-80-023, Apr. 1980, U.S. Environmental Protection Agency, Cincinnati, Ohio.

An experimental protocol for measuring the activated carbon adsorption isotherm was developed and applied to a wide range of organic compounds. Methods for treatment of the isotherm data and a standard format for presentation of results are shown. In the early phase of the study selection of compounds for testing in the experimental program presented a formidable task. Initial selections were based on the following criteria: (1) annual quantity produced, (2) critical concentration required to produce an adverse environmental effect, (3) probability of occurrence in water or wastewater, (4) persistence in the water environment, and (5) solubility. During the course of the study the Occupational Safety and Health Administration's

(OSHA) list of regulated carcinogens and the U.S. Environmental Protection Agency's Consent Decree list of priority pollutants were developed. These compounds were added to those previously selected for the experimental phase of the study.

Dowiak, M.J., et al., "Selection, Installation, and Post-Closure Monitoring of a Low Permeability Cover Over a Hazardous Waste Disposal Facility", *Proceedings of the National Conference on Management of Uncontrolled Hazardous Waste Sites*, 1982, Hazardous Materials Control Research Institute, Silver Spring, Maryland, pp. 187-190.

A closure plan, recently designed and implemented for a hazardous waste disposal facility, included the installation of a low-permeability cover over the wastes. This facility, located in western Pennsylvania, consisted of a lime neutralization plant and a 26-acre sludge impoundment. The facility started operations prior to the existence of strict hazardous waste disposal regulations and was not constructed with a liner system to protect ground water. As part of the closure requirements of the Pennsylvania Department of Environmental Resources (PADER), a low-permeability cover as well as a leachate collection and treatment system had to be installed. In addition, the facility owner is responsible for monitoring aquifers beneath the site to determine changes in water quality and the effectiveness of the liner. Presented in this paper are the hydrogeologic conditions, the considerations that went into cover selection and design, the cover installation procedure and cost, and the ground water monitoring plan.

Duffala, D.S. and MacRoberts, P.B., "Implementation of Remedial Actions at Abandoned Hazardous Waste Disposal Sites", *Proceedings of the National Conference on Management of Uncontrolled Hazardous Waste Sites*, 1982, Hazardous Materials Control Research Institute, Silver Spring, Maryland, pp. 289-290.

In this paper, the authors describe current efforts to study, design, and implement remedial actions at abandoned hazardous waste disposal sites in USEPA Regions VI through X. They also discuss the legislative background and enabling authorities to provide such remedial activities, issues related to the objectives of Superfund, and investigations and feasibility studies at an abandoned site.

Ehrenfeld, J. and Bass, J., "Handbook for Evaluating Remedial Action Technology Plans", EPA-600/2-83-076, Aug. 1983, U.S. Environmental Protection Agency, Cincinnati, Ohio.

There are four major exposure pathways for uncontrolled hazardous waste disposal sites: (1) ground water/leachate; (2) surface water; (3) contaminated soils and residual waste; and (4) air. Remedial action technologies are designed to reduce exposure to humans and the environment to acceptable levels by either containing hazardous materials in place or removing the intrinsic hazard by decontaminating or physically removing the hazardous substances. This report contains information on over 50 remedial action technologies. A brief description, status, factors for determining feasibility and reliability, principal data requirements, and basic information for cost review are given for each technology. In addition a general discussion of the major pathways and associated remedial approaches and of monitoring techniques has been included.

Ehrlich, G.G., et al., "Degradation of Phenolic Contaminants in Ground Water by Anaerobic Bacteria: St. Louis Park, Minnesota", *Ground Water*, Vol. 20, No. 6, Nov.-Dec. 1982, pp. 703-710.

Coal-tar derivatives from a coal-tar distillation and wood-treating plant that operated from 1981 to 1972 at St. Louis Park, Minnesota contaminated the near-surface ground water. Solutions of phenolic compounds and a water-immiscible mixture of polynuclear aromatic compounds accumulated in wetlands near the plant site and entered the aquifer. The concentration of phenolic compounds in the aqueous phase under the wetlands is about 30 mg/l but decreases to less than 0.2 mg/l at a distance of 430 m immediately downgradient from the source. Concentrations of naphthalene (the predominant polynuclear compound in the ground water) and sodium (selected as a conservative tracer) range from about 20 mg/l and 430 mg/l in the aqueous phase at the source to about 2 mg/l and 120 mg/l at 430 m downgradient, respectively. Phenolic compounds and naphthalene are disappearing faster than expected if only dilution were occurring. Sorption of phenolic compounds on aquifer sediments is negligible but naphthalene is slightly sorbed. Anaerobic biodegradation of phenolic compounds is primarily responsible for the observed attenuation. Methane was found only in water samples from the contaminated zone (2-20 mg/l). Methane-producing bacteria were found only in water from the contaminated zone. Methane was produced in laboratory cultures of contaminated water inoculated with bacteria from the contaminated zone. Evidence for anaerobic biodegradation of naphthalene under either field or laboratory conditions was not obtained.

Emig, D.K., "The Department of Defense's Installation Restoration Program", *Proceedings of the National Conference on Management of Uncontrolled Hazardous Waste Sites*, 1982, Hazardous Materials Control Research Institute, Silver Spring, Maryland, pp. 128-130.

The Department of Defense (DoD) began its Installation Restoration (IR) program in 1975. The IR program is a comprehensive program to identify and evaluate past DoD hazardous waste disposal sites on DoD installations, and to control the migration of contamination resulting from such operations. The IR program also is applicable to property that is excess to DoD requirements and which might be made available for other public or private use. The objectives of the DoD restoration program are: to identify and evaluate past hazardous material disposal sites on DoD facilities, to control contamination migration, and hazards to health and welfare; and to review and decontaminate as necessary land and facilities excess to DoD's mission. DoD has required that the military departments and the Defense Logistics Agency establish and operate installation restoration programs, and complete records searches at every installation listed on their priority lists by the end of FY 1985. They have also been required to develop and maintain a priority list of contaminated installations and facilities requiring remedial action. In the IR program, priority is given to control of migrating contamination that may pose a threat to the public health and welfare of surrounding communities or to on-post personnel.

Emrich, G.H. and Beck, W.W., "Top-Sealing to Minimize Leachate Generation: Case Study of the Windham, Connecticut Landfill", *Proceedings of the National Conference on Management of Uncontrolled Hazardous Waste Sites*, 1980, Hazardous Materials Control Research Institute, Silver Spring, Maryland, pp. 135-140.

It has been estimated that 15,000 landfills are now operating in the United States. The Resource Conservation and Recovery Act (RCRA) of 1976 requires that landfills not meeting criteria for solid waste disposal facilities and practices be closed. There are numerous remedial action alternatives that can be used during or after the closure of landfills and dumps. Some of these are passive requiring little or no maintenance after being emplaced. Others are active and require a continuing input of manpower and/or electricity. Techniques for the reduction or elimination of water movement into landfills may be considered in five categories: (1) surface water control; (2) passive ground water management; (3) active ground water or plume management; (4) chemical immobilization of wastes; and (5) excavation and reburial.

Emrich, G.H., Beck, W.W. and Tolman, A.L., "Top-Sealing to Minimize Leachate Generation", undated, SMC-Martin, King of Prussia, Pennsylvania.

After evaluation of more than 400 landfills, the Windham, CT landfill was selected for the implementation of a remedial action program. For the present study, SMC-Martin drilled additional wells to determine the thickness of the refuse, the depth of the water table under the landfill, the depth to bedrock and ground water quality. An electrical resistivity survey was conducted in order to define the areal extent of the leachate plume. Remedial action alternatives evaluated for this site included regrading, revegetation, surface sealing, ground water cutoff, and plume management. The first three methods proved to be the most cost effective. Suction lysimeters, pan lysimeters, staff gages, and monitoring wells were installed to determine water movement into the landfill and the effectiveness of the remedial measures. During the summer of 1979, a remedial program was implemented. The site was regraded, gas vents were installed, and 15 cm (6 in) of sand and gravel were emplaced with an additional layer of 10 cm (4 in) of fine-grained sand washings to protect the surface seal. A 20-mil PVC membrane seal was emplaced and covered with 46 cm (18 in) of sand and gravel into which composted sewage sludge and leaves were disced. Revegetation will be accomplished by hydroseeding with a mixture of grasses. Monitoring is being conducted to determine the effectiveness of the membrane seal.

Environmental Research and Application, Inc., "Concentrated Mine Drainage Disposal into Sewage Treatment Systems", Sept. 1971, Wilton, Connecticut.

Studies were undertaken on a small scale to determine the effect liquid waste artificial iron-rich acid brines had on municipal sewage treatment processes. The brines were devised to simulate concentrates from treatment of acid mine drainage. At very high concentrations, the brines neutralized with lime give virtually complete removal of phosphate from primary effluent, activated sludge effluent, or anaerobic sludge digester decantate. The cost of the iron-rich acid brine produced from acid mine drainage by the reverse osmosis membrane treatment is estimated. Costs of transportation by rail, truck, and pipeline are also shown.

Ess, T.H. and Shih, C.S., "Multiattribute Decision Making for Remedial Action at Hazardous Waste Sites", *Proceedings of the National Conference on Management of Uncontrolled Hazardous Waste Sites*, 1981, Hazardous Materials Control Research Institute, Silver Spring, Maryland, pp. 230-237.

Many modern technology related problems, especially those involving man created risks, are extremely complex and uncertain. Large volumes of data and multiple conflicting objectives lead to the requirement for incorporating subjective judgments into the decision process. If the existence of adversary positions (i.e., a politicized conflict) is involved, then something other than just intuition must be relied on to find generally acceptable solutions. The problem area of hazardous waste management is just such a situation. Multiattribute decision analysis is an apparently effective tool for this type of problem. If is highly flexible, incorporates methods to handle uncertainty, multiple objectives, etc. and is a well developed technique. However, it is not without faults. The most glaring is its inability to properly treat the public's subjective perception of risk. In addition, conventional decision analysis can become extremely cumbersome if the problem being analyzed is complex and no readily available means of tree "pruning" exists.

An obvious answer to this is to somehow integrate multiattribute decision analysis with a quantitative risk assessment technique which addresses public perception. The basic steps entailed in such a unified approach are delineated below:

(1) Construct a decision tree for the potential course of remedial actions.
(2) Complete a detailed risk analysis.
   a. Determine the objective risk of each decision branch. (Note: According to some authorities this would be termed modeled risk.)
   b. Determine the appropriate risk referents.
   c. Use an objective risk vs. risk referent comparison to determine if a decision branch requires modification or should be eliminated from consideration (i.e., pruned).
(3) Conduct a sensitivity analysis of the risk comparison to determine which branches are only "marginally" acceptable.
(4) Complete the multiattribute decision analysis with the pruned tree.
(5) Conduct a sensitivity analysis of the "solution".

This paper explores this integrated technique in more detail.

Evans, J.C. and Fang, H.Y., "Geotechnical Aspects of the Design and Construction of Waste Containment Systems", *Proceedings of the National Conference on Management of Uncontrolled Hazardous Waste Sites*, 1982, Hazardous Materials Control Research Institute, Silver Spring, Maryland, pp. 175-182.

Disposal of hazardous and toxic wastes in the subsurface environment has resulted in the widespread application of geotechnology to the design and construction of waste containment systems. The adaptation of conventional passive ground water and surface water barriers to waste containment requires certain special considerations. In this paper, the authors present a systematic engineering approach to passive containment alternatives designed to mitigate contaminant migration. Certain design and construction aspects of the application of geotechnology to these systems are also presented. The three major passive containment components which are covered in detail are: 1) top seals, including covers of native clay, processed clay and polymeric (synthetic) membranes, 2) barrier walls, including soil-bentonite slurry trench walls, cement-bentonite slurry trench walls, and vibrating beam cutoff walls, and 3) bottom seals, including liners of native clay, processed clay, and polymeric membranes. Emphasis is placed on the geotechnical aspects of design and construction practices of waste encapsulation, including advantages and limitations. The functions of each of these components are defined. The design process is then detailed with emphasis on the difference between conventional applications versus waste containment applications (for example, slurry trench cutoff walls for waste containment must incorporate additional considerations in design as compared to their use to control ground water during excavation).

Many site investigations are conducted by hydrogeologists to obtain data for assessment of contaminant migration. Subsequently, waste containment systems may be engineered as part of a pollution abatement program. It is essential that the geotechnical engineering information required be obtained during the site investigation phase in order to allow for a thorough, yet cost-effective, engineering design. In this paper, the authors present a guide to the engineering data required to enable geologists to plan and conduct subsurface investigations which will provide geotechnical as well as hydrogeological data for design.

Fabolini, F. and Rice, P.A., "Biological Treatment of Acid Mine Water", July 1971, Syracuse University, Syracuse, New York.

A strain of Desulfovibrio Desulfuricans was used to test the feasibility of treating acid mine water biologically. Sulfate reduction rates were measured with lactate media at different temperatures, influent pH, and sulfate concentrations in batch and continuous reactors. Among the several additional organic substrates tested, digester sludge appears to be the most attractive because of its low cost and its neutralizing properties. Sulfate reduction in mixtures of acid mine water and digester sludge were demonstrated in a semi-continuous reactor and the sulfate reduction rates were measured. The controlling factors of sulfate reduction are discussed.

Fitzwater, P.L., Brassow, C.L. and Fetter, C.W., "Assessment of Ground Water Contamination and Remedial Action for a Hazardous Waste Facility in the Gulf Coast", *Proceedings of the Third National Symposium on Aquifer Restoration and Ground Water Monitoring*, 1983, National Water Well Association, Worthington, Ohio, pp. 135-141.

This paper outlines the contamination assessment and closure plan for a solid waste pit near Houston, Texas. The hydrogeologic investigation was initiated with a review of the literature and study of aerial photographs. This was followed by soil test borings and laboratory testing, installation of piezometers, and in situ testing of the hydraulic conductivity. The investigation defined the geometry of a paleochannel that ran beneath the site. This channel was the major pathway for ground water flow. The water-quality investigation determined the extent of contamination. Water samples were taken for eight indicator chemical constituents. Nested piezometers were installed at the facility to collect immiscible as well as miscible contaminants. The water-quality study concluded that contamination was restricted to areas adjacent to the earthen pit. No plume was found to extend more than a few yards away from the pit. The remedial action plan was designed in conjunction with a permanent closure plan for the earthen pit. The plan included remedial pumping adjacent to the earthen pit and the installation of slurry trenches across the three legs of the paleochannel. The earthen pit will be closed in place and the consolidation of solid waste will be accomplished by surcharging the waste with a load consisting of a permeable liner topped with 2 feet of sand. The sand blanket will allow free liquid to escape the solid waste and be collected in a perimeter drain. The final closure will include an impermeable liner and clay cap.

Fochman, E.G., "Biodegradation and Carbon Adsorption of Carcinogenic and Hazardous Organic Compounds", EPA-600/2-81-032, Mar. 1981, U.S. Environmental Protection Agency, Cincinnati, Ohio.

This research program was conducted to determine the capability of biological treatment and activated carbon adsorption to remove chemical carcinogens and other hazardous organic compounds from water and wastewater. Compounds studied were benzidine, 4-nitrobiphenyl, 3,3-'dichlorobenzidine, benzo(g,h)perylene, benzo-(a)pyrene, dibenzo(a,h)anthracene, dibenzo(a,h)anthracene, 2-acetyl-aminofluorene, benzo(k)fluoroanthene, 4-aminobiphenyl, 3,4-benzofluoranthene, and 4-dimethylaminoazobenzene. All of the compounds tested exhibited some degree of biological degradation. Carbon adsorption was also effective in removing the compounds from aqueous solution. Large polynuclear aromatic compounds exhibited reduced adsorption capacities due to exclusion from the small diameter pores of the carbon.

Ford, C.T. and Boyer, J.F., "Treatment of Ferrous Acid Mine Drainage with Activated Carbon", Jan. 1973, Bituminous Coal Research, Inc., Pittsburgh, Pennsylvania.

Laboratory studies were conducted with activated carbon as a catalyst for oxidation of ferrous iron in coal mine water. Batch tests and continuous flow tests were conducted to delineate process variables influencing the catalytic oxidation and to determine the number and types of coal mine water to which this process may be successfully applied. The following variables influence the removal of iron with activated carbon: (a) amount and particle size of the carbon; (b) pH, flow rate, concentration of iron, temperature, and total ionic strength of the water; and (c) aeration rate. Adsorption as well as oxidation are the mechanisms involved in iron removal by this process. An evaluation of this process indicated technique feasibility which would permit acid mine drainage neutralization using an inexpensive reagent, such as limestone.

Fryberger, J.S., "Rehabilitation of a Brine-polluted Aquifer", EPA-R2-72-014, Dec. 1972, U.S. Environmental Protection Agency, Washington, D.C.

A detailed investigation was made of one (among several noted) incident where a fresh-water aquifer has been polluted by accepted disposal of oil-field brine through an "evaporation" pit (an unlined earthen pit) and later a faulty disposal well. The present extent of the brine pollution is one square mile; however, it will spread to affect 4.5 square miles and will remain for over 250 years before being flushed naturally into the Red River. Detailed chemical analyses show changes in relative concentrations of constituents as the brine moves through the aquifer. Several rehabilitation methods are evaluated in detail, including controlled pumping to the Red River and deep-well disposal. None of the methods that are both technically feasible and permissible show a positive public benefit-cost ratio. Although real economic damage both present and future results from this brine pollution, rehabilitation is not now economically justified. The report emphasizes that greater effort is needed to prevent such pollution, which not only affects ground water resources but also affects water quality in interstate streams.

George, A.D. and Chadhuri, M., "Removal of Iron from Groundwater by Filtration through Coal", *Journal of the American Water Works Association*, Vol. 69, No. 7, July 1977, pp. 385-389.

Laboratory studies were made to compare the differences in coal and sand filtration for the removal of iron from ground water. The results were compared based on filter performance index (FPI) and retentivity. Water analysis was made on the basis of soluble iron and total ferrous iron. The authors report almost complete removal of iron in the coal and sand up to 26 hours. After 26 hours the sand filter effluent quality deteriorated. Further results at varying filter rates and influent iron concentrations showed the following: (1) coal filters can be operated effectively at 3 to 4 times the effective rate of sand filters; and (2) coal filters were more effective in removing soluble ferrous iron while both sand and coal were effective in removing ferric and insoluble ferrous iron.

Gibb, J.P., "Cost of Domestic Wells and Water Treatment in Illinois", *Ground Water*, Vol. 9, No. 5, Sept.-Oct. 1977, pp. 40-49.

This study provides cost information for private home ground water supply systems in Illinois. Relatively accurate cost predictions for different types and depths of wells, ranging in cost from about $150 to $2500, can be made from the graphs presented. The average cost of all wells studied is about $575. Cost data for pumping systems equipped with 10-gpm submersible pumps (approximately 50 percent of all collected data) show that the average cost of these systems is about $585, with 50 percent ranging between $400 and $680. The costs of treating water for domestic use also are summarized. Two graphs illustrate the monthly costs of softening and removing iron at varying monthly consumption rates and concentrations of hardness-forming minerals and iron. The monthly cost of continuous chlorination is calculated. Use of the data presented makes it possible to estimate the monthly costs of raw and treated water from a domestic ground water supply. Two maps show the probable costs of domestic raw water-supply systems from sand and gravel wells and bedrock wells throughout the State. For an average installation and domestic use rate in Illinois, the monthly cost of raw water is about $11.00, softened water $15.40, softened water treated for iron $22.00, and softened water treated for iron and chlorinated $25.00. Similar calculations for any type and depth of well, water quality, and treatment can be made from the information in this report. This material should provide adequate information for planning purposes and decision making in developing a desired domestic supply.

Giddings, M.T., "The Lycoming County, Pennsylvania Sanitary Landfill: State-of-the-Art in Ground Water Pollution", *Ground Water*, Vol. 15, No. 1, Jan.-Feb. 1977, pp. 5-14.

A 120-acre sanitary landfill was being developed for a six county area. A 20 mil (0.5 mm) PVC plastic liner lying between 2 protective layers of sand was used to prevent ground water contamination by landfill leachate. Sampling monitored the liner performance. A backup leachate treatment system existed on site to treat any detected contamination. The ground water flow at the site was confined by a thick glacial till within the underlying shale bedrock. The till was of low permeability and an artesian head existed within the bedrock system affording extra protection from ground water contamination. Operation of the landfill was by the area-fill method with 8 foot lifts up to a limit of 120 ft. To assure water quality a baseline monitoring program was undertaken to establish on-site conditions and off-site private well conditions. Two years of regulatory hearings allowed a good data base to be established. Four ground water quality monitoring wells were drilled to a depth of 150 feet (all penetrated at least 100 feet of bedrock). It was determined that the ground water is confined by the glacial till under artesian conditions. The artesian head varies from 9 feet (uphill) to 60 feet (downhill). Water exists at 80 feet below the ground surface. Analysis of "selected chemical parameters" was made from well samples. Monitoring of the ground water underdrain for conductivity was used to evaluate the efficiency of the liner. Should contamination be detected, the contaminated section was isolated and flow was diverted to a backup leachate treatment system.

Giddings, M.T., "The Utilization of a Ground Water Dam for Leachate Containment at a Landfill Site", *Proceedings of the Second National Symposium on Aquifer Restoration and Ground Water Monitoring*, 1982, National Water Well Association, Worthington, Ohio, pp. 23-29.

Where hydrogeologic conditions are suitable for utilization of a ground water dam for leachate containment, several advantages can be realized. Positive containment of leachate is achieved and inflow from adjacent water bodies is excluded, thereby minimizing total leachate quantity. Operating costs are limited to the pumpage cost for removal of the leachate-contaminated ground water behind the dam. Where suitable, native soils are available, construction costs can be minimized through on-site borrowing of clay to construct the dam without the need for purchasing artificial materials. Many site conditions will allow the use of standard construction equipment for excavation of the trench and for implacement of the dam so that landfill operation personnel can operate the equipment under proper engineering supervision. This case history example illustrates the utilization of a compacted clay till ground water dam from upgrading an existing landfill site to meet current design and operational requirements. On-site clay till was remolded to a permeability of 3×10^{-8} centimeters per second and keyed into a glacial till stratum underlying the landfill having an inplace permeability of 3×10^{-7} centimeters per second. Between 14 and 20 feet of sand and gravel alluvium were excavated and the dam constructed on the flood plain between the landfill and an adjacent river. Ground water quality monitoring wells located downgradient of the dam have indicated abatement of the previously existing ground water contamination due to leachate seepage from the landfill and have documented that the leachate withdrawal from a perforated pipe within a gravel envelope behind the dam is satisfactorily extracting leachate that is generated. The leachate is currently treated on an interim basis by recycling into the landfill, but will ultimately be discharged to an adjacent sanitary sewer interceptor located within the flood plain area between the ground water dam and the river.

Gillespie, D.P., Schauf, F.J. and Walsh, J.J., "Remedial Actions at Uncontrolled Hazardous Waste Sites: Survey and Case Study Investigation", *Proceedings of the Second National Symposium on Aquifer Restoration and Ground Water Monitoring*, 1982, National Water Well Association, Worthington, Ohio, pp. 369-374.

The purpose of this project was to investigate remedial actions applied at hazardous waste sites in the past and to demonstrate their success or failure before Superfund-sponsored cleanups were initiated. The specific objectives of this project were twofold. First, a comprehensive survey of uncontrolled hazardous waste disposal sites at which remedial actions had been practiced was performed. Remedial actions were found to have been implemented at many kinds of hazardous waste disposal facilities including some drum storage areas, incinerators and injection wells, but most frequently at landfills/dumps and surface impoundments (i.e., wastewater pits, ponds and lagoons). Contamination of numerous media included ground water, surface water, soil, air and the food chain. Waste types involved included mercury, arsenic, solvents, oil, tire wastes, organic and inorganic wastes and septic wastes. During the second phase of the project, nine sites were studied in detail to document typical pollution problems and remedial actions at such uncontrolled hazardous waste sites. Remedial action technologies employed at these sites would be generally classified as either 1) surface water containment measures or 2) more sophisticated active and passive ground water control measures which were used to contain further spread of contaminant materials on the ground surface. Selected other sites utilized ground water cleanup measures including two facilities which were subjected to ground water extraction for purposes of containing and treating contaminant plumes.

Gilmer, H. and Freestone, F.J., "Cleanup of an Oil and Mixed Chemical Spill at Dittmer, Missouri, April-May, 1977", *Proceedings of the 1978 National Conference on Control of Hazardous Material Spills*, 1978, Hazardous Materials Control Research Institute, Silver Spring, Maryland, pp. 131-134.

In response to neighbor complaints of odors, fire and obnoxious smoke and deteriorating creek quality, the Missouri Conservation Commission (MCC) and the Department of Natural Resources (MDNR) made a series of investigations of the Albert Harris Property located in Dittmer, Jefferson County. These investigations, which began March 24, 1977 and continued through April 1, revealed that Mr. Harris was operating an illegal solid waste disposal site in conjunction with a container reconditioning operation. As a result of exploratory trenching and site survey, several actions were decided upon, including stream by-passing, land-fill disposal, on-site treatment and analysis of the contaminated stream water, excavation and administrative support from the U.S. Coast Guard Gulf Strike Team. On Monday, April 11, the work began as personnel and equipment arrived on scene. By Tuesday, April 12, disposal arrangements were made at a chemical landfill in Illinois.

Contaminated water impounded on the stream was pumped to a carbon adsorption column built on a design provided in an EPA research publication on field improvised water treatment systems. The treatment system consists of three five-micron bag filters and a 100 ft^3 gravity flow carbon adsorption column built from a 12 ft long by 4 ft ID concrete culvert buried six feet in the ground on a reinforced concrete base. The culvert was filled to a depth of eight feet with carbon. The field-improvised treatment system was successfully designed, fabricated and made operational within nine days. One of the most difficult problems associated with long term post-spill treatment is not technical, however. The problem is defining willing parties to be responsible for continuing maintenance requirements. The long term effectiveness of the system is a direct function of the reliability of regularly scheduled maintenance by trained personnel.

Glover, E.W., "Containment of Contaminated Ground Water—An Overview", *Proceedings of the Second National Symposium on Aquifer Restoration and Ground Water Monitoring*, 1982, National Water Well Association, Worthington, Ohio, pp. 17-22.

The containment of contaminated ground water is a difficult and expensive proposition. The initial step in developing a containment plan is to conduct a detailed hydrogeologic investigation to assess the seriousness of the problem. With the hydrologic and geologic data developed from the assessment study, a remedial containment plan is developed. The concepts used in developing the hydrogeologic study plan are briefly reviewed. Using the data developed during the hydrogeological assessment, a containment plan is developed to meet the site-specific hydrogeological conditions and nature of the contaminant. This paper reviews the current technologies available for controlling contaminated ground water and presents conditions where each may be applicable. The paper also discusses the relative effectiveness of the methods and some of the more difficult problems associated with the proper installation of the various systems. The paper concludes with a case history for a typical installation.

Graybill, L., "Evolution of Practical On Site Above Ground Closures", *Proceedings of the National Conference on Management of Uncontrolled Hazardous Waste Sites*, 1983, Hazardous Materials Control Research Institute, Silver Spring, Maryland, pp. 275-279.

Most closure methods are based on surface or subsurface containment barriers. The waste, however, continues to be exposed to the same hydrodynamic environmental forces as in the case of secure landfills but without the design advantages of secure landfills. Since they are located on all kinds of geological formations, the integrity and useful life of the containment method can be expected to be much less than secure landfill. The above ground closure method discussed in this paper has successfully addressed the problems associated with typical containment approaches and can be executed at a fraction of the traditional closure methods. The above ground approach differs from traditional containment methods in the following ways:

(1) The above ground method utilizes high volume waste handling techniques developed by waste disposal firms. Use of construction methods and special pumping equipment allows the economical bulk handling, manipulation and exhumation of waste at large hazardous waste sites.

(2) The above ground method uses waste solidification processes that were also developed from operating experience. The waste solidifica-

tion processes allow the historic waste to be compacted into a load bearing material.

(3) The solidification and bulk handling capability allows the use of traditional civil engineering methods to design above ground closures that effectively address the environmental hydrodynamic forces operating against traditional landfills and on-site containments.

Above ground closure also offers the following advantages of conventional surface and subsurface approaches: (1) there is essentially zero risk of contaminating the uncontrolled medium of soil and ground water because the entire system is above ground, with buffer zones and failure detection systems between the waste and ground water; (2) easier access for maintenance; (3) top, walls and effluent can be visually inspected; (4) provides opportunity to detect any errant leachate prior to ground water contamination; (5) provides easy access for future manipulation of the waste for resource recovery and new treatment methodology; (6) versatility – the above ground cells can be used over low and high water tables and over clay and non-clay strata; (7) the design is in harmony with environmental, hydrodynamic pressure; therefore, no leachate collection system is necessary for pumping leachate up from the bottom of a landfill; (8) risk of ground water contamination is reduced in the event of geological action such as fault slippage, slumping or any other massive earth movement. Remedial action is easier, quicker and less costly. The above advantages, along with the long-term security, economics and widespread application make the above ground closure approach a powerful new tool for addressing many of the pressing hazardous waste problems in this country.

Gregg, D.O., "Protective Pumping to Reduce Aquifer Pollution, Glynn County, Georgia", *Ground Water*, Vol. 9, No. 5, Sept.-Oct. 1971, pp. 21-29.

Water-level declines in a principal artesian aquifer have created a head imbalance between the aquifer and an underlying brackish-water zone containing up to 4,550 mg/l chloride. The brackish-water zone leaks brackish water into the aquifer through several breaks in a confining unit. A relief well tapping the brackish-water zone was drilled near a suspected break and pumped at about 3,000 gpm to lower the potential in the zone and bring it into hydrostatic equilibrium with the aquifer. The pumping apparently succeeded in decreasing the rate of brackish-water leakage into the aquifer. Successive samples of water from a well tapping the aquifer and downgradient from the relief well showed a decrease in the chloride content. Several more relief wells may be necessary to ultimately control chloride contamination of the aquifer.

Gruber, P., "Evaluation of Ground-Water Contamination Associated with the Use of Organic Solvents at Romansville, Pennsylvania", *Proceedings of the Third National Symposium on Aquifer Restoration and Ground Water Monitoring*, 1983, National Water Well Association, Worthington, Ohio, pp. 69-75.

A new residential well in Romansville, Pennsylvania, produced water with an odd taste and smell. Analysis by Pennsylvania Department of Environmental Resources (PADER) revealed organic chemical concentrations ranging from 1 ppb to 11,000 ppb. Of 11 neighboring wells also tested, six showed concentrations of organic chemicals in excess of 100 ppb. The Richard M. Armstrong Co., which used trichloroethane as a degreaser, was identified as the probable source of contamination. Armstrong retained ERM to define the extent of the ground water degradation and develop an economical and technically feasible remedial action plan. ERM's work effort was designed in consultation with the PADER and Armstrong. It consisted of: an assessment of existing site data; on-site field mapping; collection of hydrological and climatological data (temperature, rainfall), and surface data (slope, percent vegetation) necessary to establish a water balance; excavation of trenches to bedrock in order to observe and describe the subsurface deposits and evidence of soil contamination; mapping of fracture traces from stereoscopic aerial photographs; drilling a new well to provide additional subsurface geological control and water quality monitoring points as well as to serve as a recharge/recovery well during remedial measures; and sampling and analysis of 58 neighboring house wells for selected organic chemicals. The remedial actions consisted of: carbon filter installation and maintenance; well purging; contaminated soil removal; and site renovation and monitoring. In addition to a carbon filter installed at the Armstrong plant, 12 carbon filters were installed in residences surrounding the facility. The treated water is monitored on a bimonthly basis. When the organic chemical concentration of the treated water exceeds the PADER recommended limits, all filters are recharged. The new well on the Armstrong site and the existing water supply wells are pumped to create a local cone of depression and prevent continued migration of ground water. Water quality monitoring has continued for nearly 18 months and has shown a slow but continual improvement in water quality in the affected wells.

Gushue, J.J., Ayres, J.E. and Snyder, A.J., "Hazardous Waste Site Investigation Sylvester Site, Nashua, New Hampshire", *Proceedings of the National Conference on Management of Uncontrolled Hazardous Waste Sites*, 1981, Hazardous Materials Control Research Institute, Silver Spring, Maryland, pp. 359-370.

In this paper, the authors describe the results of an engineering and hydrogeological investigation of the Sylvester Site. The investigation of the Sylvester Site and adjoining properties was designed to: (1) define the extent of soil, surface water and ground water contamination; (2) develop an engineering assessment of the hydrogeological and hydrological conditions governing contaminant movement from the site; and (3) evaluate alternative remedial actions on the basis of technical feasibility, estimated costs and environmental implications. The evaluation of remedial actions included a laboratory-scale assessment of the feasibility of removing contaminants from ground water in an on-site, above-ground water treatment plant. The final project report, including recommendations for remedial action, provided the basis for the first Federal/State cooperative agreement under Superfund, with $2.3 million being awarded to the State of New Hampshire in August 1981 for design and construction of remedial measures at the site.

Hager, D.G. and Loven, C.G., "Operating Experiences in the Containment and Purification of Groundwater at the Rocky Mountain Arsenal", *Proceedings of the National Conference on Management of Uncontrolled Hazardous Waste Sites*, 1982, Hazardous Materials Control Research Institute, Silver Spring, Maryland, pp. 259-261.

The Rocky Mountain Arsenal (RMA) was established by the United States Department of the Army in 1942 on approximately 20,000 acres northeast of Denver. Its mission has been to manufacture chemicals needed by the military branches of the U.S. Government. In more recent years the facilities have been used to destroy and detoxify obsolete chemical weapons and contaminated hardware as well. Since 1946, some of the facilities have been leased to a private contractor who has manufactured various types of insecticides and herbicides for commercial agricultural applications. In 1974, a portion of the RMA site was taken over by Stapleton Airport and additional land acquisition by the airport is expected in the near future. As a result of these manufacturing and demilitarization activities both soil and water have become contaminated with chemicals which are both inorganic and organic in nature. Many of the materials that were processed were either toxic and/or suspected carcinogens.

In 1974, it was discovered that several of the chemicals were being transported off site by ground and surface waters. To address this and other potential problems, a Program of Installation Restoration was established at RMA. The permanent facility for ground water containment at the North Boundary included a 6800 ft long clay slurry wall which was installed 200 ft inside the RMA perimeter. Dewatering and reinjection wells along the clay barrier remove water for treatment at a central facility and deliver the purified water back to the ground flow. The dewatering system is divided into three separate water collections systems which have waters of varying organic composition. It was anticipated that three adsorption process trains would provide added treatment flexibility than that afforded by treating a mixed influent. A pulsed bed adsorption system was selected for the permanent RMA North Boundary treatment facility. The system is comprised of three adsorbers in parallel, cartridge filters ahead of and following the adsorbers, storage vessels for both fresh and exhausted carbon and two pressure transfer vessels to move carbon between the equipment in a measured slurry volume.

Hager, D.G., et al., "Ground-Water Decontamination at the Rocky Mountain Arsenal", *Proceedings of the Third National Symposium on Aquifer Restoration and Ground Water Monitoring*, 1983, National Water Well Association, Worthington, Ohio, pp. 123-134.

Previous chemical production and demilitarization activities at the Rocky Mountain Arsenal near Denver, Colorado, have resulted in soil and ground water contamination at the site. Ground water flow from the arsenal now threatens nearby potable water wells. Since the initiation of an Installation Restoration Program in 1974 the migration of organic contaminants off-site has been mitigated by the installation of boundary ground water containment systems combined with granular activated carbon treatment of the ground water. Both hydraulic and slurry wall barriers have been installed to contain the flow of ground water. The water is then extracted, purified and recharged to the ground. These systems are described and operating results are presented. Ground water below the chemical production area lies in a low transmissivity zone and contains a variety of organic solvents including chloroform, trichloroethylene, carbon tetrachloride and other low molecular weight chlorinated hydrocarbons. One source control alternative under consideration in the Installation Restoration Program is to dewater, purify and recharge the ground water in the production area. To evaluate this approach a dynamic flow pilot system has been designed to examine several purification techniques for volatile organic contaminant removal. The pilot system will include aeration, catalyzed chemical oxidation and adsorption. In addition to these organic removal processes, several inorganic treatment processes will be included to assess the removal of heavy metals, hardness and silica. Preliminary studies have been conducted to provide design data for the 20 gpm pilot system. The results of these studies are reviewed and the design rationale for the pilot system is presented.

Hallberg, R.O. and Martinelli, R., "Vyredox—In Situ Purification of Ground Water", *Ground Water*, Vol. 14, No. 2, Mar.-Apr. 1975, pp. 88-93.

The abundance and relative purity of ground water guarantees its increase in usage. In some localities, the content of iron and manganese in ground water is so high that these metals must be removed before the water can be used for drinking or industrial purposes. Iron occurs in two states of oxidation in nature—the divalent (ferrous) and trivalent (ferric) forms. The Vyredox method developed in Finland and used now also in Sweden and some other countries oxidizes the ferrous ion, which is soluble in water, to the ferric ion, which is insoluble before the water enters the well. The Vyredox method achieves a high degree of oxidation in the strata around the well. The method makes use of iron-oxidizing bacteria and aeration wells. A number of aeration wells are placed in a ring around the supply well. Water is forced down the aeration wells but first it is degassed and then enriched with oxygen. The oxygen-rich water provides a suitable habitat for the iron-oxidizing bacteria which assist in the oxidation of ferrous iron. The process must be repeated at specific time intervals to avoid further increases of iron content. The process of precipitating iron in the aquifer has only a slight effect on aquifer permeability. Cloggage of the aquifer surrounding the well should not occur for a period many times longer than the life span of a typical well.

Halliburton Company, "Feasibility Study on the Application of Various Grouting Agents, Techniques and Methods to the Abatement of Mine Drainage Pollution. Part II. Selection and Recommendation of Twenty Mine Sites", Aug. 1967, Duncan, Oklahoma.

The study involved a survey of all drift mine openings in the Upper West Fork River sub-basin and the development of alternate remedial methods for the abatement of mine drainage pollution. Twenty mine sites were selected from the two hundred twenty-eight mines surveyed in the initial phase. These twenty sites are described in detail herein.

Halliburton Company, "Feasibility Study on the Application of Various Grouting Agents, Techniques and Methods to the Abatement of Mine Drainage Pollution. Part IV. Additional Laboratory and Field Tests for Evaluating and Improving Methods for Abating Mine Drainage Pollution", Aug. 1967, Duncan, Oklahoma.

The initial phase of this project involved the studies and investigations performed in the exploration of drift mines in the Upper West Fork River Sub-basin, West Virginia. Included as a part of the study was the application of various grouting agents and techniques to the abatement of mine drainage pollution. Part IV of the project is a modification of the original contract for the purpose of conducting additional field tests on methods and materials not previously evaluated in this project. The Part IV program includes an experimental test program, a seismic survey and study, and field construction tests and evaluation. The experimental test program was conducted for the purpose of determining improvements in techniques and materials previously used for remedial applications as well as consideration of different techniques which might be evaluated in field tests. The seismic survey and study included was for the purpose of locating hidden mine openings or thin high wall sections by seismic methods.

Halliburton Company, "Feasibility Study on the Application of Various Grouting Agents, Techniques and Methods to the Abatement of Mine Drainage Pollution. Part III. Plans, Specifications and Schedules for Remedial Construction at Mine No. 12-007A, Mine No. 62-067, Mines No. 64-014, 64-016, and 64-017", Nov. 1967, Duncan, Oklahoma.

The report is a continuation of previous reports concerning the studies and investigations performed in the exploration of drift mines and feasibility studies into the application of materials and material placement techniques toward the abatement of mine drainage pollution. Twenty drift mine sites were selected previously. Report II studies them. In the present report, 3 of those 20 mines were further studied. Detailed plans, specifications, and schedules for contracting and construction of pollution abatement remedial measures for each of the three selected sites is presented. This includes a general description of the three selected mines, the schedule, cost estimate and recommendations for each mine site, and the plans and specification sections for each mine.

Halliburton Company, "New Mine Sealing Techniques for Water Pollution Abatement", Mar. 1970, Duncan, Oklahoma.

The purpose was to develop and field test new concepts for watertight mine seal and bulkhead construction applicable to abatement of acid mine water pollution. Laboratory research determined proper materials, equipment and techniques for constructing mine seals or bulkheads. Field testing was conducted on remedial grouting techniques for constructing mine seals or bulkheads. Two new processes were developed. One involved a technique of placing a plug or graded limestone aggregate in a mine drift or portal to neutralize an acid mine water discharge until a seal was effected. The second process consisted of remotely constructing a mine seal including rear and front bulkheads of a self-supporting, quick-setting sodium silicate cement specifically developed for this application. A filler material of expansive cement was used between the bulkheads to complete the seal. Field testing in West Virginia substantiated the feasibility of both processes when two aggregate and two bulkhead type seals were placed in abandoned mines which had drainage flows up to 58 gallons per minute. Cost of the seals are reported.

Hammond, J.W. and Metry, A.A., "An Evaluation of In-Situ Groundwater Ion Exchange Barriers for Control of Low-Level Radioactive Wastes", undated, Roy F. Weston, Inc., West Chester, Pennsylvania.

This paper presents a methodology for control of ground water contamination resulting from shallow burial of low-level radioactive waste. Such sites fall in one of the following categories: (1) Formerly Utilized Site Remedial Action Project (FUSRAP); (2) Uranium Mill Tailings Remedial Action

Project (UMTRAP); and (3) Remedial Action of Inactive Commercial Low-Level Radio-active Waste Disposal Sites. In the past few years a great deal of experience has been gained in the design of low-level radioactive residuals disposal facilities and remedial action rehabilitation plans for existing low-level radioactive facilities and waste disposal sites. During many projects, the following have been identified as feasible techniques to protect public health and the environment: (1) use of ion-exchange media as part of the liner or barriers for radionuclides attenuation; (2) use of a cover system for radon gas control and attenuation; (3) containment and isolation of the site using a surface cap; (4) use of a liner system beneath waste material for containment and encapsulation; (5) hydrogeologic isolation of affected aquifer by bentonite slurry walls; (6) in situ grouting, solidification, conditioning, and chemical or thermal stabilization; (7) use of passive ground water treatment systems; (8) application of overall site management and closure techniques; and (9) environmental management, surveillance, and monitoring. For given projects, the optimum design may include various combinations of these techniques, depending on site-specific characteristics. The proper selection and specification of a technique requires an understanding of the principal involved and methodology used for evaluation. This paper presents a discussion of ion exchange properties of barriers or liners as passive means of attenuation of radionuclides and protection of ground water quality.

Hanson, J.B., et al., "Hazardous Substance Response Management Model", *Proceedings of the National Conference on Management of Uncontrolled Hazardous Waste Sites*, 1981, Hazardous Materials Control Research Institute, Silver Spring, Maryland, pp. 198-200.

The Environmental Protection Agency (EPA) is now using on an interim basis a Hazardous Substance Response Model as a management tool to lead governmental project managers through the activities related to cleaning uncontrolled releases of hazardous substances. Response funds are available primarily from the Comprehensive Environmental Response Compensation and Liability Act of 1980, P.O. 96-510 ("CERCLA" or "Superfund"). The Model presents in Flow Chart form a suggested master network of activities designed to orchestrate a cleanup project in a manner that approximates the process of response currently being used on an interim basis by the EPA. The Model, in its completed form, is currently being developed and will include management-related activities, community relations, government/agency coordination, contractor procurement, contractor monitoring and technical decision-making. The goals of the Model are: (1) to describe the activities necessary to manage cleanup of an uncontrolled hazardous waste site; and (2) to present this information in a form that is easy to understand, and which will assist those responsible for the day-to-day management of a project. In short, it is comprehensive in scope and designed for practical use.

The Model is a vehicle incorporating the requirements of CERCLA with the experience of the people who are well-versed in evaluating, planning and cleaning releases.

Harris, M.R. and Hansel, M.J., "Removal of Polynuclear Aromatic Hydrocarbons from Contaminated Groundwater", *Proceedings of the National Conference on Management of Uncontrolled Hazardous Waste Sites*, 1983, Hazardous Materials Control Research Institute, Silver Spring, Maryland, pp. 253-265.

Reilly Tar and Chemical Corporation operated a coal tar distillation and wood preserving plant in St. Louis Park, Minnesota, between 1918 and 1972. Reilly disposed of wastes from the operation consisting of a mixture of many compounds, including a class of organic compounds known as Polynuclear Aromatic Hydrocarbons (PAH), some of which are carcinogenic, in a network of ditches that discharged to an adjacent wetland. In the mid-1970's concerns were raised relative to PAH in ground water in the area. This paper presents bench-scale and pilot-scale testing data for a variety of water treatment technologies, illustrating the degree of treatment achievable with each technology. Influent and effluent concentrations are reported for 34 PAH and other coal tar derivative compounds. Three of the technologies tested (GAC, O_3/US and H_2O_2/UV) achieved compliance with project specific treatment goals and provided effluent water quality adequate for use in a potable water distribution system.

Harsh, K.M., "In Situ Neutralization of an Acrylonitrile Spill", *Proceedings of the 1978 National Conference on Control of Hazardous Material Spills*, 1978, Hazardous Materials Control Research Institute, Silver Spring, Maryland, pp. 187-189.

At 2:52 p.m., June 11, 1974, an east bound Chessie System Freight train hit an obstruction placed on the tracks by vandals and derailed about 6 miles east of Dayton, Ohio. A tank car containing 133,300 pounds of Acrylonitrile was punctured and caught fire. The plume from the fire was visible for 6-10 miles. Local fire departments and disaster agencies responded quickly and put out the fire by about 4:30 p.m. By 1:00 p.m., on June 14, the Ohio EPA had decided that the acrylonitrile would be neutralized by chemical action upon the cyanide portion of the acrylonitrile molecule. The oxidation of the cyanide/acrylonitrile would be accomplished by first raising the pH of the acrylonitrile contaminated area above 10, with lime and then spraying chlorine over the area. On Monday, June 17, and Wednesday, June 19, monitor wells were drilled on all sides of the spill. Samples were also taken of the affected area to determine if the neutralization had been successful. Samples taken for an extended time period from the monitor wells showed no acrylonitrile, and therefore the operation was a success.

Headworth, H.G. and Wilkinson, W.B., "Measures for the Protection and Rehabilitation of Aquifers in the United Kingdom", *Proceedings of Conference on Groundwater Quality: Measurement, Prediction and Protection*, 1976, Water Research Centre, University of Reading, Berkshire, England, pp. 760-817.

At present about 30% of public water supply in the United Kingdom is derived from ground water sources, and regional schemes for future development are now being planned and implemented. However, there is an increasing demand for these aquifers to fill a multiple role in water supply, waste disposal, and as sources of construction material. It is, therefore, appropriate that authorities concerned with water supply and waste disposal, and the associated organizations, are now directing their attention towards the best use of these aquifers while attempting to resolve the problems arising from their conflicting functions. Water authorities have a statutory duty and are currently engaged in reviewing the use and management of their water resources and in drawing up future plans. Thus there is an opportunity to prepare a long-term policy for the protection of ground water quality. This paper firstly categorizes the principal sources of pollution and the ground water development conditions that will lead to contamination of underground resources. The control and rehabilitation measures that may be adopted for existing and potential pollution situations are described and illustrated by a number of case histories.

Hemsley, W.T. and Koster, W.C., "Remedial Technique of Controlling and Treating Low Volume Leachate Discharge", *Proceedings of the National Conference on Management of Uncontrolled Hazardous Waste Sites*, 1980, Hazardous Materials Control Research Institute, Silver Spring, Maryland, pp. 141-146.

A landfill was ordered to be closed by the Pennsylvania Department of Environmental Resources. The landfill had an uncontrolled leachate discharge which the State ordered to be remedied. The problem of collecting leachate after-the-fact from certain areas of the landfill was strictly hydrogeological in nature. How much leachate was generated had to be determined for the design of a treatment system of adequate capacity. This was a question that could only be solved through hydrogeological analysis. It was determined that the peak discharge rate of leachate was 7.9 gpm/acre. Treatment of small volumes of highly concentrated leachate is by chemical addition, settling, and activated carbon filtration. This paper discusses the procedure of the hydrogeologic methods employed, and the treatment plant design including its capital and operating expenses.

Hill, R., Schomaker, N. and Wilder, I., "Uncontrolled Hazardous Waste Site Control Technology Evaluation Program", *Proceedings of the National Conference on Management of Uncontrolled Hazardous Waste Sites*, 1982, Hazardous Material Control Research Institute, Silver Spring, Maryland, pp. 233-236.

In anticipation of the passage of the Comprehensive Environmental Response, Compensation, and Liability Act of 1980 (CERCLA or Superfund), the Office of Research and Development, of the USEPA began a program in 1980 to support the Agency's activities concerned with uncontrolled hazardous waste sites. Since CERCLA only provided for a five year program for the uncontrolled hazardous waste site problem, time was not available to establish a fundamental research and development program. The approach taken by the Agency for the Office of Research and Development was one of technical support to the Office of Emergency and Remedial Response. Technologies that had been developed under the Clean Water and Solid Waste programs were adapted to the uncontrolled hazardous waste site situations. In addition, construction techniques, e.g., slurry trench cutoff walls, injection grouting, and chemical stabilization, that had been used for other purposes, were evaluated to determine their applications to uncontrolled sites. Also there were very limited data available on the cost and effectiveness of various remedial techniques. The task of collecting and analyzing the available data was initiated. Once CERCLA became law, the Office of Research and Development developed, in consort with the Office of Emergency and Remedial Response, a five year support strategy which is updated each year. The srategy outlined a program with peak funding in the early years to meet the immediate needs of the program; the latter years concentrates on technical assistance.

Holmes, J.G. and Kreusch, E.G., "Acid Mine Drainage Treatment by Ion Exchange", Nov. 1972, Culligan International Co., Northbrook, Illinois.

In areas where acid mine drainage has diminished the available water supplies, processes which are capable of producing potable water from this drainage are of interest. Ion exchange processes have this capability. The investigation was intended to study several conventional ion exchange processes, using commercially available materials and to determine if any of these processes could be used to produce potable water from acid mine drainage. The three best systems were found to be: strong acid cation exchanger regenerated with sulfuric acid; weak base anion exchanger regenerated with caustic soda; and weak base anion exchanger regenerated with sodium hydroxide and carbon dioxide (modified Desal process). Treatment plants in three sizes were designed for each system so that cost estimates could be established.

Horizons, Inc., "Foam Separation of Acid Mine Drainage", Oct. 1971, Cleveland, Ohio.

Laboratory studies of continuous flow foam separation were conducted to determine the optimum conditions for maximum extraction of dissolved metal cations (Fe, Ca, Mg, Mn, and Al) from acid mine drainage. Foaming experiments were conducted in a 6 in.-diameter glass column capable of liquid flow rates of 3-12 gal. per hour. The approach to foam separation taken was the production of the persistent foams which allowed protracted foam drainage to reduce liquid carry-over in the foam. The effects of pH, chelate addition, surfactant type and concentration, air sparging rate, metal concentration and foam drainage were investigated in relation to metal extration. Results show that sewage foamability is too low for foam separation alone to be a feasible sewage treatment method.

Huibregtse, K.R. and Kastman, K.H., "Development of a System to Protect Groundwater Threatened by Hazardous Spills on Land", EPA-600/2-81-085, May 1981, U.S. Environmental Protection Agency, Cincinnati, Ohio.

Treatment of hazardous materials spills onto land that create threats to the quality of ground water is frequently limited to excavation or flushing of the area with water. The purpose of this project was to establish an alternative approach to the solution of the problem. The objective was to develop, design and construct a vehicle capable of containing about a 40 m³ (10,000 gal) hazardous spill on contaminated land and treating the polluted soils by oxidation/reduction, neutralization, precipitation or polymerization. The approach was to use direct injection of grout into the soil around the contaminated area to envelop the spill and isolate it from the ground water, nd detoxify the spill area by injection of treatment agents. Since information regarding this approach was minimal, both laboratory and pilot-scale tests were performed before the unit was designed, constructed and delivered to the U.S. Environmental Protection Agency in Edison, New Jersey.

Treatability tests were performed on four soil types (gravel, sand, silt and clay) with three reaction types (oxidation/reduction, neutralization and precipitation). Flow-through tests were performed to simulate conditions in gravel, sand and silt while sealed tests were performed to approximate conditions in clay soils. Detoxification in flow-through tests exceeded 99 percent of the initial contaminant added and appeared to be the result of both displacement and chemical reaction. Sealed test detoxification efficiencies in clays ranged from 56.5 to 99.5 percent. Formation of insoluble products did not adversely affect detoxification. A simultaneous evaluation of grout formulations and injection devices indicated that grouting was most applicable in gravel, sand and some silty soils. The direct driving of heavy duty pipe is recommended to minimize short-circuiting and to lower costs. Two types of grout (bentonite/cement and sodium silicate) were selected as most applicable.

A mobile in situ containment/treatment unit was designed, constructed and mechanically tested before it was delivered to EPA. Major components located on a 12.1 m (43 ft) drop deck trailer include: a diesel electric generator, an air compressor, two 2.46 m3 (650 gal) lined tanks with mixers, four 0 to 30 1pm (0 to 8 gpm) progressive cavity pumps, four 0 to 76 lpm (0 to 20 gpm) air diaphragm pumps, a control panel, flow controls and a solids feed conveyor. Decision guidelines for unit mobilization and an operations and maintenance manual were provided with the containment treatment unit.

Intorre, B.J., et al., "Evaluation of Ion Exchange Processes for Treatment of Mine Drainage", Jan. 1973, Burns and Roe Construction Corporation, Paramus, New Jersey.

> Laboratory and pilot plant research work was conducted on a modified Desal, Sul-biSUL, and conventional ion exchange processes to evaluate their performance on acid mine drainage feed waters. The study included process variables, analytical techniques, daily monitoring of acid mine drainage source, and economics of potable water production. This work was carried out at the Acid Mine Drainage Demonstration Plant at Hawk Run, Pennsylvania. Disposal of wastes was explored briefly.

James, S.C., Shuckrow, A.J. and Pajak, A.P., "History and Bench Scale Studies for the Treatment of Contaminated Groundwater at the Ott/Story Chemical Site, Muskegon, Michigan", *Proceedings of the National Conference on Management of Uncontrolled Hazardous Waste Sites*, 1981, Hazardous Materials Control Research Institute, Silver Spring, Maryland, pp. 288-293.

> Hazardous leachates and contaminated ground and surface waters are often associated with unsecured industrial waste storage and disposal sites. Numerous problems are encountered in the cleanup of such a site. One major problem is identifying the most effective treatment technology for the contaminated stream. Contributing to this problem are: (1) the inability to characterize completely the contaminated stream due to technical and economical limitations; (2) the paucity of information on the effectiveness of techniques for treating the broad spectrum of organic and inorganic compounds frequently present in these streams. Therefore, the U.S. Environmental Protection Agency began a project to evaluate techniques for concentrating hazardous constituents of aqueous waste streams. A literature review, desktop evaluation and laboratory bench scale experimental studies form the basis for judging the potential of numerous candidate technologies. During the course of this project, a part of the laboratory experimental work was carried out at Cordova Chemical Company, present owners of the Ott/

Story Site, using ground water contaminated by prior operations. This site provided an opportunity to interface research activities with development of a treatment system for a high priority problem site. In this paper, the authors provide a brief background on the site history, a method selection of candidate treatment technologies and summary of results of the experimental studies.

Jhaveri, V. and Mazzacca, A.J., "Bio-reclamation of Ground and Groundwater Case History", *Proceedings of the National Conference on Management of Uncontrolled Hazardous Waste Sites*, 1983, Hazardous Materials Control Research Institute, Silver Spring, Maryland, pp. 242-247.

In Aug. 1975, contamination was observed in a small creek that discharges into Allendale Brook in New Jersey. A storm sewer line that discharges into the stream was a point source of pollution. Data obtained in Nov. 1975 indicated the contamination was coming from Biocraft Laboratories, Inc., a semi-synthetic penicillin manufacturing plant located nearby. Biocraft's investigation revealed a leak in an underground process line that carried a mixture of methylene chloride, acetone, n-butyl alcohol and dimethyl aniline. This mixture infiltrated into the storm sewer line in front of the building which was the point source of contamination.

On Nov. 24, 1975, all underground lines were disconnected and above ground lines installed. The investigation failed to determine when the leak started. Based on daily tank inventory readings from June 1972 when the plant started up, it was estimated that the following amounts of substances may have leaked into the subsurface: methylene chloride—181,500 lb; n-butyl alcohol—66,825 lb; dimethyl aniline—26,300 lb; and acetone—10,890 lb. The biostimulation-decontamination system consists of the following major components: (1) a ground water collection system down gradient of the contamination area; (2) a four-tank, dual aerobic biological treatment system; (3) a series of in situ aeration wells along the path of ground water flow; (4) effluent injection trenches up gradient of the contaminated area; and (5) a control room. The biostimulation process implemented at the Biocraft site is reducing pollutant concentrations in the ground and ground water. Chemical analyses of ground water for selected pumping and monitoring wells prior to the startup of the continuous process and after two years of operation strongly indicate a gross reduction of contamination in ground water. This demonstrates that biostimulation can be successful not only in relatively permeable but also in low permeable formations.

Jones, A.K., et al., "Coverings for Metal Contaminated Land", *Proceedings of the National Conference on Management of Uncontrolled Hazardous Waste Sites*, 1982, Hazardous Materials Control Research Institute, Silver Spring, Maryland, pp. 183-186.

The information derived from this project has shown that the measures used for restoration of metal contaminated land have been successful. In general, where covering has been carried out in accordance with recommendations there is no evidence that regression due to reappearance of toxicity will occur. Toxicity problems still exist at some sites due to areas of contaminated material remaining uncovered. At some sites the provision of an insufficient depth of cover has resulted in toxic concentrations of metals within plant rooting zones causing adverse effects and in some cases, death of vegetation. These investigations at reclaimed sites have produced no evidence of upward migration of metals into covering layers through transport of metals in solution in capillary water under field conditions. At some sites elevated metal concentrations in surface covers have occurred through physical incorporation of contaminated materials into the amendments during restoration work. At two sites where chromium is present in a soluble anionic form, contamination of the surface covers with resulting toxicity to vegetation has occurred where water enters waste mounds, becomes contaminated, and re-emerges through the covering layers at lower levels. At none of the sites was there any evidence of similar movement of cationic metal ions through soil covers.

Kaschak, W.M. and Nadeau, P.F., "Remedial Action Master Plans", *Proceedings of the National Conference on Management of Uncontrolled Hazardous Waste Sites*, 1982, Hazardous Materials Control Research Institute, Silver Spring, Maryland, pp. 124-127.

The Remedial Action Program has been developed to respond to releases of hazardous substances from the 400 sites comprising the National Priorities List under the commonly known Superfund Program. These sites may require long-term cleanup actions to provide adequate protection of public health, welfare and the environment. The Interim Priority List, which identified 115 sites eligible for remedial funding, was published on Oct. 23, 1981. This list was expanded to 160 sites on July 23, 1982. As the program expands and the National Priorities List of 400 sites is published, an effective long-range planning mechanism becomes essential. An effective technical and financial planning document which has been developed to assist with the long-range planning needs of the USEPA is the Remedial Action Master Plan or RAMP. In this paper, the authors discuss the development of the RAMP and its current and future uses as a planning tool for conducting remedial activities at uncontrolled hazardous waste sites, and outlines the basic structure and contents of the RAMP with the use of an illustrative example.

Kastman, K.H., "Remedial Actions for Waste Disposal Sites", *Proceedings of Fourth Annual Madison Conference of Applied Research and Practice on Municipal and Industrial Waste*, 1981, Department of Engineering and Applied Science, University of Wisconsin-Extension, Madison, Wisconsin, pp. 131-150.

> The current problems associated with waste disposal sites which are polluting the environment have received much public attention in the last few years. The situations can be extremely complex and many of the techniques for mitigating the contaminant effects are marginally effective, or extremely costly. Also, those faced with solving specific problems may not have sufficient information available to determine the best type of remedial action which may be implemented. This paper provides a state-of-the-art review of specific remedial actions which are available to reduce or eliminate ground water and subsurface contamination problems at existing waste disposal sites.

Kaufmann, H.G., "Cost-Effective Treatment of Organically Contaminated Potable Well Waters", *Proceedings of the Second National Symposium on Aquifer Restoration and Ground Water Monitoring*, 1982, National Water Well Association, Worthington, Ohio, pp. 94-98.

> During the past several years, more and more emphasis has been placed on the quality of ground water that is being used as a source of municipal potable drinking water. The identification of large numbers of chemical, sanitary and hazardous dump sites throughout the country illustrate the potential, and in many cases, actual contamination of ground water. Evidence of contamination with synthetic organic chemicals that were disposed of in prior years is now being found. Once organic chemical contamination of a municipal well water is detected, several options are immediately exercised: close down the well; buy bottled water; buy water from adjacent communities; do nothing; drill new wells upgradient of the source of contamination; or treat the contamination. This paper deals with the option of cost-effectively treating the municipal well water before it is pumped to the distribution system. The paper discusses municipalities which reviewed the available options. Granular activated carbon systems were installed by these municipalities for the removal of the organic chemical compounds. Data is presented illustrating the removal efficiencies of the granular carbon systems. The capital and operating economics of the systems are presented. Economics illustrating treatment costs in the range of 40 cents to $1 per capita per month utilizing the service system approach are presented.

Kelleher, D.L. and Stover, E.L., "Investigation of Volatile Organics Removal", Presented at the New England Water Works Association Meeting, Jan. 1980, Randolph, Massachusetts.

Organic contaminations were detected from the well supply in the town of Norwood, Massachusetts. This prompted the Dedham Water Company to consider the possible interaction of ground water between the Norwood wells and its own White Lodge well field in Westwood. The Company's well field consisting of four conventional gravel-packed wells with depths ranging from 70-76 feet, has an estimated safe yield of 4.5 mgd. Organic analyses indicated the presence of more than 35 ppb, suggested maximum contaminant level, of 1,1,1, trichloroethane (TCEA) in two of the wells while the other two contained less than 35 ppb.

Pilot plant tests were conducted to evaluate methods for organic removal, particularly TCEA and iron and manganese removal which were capable of meeting not only current regulations but also any more stringent regulation not yet mandated. The treatment alternatives were: air stripping for volatile organic removal and possible oxidation of manganese and subsequent removal, activated carbon for volatile organics removal, and ion exchange for manganese removal. Parallel air stripping and carbon adsorption pilot studies were conducted for volatile organics removal. The system consisted of 4-inch diameter glass columns packed with 6 mm glass Raschig rings, supported by a stainless steel screen. The system was allowed to stabilize at each operating condition (various air and water flow rates) prior to collecting samples to ensure representative samples and to eliminate start-up effects. Analyses of the volatile organics were conducted with a gas chromatograph utilizing the purge and trap method. The carbon pilot system conducted consisted of 4 glass columns of 4 inch inside diameter by 5 feet in length. These columns were operated in series in the downflow mode during the 72 day test period. The carbon exhaustion rates and required contact time to achieve the desired effluent quality were also determined. Volatile organics were analyzed for the well water before and after treatment. The feasibility of employing aeration and glauconite or greensand were evaluated for iron and manganese removal. The pH of the well water was raised to 9.0 with caustic and vigorously aerated. The manganese greensand was formed by treating glauconite with manganese salt. The exhausted media was regenerated by applying potassium permanganate to oxidize the adsorbed material. An intermittent aeration technique, operating under constant head, was also employed. Samples were analyzed for manganese, iron, pH and turbidity.

TCEA removal efficiencies by air stripping were 87 percent and 74 percent with an air-to-liquid ratio of 30.1 to 1.0 for the two contaminated wells. A composite mix of water from both the wells at an air to liquid ratio of 26.5 to 1.0 gave an 82 percent removal efficiency. The four carbon adsorption pilot columns were able to reduce TCEA and TOC to less than 1.0 ppb and 0.5 ppm, respectively. The carbon exhaustion rate and the carbon retention time required for design was determined by comparing the TCEA carbon adsorption loading rate to the retention time of specific effluent quantities. A retention time of around 15 to 20 minutes yielded the highest carbon adsorption loading rate and lowest carbon exhaustion rate. Hence, approximately 1.4 pounds of carbon per 1,000 gallons of water is required for an influent of about 100 ppb. Aeration did not prove to be a feasible method of manganese removal, but the adsorption technique by greensand filter was found to be effective. The least expensive alternative evaluated was air stripping and manganese greensand filtration whereby TCEA concentration could be effectively stripped to less than 10 ppb.

Kelly, W.E., Powers, M.A. and Virgadamo, P.P., "Control of Groundwater Pollution at a Liquid Chemical Waste Disposal Site", *Studies in Environmental Science: Quality of Groundwater*, Vol. 17, 1981, Elsevier Scientific Publishing Company, Amsterdam, The Netherlands, pp. 273-278.

Waste chemicals dumped in open pits have polluted ground water and nearby domestic wells. The existing situation has been evaluated by field studies and predictions tested with analytic dispersion models. A renovation scheme involving recovery of contaminated ground water, air stripping and vapor recovery, and recharge to the ground water has been developed and approved for final design.

Kerfoot, W.B., "Direct Measurement of Gasoline Flow", *Proceedings of the Third National Symposium on Aquifer Restoration and Ground Water Monitoring*, 1983, National Water Well Association, Worthington, Ohio, pp. 396-401.

Direct measurement of gasoline flow on top of ground water provides rapid information on direction and rate of movement. Two procedures can be used for calibration: directly if a sufficient volume of strata is available; or indirectly, utilizing a conversion of the hydraulic conductivity of ground water versus petroleum product. Careful selection of well screen should be followed if measurements are made inside monitoring wells. Swelling of PVC can cause closure of slit widths less than 20/1,000 width. During recovery of gasoline product, the efficiency of collection can be monitored to assure efficient operation. Mapping of the zone of depression provides on-site information on rate of withdrawal and intercepted zone of flow. Several useful formulas are given for estimating time to recovery and anticipated volume.

Knox, R.C., "Effectiveness of Impermeable Barriers for Retardation of Pollutant Migration", *Proceedings of the National Conference on Management of Uncontrolled Hazardous Waste Sites*, 1983, Hazardous Materials Control Research Institute, Silver Spring, Maryland, pp. 179-184.

One of the most popular current methods of containing contaminated ground water is through the use of subsurface impermeable barriers. These barriers can take one of three forms: slurry walls; grout curtains; or steel sheet-piles. Successful operation of these barrier systems depends upon three basic criteria. First, the barrier must be truly impermeable and remain so even upon exposure to the contaminated ground water. Second, there must exist an underlying impermeable formation, at a reasonable depth, to which the barrier can be connected. Third, an adequate connection between the barrier and the underlying formation must be assured. In this paper, the author presents the results of the analysis of the movement of contaminated ground water under or through an imperfect barrier. The first phase of the analysis consists of the development of an analytical solution for the flow of ground water under a barrier and a simple numerical integration technique for developing concentration breakthrough curves. This simple solution algorithm was applied to the cases of variable recharge rates and lengths, variable depths of penetration of the barrier and anisotropic soils. The second phase of the analysis involves applying a numerical solute transport model to analyze the performance of a barrier with and without the effects of hydrodynamic dispersion, in the presence of a layered soil and a fully penetrating but partially permeable barrier.

Kosson, D.S. and Ahlert, R.C., "Treatment of Hazardous Landfill Leachates Utilizing In-Situ Microbial Degradation", *Proceedings of the National Conference on Management of Uncontrolled Hazardous Waste Sites*, 1983, Hazardous Materials Control Research Institute, Silver Spring, Maryland, pp. 217-220.

Industrial wastes have been frequently placed in landfills, lagoons and uncontrolled dump sites, subsequently producing liquid residuals as the result of compaction, rainfall runoff and infiltration, and biodegradation. These liquids can leach from a disposal site and contaminant underlying ground water and adjacent surface waters. An in situ treatment process for treatment of aqueous landfill leachates has been developed. Transport of hazardous materials is avoided, offering safety and economic advantages. A combination of physical, chemical and biological processes results in effective ultimate disposal. Landfill leachate is pretreated to remove suspensions, dispersed oil and heavy metals. Subsequently, treatment is effected with a land based biological process. Utilizing aerobic and anaerobic degradation processes associated with mixed microbial populations, hazardous waste treatment is viewed as proceeding either in the landfill structure or in a neighboring soil system.

Kraus, D.L. and Dunn, A.L., "Trichloroethylene Occurrence and Ground-Water Restoration in Highly Anisotropic Bedrock: A Case Study", *Proceedings of the Third National Symposium on Aquifer Restoration and Ground Water Monitoring*, 1983, National Water Well Association, Worthington, Ohio, pp. 76-81.

A study was conducted to determine the extent of ground water contamination by trichloroethylene (TCE) and possible aquifer restoration methodologies at a manufacturing facility in southeastern Pennsylvania. Analysis of soil samples collected from the site showed that the contamination originated in an area near the plant machine shop. Approximately 30 cubic yards of soil containing up to 68,000 ppb TCE was subsequently excavated and transported to a state-approved disposal site. Initial ground water samples collected from residential and commercial wells in the area indicated that the contamination appeared to be confined to the plant property. To more accurately define the extent of TCE occurrence, a monitoring well network was installed on plant property. A recovery well was drilled at the TCE source, with its discharge routed through a counter-current air stripper for TCE removal. Pumping tests showed a very strong hydraulic connection between the recovery well and one of the downgradient monitor wells, while the remaining monitor wells showed little, if any, effects from pumping. Shortly after recovery began, the hydraulically-connected downgradient well showed significant increases in TCE concentrations, indicating that the plume had extended beyond this well and was being drawn back toward the recovery well. This indicated that TCE transport was possibly fracture-controlled and that the ground water recovery program would be an effective method for aquifer restoration.

Kufs, C., et al., "Alternatives to Ground Water Pumping for Controlling Hazardous Waste Leachates", *Proceedings of the National Conference on Management of Uncontrolled Hazardous Waste Sites*, 1982, Hazardous Materials Control Research Institute, Silver Spring, Maryland, pp. 146-149.

Many techniques are available for controlling ground water contamination from hazardous waste management facilities including pumping, removal (excavation), subsurface drains, low permeability barriers, and in situ treatment. Ground water pumping systems are very commonly used because of their large range of applicability and low installation costs. Pumping systems can be designed to perform almost any function such as adjusting potentiometric surfaces, containing leachate plumes to prevent further migration, and removing contaminated ground water. The chief technical advantages of pumping systems are: applicability—usable in confined and unconfined aquifers of rock or unconsolidated materials under any conditions of homogeneity and isotropy; design flexibility—usable for injection as well as extraction regardless of the depth of contamination; construction flexibility—able to be installed using a variety of readily available materials by most qualified well drillers; and operational flexibility—easy to repair most systems components or modify the system or its operation as site conditions dictate. Although there are many advantages to using ground water pumping techniques, some distinct disadvantages to their use exist. The main technical disadvantages of pumping are that it requires: extensive design data—on the site's hydrogeology and leachate characteristics so that the system's components and operating conditions can be designed or selected properly; effluent treatment—or some other means of managing well discharges; continuous monitoring—to verify adequacy of system design and function and to safeguard against component (e.g., pump) failure which can result in contaminant escape; permeable aquifers—otherwise the system cannot function efficiently or effectively; leachate compatibility—in terms of the leachate's ability to move in ground water to the wells and no deteriorate well materials; and dilution—leachate may significantly dilute with ground water making it necessary to pump larger volumes of ground water to remove contaminant. Of these requirements, aquifer permeability and leachate compatibility are the most common reasons for seeking an alternative to ground water pumping.

Probably the single most significant drawback to using ground water pumping is the high operation and maintenance (O&M) costs typically associated with the system after installation. Many times this is overlooked in remedial action planning especially when capital (installation) costs are much higher for other options. Minimizing long-term costs is particularly crucial to site remediation under Superfund, because states must pay all O&M costs. Key O&M costs for pumping systems include: treating the contaminated discharge, providing electricity for the pumps, maintaining and repairing system components, and monitoring system performance. While other remedial actions also have these O&M costs associated with them, pumping costs are frequently higher because system operation is more demanding. For example, a subsurface drain may require the same amount of effluent treatment, maintenance, and monitoring as a pumping system. However, electrical requirements will be lower because the leachate is collected by gravity flow. Low-permeability barriers will have very low O&M costs, especially if leachate treatment is not required. Consequently, these techniques may be more cost effective than pumping in the long run even though their installation costs are higher. In the remaining sections of this paper, the authors describe alternate technologies for controlling ground water contamination in cases where pumping is inappropriate. Also discussed are the conditions under which an alternate technology might be a more effective remedial action than pumping.

Lafornara, J.P., et al., "Soil Surface Sealing to Prevent Penetration of Hazardous Material Spills", *Proceedings of the 1978 National Conference on Control of Hazardous Material Spills*, 1978, Hazardous Materials Control Research Institute, Silver Spring, Maryland, pp. 296-302.

> The problem of control of chemical spills on land has recently received a great deal of attention from the public, in general and the technical community, in particular, due at least in part to several incidents throughout the country where ground water has been contaminated. Most of the scientific activity in this area has been in the monitoring and health effects areas but little has been done to develop methods to alleviate or mitigate the problem of the contaminated ground water. Presently, mitigating action involves either the physical removal of the soil from the spill-impacted area and disposing of it at a secure landfill or the drilling of "cone of influence" wells followed by the removal of the pollutant from the water by conventional treatment technology. These techniques are both time consuming and expensive.
>
> It is the purpose of this program to develop an alternative method to mitigate the effects of land spills and thereby protect the nation's ground water resources. The main thrust of this alternative will be to prevent a spilled chemical from reaching an aquifer in the first place. This will be accomplished by soil surface sealing.

Lakshman, B.T., "Is it Feasible to Clean the Aquifer Once it is Contaminated by Landfill Leachate?", *Studies on Environmental Science: Quality of Groundwater*, Vol. 17, 1981, Elsevier Scientific Publishing Company, Amsterdam, The Netherlands, pp. 1065-1072.

> In this paper an attempt is made to critically examine the implications and experiences of ground water contamination due to landfill leachate in the State of Delaware in particular, and the United States of America in general. The feasibility of cleaning such contaminated aquifers through abatement programs has also been weighed and analyzed. Also discussed in general are other sources of contamination contributing to the impairment of aquifers used for potable water supplies.

Lamarre, B.L., McGarry, F.J. and Stover, E.L., "Design, Operation and Results of a Pilot Plant for Removal of Contaminants from Ground Water", *Proceedings of the Third National Symposium on Aquifer Restoration and Ground Water Monitoring*, 1983, National Water Well Association, Worthington, Ohio, pp. 113-122.

> The ground water at the Gilson Road Hazardous Waste Site in Nashua, New Hampshire, was contaminated with several million gallons of volatile organic solvents, extractable organics, various heavy metals sludges and leachate from domestic refuse. A detailed analysis of the site conditions and

the likely effects of the surrounding environment identified a two-part remedial action program as the most cost-effective solution. The plume of contaminated ground water has been partially isolated by surrounding it with a
slurry wall and membrane surface cap. Depth to bedrock varied from 30 to
110 feet around the 20-acre site. Bedrock beneath the site is highly fractured
and the glacial till discontinuous, leading to the possibility of significant
leakage from the contained area. Also the potential for eventual leakage
through the slurry wall due to chemical degradation is high. The concentration of organic solvents in the ground water is approximately 0.2 percent by
weight. A ground water treatment system was proposed to reduce the contaminant concentration by approximately 90 percent, based on bench scale
treatability studies. A 20 gpm pilot plant was designed, built and operated to
establish the cost-effectiveness of several process alternatives and to establish firm design parameters for the full scale (300 gpm) plant. Processes
piloted included metals removal by chemical neutralization and precipitation, high temperature air stripping, biological treatment, distillation and
incineration. The key process and economic factor is disposal of the concentrated organic solution.

Landon, R.A. and Sylvester, K.A., "Delineation of Subsurface Oil Contamination and Recommended Containment/Cleanup, U.S. Naval Station, Norfolk, Virginia", *Proceedings of the Second National Symposium on Aquifer Restoration and Ground Water Monitoring*, 1982,
National Water Well Association, Worthington, Ohio, pp. 335-338.

A hydrogeologic investigation was conducted at the Norfolk, Virginia,
Naval Station to delineate the subsurface occurrence of fuel oil and associated hydrocarbon products at two bulkhead locations. The initial phases of
this investigation entailed the drilling of test borings and the excavation of
numerous backhoe pits to delineate the areal extent of oil products. Small-
diameter monitoring wells were constructed to afford periodic monitoring
of the occurrence and thickness of oil products on a fluctuating ground
water table. Numerous water/oil measurements were made to determine and
assess the effects of recharge from rainfall, normal tidal fluctuations and
storm tidal fluctuations relative to locally significant changes in the occurrence and thickness of oil. A French drain interceptor trench was recommended for oil recovery and cleanup. This trench was installed in conjunction with bulkhead repair and reconstruction. An additional use of the
trench would be to monitor and recover any future oil leaks or spills.

Larson, R.J. and Ventullo, R.M., "Biodegradation Potential of Ground-
Water Bacteria", *Proceedings of the Third National Symposium on
Aquifer Restoration and Ground Water Monitoring*, 1983, National
Water Well Association, Worthington, Ohio, pp. 402-409.

The biodegradation potential of bacteria in ground water was examined using representative natural and xenobiotic compounds such as carbohydrates, amino acids, industrial chemicals and pesticides. Trace amounts (ppb level) of ^{14}C-labeled organic compounds were added to replicate samples and the amount of carbon assimilated (^{14}C into biomass) and/or respired to [14]CO_2 ([14]CO_2 evolved) was measured. Turnover times, the amount of time required to remove the in situ level of substrate present, maximum velocities (V_{max}) and biodegradation rate constants were calculated from these data. The abundance of bacteria was assessed by plate counts (PC) and acridine orange direct counts (AODC). Overall, the abundance and activity of the ground water microbial communities were comparable to values obtained in oligotrophic surface water. AODC were on the order of 10^6 ml[-1] and PC, 10^3-10^4 ml[-1]. Turnover time values for natural compounds ranged from 90-200 hr, whereas the values for the xenobiotic compounds were two- to 30-fold higher. Surfactants exhibited the most rapid rates of degradation of the industrial chemicals tested ($V_{max} \sim 120$ ng/l/hr), whereas pesticides exhibited the slowest rates ($V_{max} \sim 2$ ng/l/hr). For most of the ^{14}C-pesticides tested, mineralization of the carbon skeleton to [14]CO_2 was not observed and utilization of these compounds as carbon and energy sources by ground water bacteria could not be demonstrated.

Lee, G.W., Jr., et al., "Investigation of Subsurface Discharge from a Metal Finishing Industry", *Proceedings of the National Conference on Management of Uncontrolled Hazardous Waste Sites*, 1983, Hazardous Materials Control Research Institute, Silver Spring, Maryland, pp. 346-351.

An order was issued to a metal finishing and manufacturing facility in Aug. 1981. The document required that a complete hydrogeological investigation pertaining to the subsurface discharge of 1,1,1-trichloroethane which allegedly migrated through the ground water system to a nearby Town water supply well be performed. An investigation was performed which included the installation of ground water monitoring wells, soil and ground water sampling and research oriented work to confirm or refute the allegations and to address the concerns of the consent order. It was found that the metal finishing and manufacturing facility was responsible for the contamination and that remedial work was in order. Alternatives were developed and evaluated. The work completed to date indicates that the development of a new water supply, with the possible treatment of the ground water aquifer, is the most reasonable remedial alternative. Investigations are ongoing to determine the level of treatment, if any, required for the clean up of the ground water aquifer.

Lindorff, D.E. and Cartwright, K., "Ground Water Contamination: Problems and Remedial Actions", EGN 81, May 1977, Illinois State Geological Survey, Urbana, Illinois.

> Case histories of 116 ground-water contamination incidents reveal that remedial action is usually complex, time consuming, and expensive. Quick recognition of and response to an emergency involving ground water are essential to minimize the spread of contamination and make only minor remedial action necessary. However, the hydrogeologic setting and the pollution hazard at the specific locality should be assessed before corrective action is taken. The remedies most frequently used to renovate the subsurface environment include pumping and treating ground water, removing soil, draining away ground water via trenches, and removing the source of contamination. A combination of these techniques also has been used. Guidelines for strategies that consider the hydrogeologic framework and administrative structure of Illinois are presented. A committee or task force should define the scope of the problem and develop a formal procedure for dealing with various kinds of land and ground water pollution emergencies.

Lippitt, J., et al., "Worker Safety and Degree-of-Hazard Considerations on Remedial Action Costs", *Proceedings of the National Conference on Management of Uncontrolled Hazardous Waste Sites*, 1982, Hazardous Materials Control Research Institute, Silver Springs, Maryland, pp. 311-318.

> In this paper, the authors review two projects undertaken to determine remedial action costs. The basic approach to developing costs was to identify the remedial action unit operations which may be combined to provide a remedial action plan. Component cost items were identified which contribute to each unit operation. By determining the cost associated with the component cost items, the total unit operation cost can be developed. A completed remedial action plan for a site can be evaluated by summing the costs of each of the remedial action unit operations involved.

Lochrane, T.G., "Ridding Groundwater of Hydrogen Sulfide — Part I", *Water and Sewage Works*, Vol. 126, No. 2, Feb. 1979, pp. 48-50.

> Hydrogen sulfide (H_2S) can be removed from water by aeration via gas transfer. The main function of the aerator is to speed up the reaction rate of H_2S releasing it to the atmosphere and reducing the overall chlorine need. As H_2S decreases in content, its ionic forms, HS^- and S^- increase. If CO_2 is present in significant amounts it will be stripped faster than H_2S. This causes the pH to rise and decreases the removal efficiency. Ionized forms of H_2S, HS^- and S^- can be removed by chlorination. Alkaline sulfide forms are not

as simple to remove. Colloidal sulfur and polysulfides are converted to sulfates by adding sulfite. Thiosulfate can be converted to sulfate by rechlorination.

Lord, A.E., Jr., "The Hydraulic Conductivity of Silicate Grouted Sands with Various Chemicals", *Proceedings of the National Conference on Management of Uncontrolled Hazardous Waste Sites*, 1983, Hazardous Materials Control Research Institute, Silver Spring, Maryland, pp. 175-178.

Grouting can be used to decrease permeability with special emphasis on stopping the flow of various hazardous chemicals moving within leachate or the ground water flow regime. Indeed, there is a very strong effort currently underway toward mitigating the effect of contamination of ground water, rivers and lakes by hazardous materials leaking from waste disposal sites. Cutoff walls of various types are being used to intercept the flow between the landfill site and the potentially pollutable water. Types of cutoff walls used in geotechnical engineering and heavy construction work, which are usually adopted to make seepage cutoff walls, are the following: cement grout type wall — cement, cement/clay, cement/fly ash and cement/clay/fly ash; chemical grout type wall — silicate, acrylamide and polyacrylamides, phenoplast, lignosulfite, aminoplast, some combinations of above, and new products (e.g., acrylate); slurry walls — soil/bentonite backfilled trench, and geomembrane liner in a backfilled trench; and sheet pile walls — interlocking steel sheets, tongue and groove precast concrete panels, and tongue and groove lumber.

The time dependent reaction of these cutoff wall materials with various chemicals is, in general, unknown. It is thought (usually without experimental justification) that a number of these materials are quite durable and impermeable to most chemical agents and environmental conditions. However, there has been very little work on the durability of these wall materials when in a possibly corrosive chemical environment. In retrospect, however, it was also thought that the clay materials commonly used at sanitary landfill sites for liners would be impervious to most chemicals. However, some very significant recent work showed this thinking to be quite false. In this work, it was shown that most organic solvents (acetone, xylene, methanol, etc.) will permeate readily through clay liners. The object of this study is to see if a similarly adverse behavior also occurs in chemically grouted walls, and the results of this study are presented.

Lu, J.C.S., Morrison, R.D. and Stearns, R.J., "Leachate Production and Management from Municipal Landfills: Summary and Assessment", *Land Disposal: Municipal Solid Waste — Proceedings of the Seventh Annual Research Symposium*, 1981, U.S. Environmental Protection Agency, Cincinnati, Ohio, pp.1-17.

This report is a review, analysis, and evaluation of current literature and information on municipal landfill leachates. The overall objective is to provide documentation of practical information, current techniques, comparative costs, and additional research needs on the generation, composition, migration, control, and monitoring of leachates from municipal landfills. Due to the sizeable amount of information available on the subject matter only the conclusions are presented in this report.

Lundy, D.A. and Mahan, J.S., "Conceptual Designs and Cost Sensitivities of Fluid Recovery Systems for Containment of Plumes of Contaminated Groundwater", *Proceedings of the National Conference on Management of Uncontrolled Hazardous Waste Sites*, 1982, Hazardous Materials Control Research Institute, Silver Spring, Maryland, pp. 136-140.

Removal and treatment of contaminated fluids from wells or drains is the most commonly used method for controlling the movement of a plume of contaminated ground water. Other methods, such as fluid encapsulation within subsurface physical barriers and in situ treatment or fixation, are being researched and applied, but those involving fluid removal probably will continue to be the most popular on the basis of their cost and reliability. This paper: (1) describes simple hydraulic models for making conceptual designs and cost estimates for recovery-well systems, and (2) examines the general sensitivities of recovery system costs to selected plume and aquifer characteristics. The costs presented are for simplified scenarios of contamination, which represent a foundation upon which more complex models can be built. Although the more complex models are increasingly being applied in the detailed design and evaluation of systems of recovery wells, there is and will continue to be a place for simpler mathematical models that can provide rough estimates of system discharge and number/size of recovery wells. The costs apply only to engineered structures and related services for a containment system, with other cost elements such as source removal, land purchase, legal services, etc., not included. Also, the paper is concerned not only with containment and treatment of contaminated ground water, but also with total cleanup of the site and the aquifer.

Lycott Environmental Research, Inc., "Final Report, Ground Water Resources Evaluation and Aquifer Protection Plan, Phase II", E-007-79, undated, Southbridge, Massachusetts.

Because Acton relies solely upon ground water for its water supply, it is extremely important for the Town to have a comprehensive inventory and evaluation of its ground water resources. This evaluation provides a basis for the Town's aquifer protection plan. Part of this evaluation is a ground water monitoring well network designed to provide an "early warning" system for the quality of water in Acton's aquifers. Analysis of water budgets for the Town's aquifer indicates that considerably more ground water is available for municipal development. The most favorable areas for additional water development appear to be the lower reaches of Fort Pond Brook, North Acton sites near Nashoba Brook and Butter Brook, west of Ice House Pond, and south of Mt. Hope Cemetery. The Sinking Pond aquifer is currently under study for rejuvenation (it is not currently being used for municipal drinking water supply because of organic chemical contamination), and this aquifer also may be able to sustain greater withdrawals once the contamination problems are solved. Modeling of nitrogen and sodium loading of the Town's aquifers will be completed as part of the aquifer protection plan. This plan will include recommendations for road-salting policies as well as other measures designed to protect the Town's aquifers.

Malone, P.G., Francingues, N.R. and Boa, J.A., Jr., "The Use of Grout Chemistry and Technology in the Containment of Hazardous Wastes", *Proceedings of the National Conference on Management of Uncontrolled Hazardous Waste Sites*, 1982, Hazardous Material Control Research Institute, Silver Spring, Maryland, pp. 220-223.

Uncontrolled hazardous waste sites include a wide variety of waste disposal locations where toxic wastes from manufacturing or mining activities have been discarded or dumped in a manner that poses a threat to human health and the environment. The potential pollutants vary from highly toxic organics to inorganic weathering products from mine spoils. In the superfund list of 160 problem sites, 134 sites involve significant pollution from toxic organics and 26 involved inorganic industrial wastes (such as battery or plating wastes), mine tailings or mine drainage or slag. The wastes at the disposal sites can be in tanks, lagoons, drummed storage or waste piles. In most cases, soil and ground water contamination from leakage or intentional discharge is already evident. Remedial actions at the sites usually take the form of removing waste to a safer location and/or developing a strategy for on-site containment by installing barriers to control waste movement or by immobilizing toxic constituents in place. In this paper, the authors discuss grouting technology and grouting equipment in remedial actions other than diversion of uncontaminated ground water. While grouts have long been used to stabilize soils for foundations or to produce barriers to control ground water movement in tunnelling or excavation, there are other aspects of grout chemistry that can be used to directly restrict chemical transport from waste materials.

Malone, P.G., Larson, R.J. and Meyers, T.E., "Stabilization/ Solidification of Waste from Uncontrolled Disposal Sites", *Proceedings of the National Conference on Management of Uncontrolled Hazardous Waste Sites*, 1980, Hazardous Materials Control Research Institute, Silver Spring, Maryland, pp. 180-183.

> Recent U.S. Environmental Protection Agency reports have estimated the number of unauthorized or uncontrolled hazardous waste disposal sites in the United States to be 2,000. At most of these sites, the wastes are liquids and are held in 208-1 (55-gal) drums in open-air surface storage or stored in bulk in open impoundments. Cleanup of such sites involves identifying the wastes, attempting to treat or contain the wastes, and preparing the wastes to be shipped to a site for final disposal. In incidences where leakage has occurred, it is necessary to collect the contaminated soil also, for treatment or transportation to a secured landfill. The magnitude of the recontainerization or treatment effort is often so great that it involves building a waste disposal operation at major sites. The safety and security of cleanup and disposal operations can be improved if the wastes are converted into less reactive solid forms through processes termed stabilization/solidification. The purpose of this paper is to explore the possible application of stabilization/solidification technology to waste cleanup at dangerous abandoned or uncontrolled dump sites. Stabilization/solidification can be employed as a step to facilitate safer transportation and disposal.

Mathes, G.M., "Recovery of Spilled Petroleum Products from Fine Alluvium in the Mississippi River Aquifer", *Proceedings of the Conference on Oil and Hazardous Material Spills: Prevention, Control, Cleanup, Recovery, Disposal*, 1979, Hazardous Materials Control Research Institute, Silver Spring, Maryland, pp. 164-174.

> This paper summarizes the results of Phases II and III of an on-going study aimed at identifying, locating, and dealing with the source(s) of gas odors experienced periodically since 1966 at many widespread locations in the city of Hartford, Illinois. The initial phase of this study identified multiple sources for the varied odors experienced. A major source for many of the odors was identified to be an underground accumulation of petroleum products trapped above the ground water beneath the city. Several house fires in the city have been attributed to the presence of air-gas mixtures. Two potential systems for providing high capacity depression wells and low capacity recovery wells were considered feasible. These may generally be described as follows:
>
> (1) A single larger diameter well system with a 60 inch diameter hole and a 40 inch diameter screen where both the large capacity well and product recovery systems are housed in the same well. An estimated well depth on the order of 100 feet would be required to provide the necessary flow.

(2) A double well system where a different well is used to create a cone of depression than is used to recover the product. In this case a 28 inch diameter hole with 12 inch screen approximately 100 feet in depth would be required for the high capacity well and a 48 inch diameter hole with 30 inch screen to a depth of approximately 50 feet would be installed for recovery.

The first recovery system was installed in July, 1978 and the second system was installed in March, 1979. The first well was installed intersecting the largest of the three areas of product concentration. During the late summer, fall, and early winter, this well achieved production in the range of 500 to 1,400 gallons of product per day. Starting in late winter and early spring and extending through midsummer 1979, low production was obtained from both wells. This is attributed primarily to corresponding periods of rapid fluctuation in the aquifer due to precipitation and river level change. Monthly production figures and well hydrograph information were not available at the time of this paper to allow more detailed comparison. It has been reported that total monthly production has ranged from 6,000 to 30,000 gallons during the history of system operation. Through October, 1979, a total of approximately 192,000 gallons of product were recovered from the first well, and 40,000 gallons from the second well.

Matis, J.R., "Petroleum Contamination of Ground Water in Maryland", *Ground Water*, Vol. 9, No. 6, Nov.-Dec. 1971, pp. 57-61.

Petroleum contamination of ground water is a widespread problem that plagues many areas. Historical data collected in Maryland indicates that most counties in the state record cases of this contamination problem annually. Cases in the "hard-rock areas" west of the Fall Zone have the highest frequency of occurrence, in contrast to the Coastal Plain geologic province to the east. In both areas, the problems have been localized. It is difficult to handle petroleum contamination cases, and the legal implications are very complex. Many petroleum fuels do not deteriorate in the ground water system. Further, the identification of the specific petroleum products in ground water is generally not possible with present techniques. An investigation of a particular complaint can often be split into a preliminary phase and a detailed site investigations phase. Once a source of contamination is located, it must be stopped or removed. Since it is virtually impossible to remove the contamination from the ground water, legal and regulatory problems continue on for months or years after an original complaint.

Matthess, G., "In Situ Treatment of Arsenic Contaminated Groundwater", *Studies in Environmental Science: Quality of Groundwater*, Vol. 17, 1981, Elsevier Scientific Publishing Company, Amsterdam, The Netherlands, pp. 291-296.

Ground water in a sand and gravel aquifer was contaminated by arsenic compounds. The extent and the As concentration of the polluted ground water plume decreased from 1971 to 1975, whereas the content of free dissolved oxygen increased. High As concentrations (>1 mg/l) occurred in ground water with typical characteristics of a "reduced" water with negative Eh values and high concentrations of dissolved iron (up to 140 mg/l in 1971). When plotted into an As stability field diagram, the higher values (>1 mg As/l) coincided with the fields of trivalent As species, whereas the lower value (<0.1 mg As/l) fitted to the fields of the pentavalent arsenic species. Therefore it was concluded that an improvement of the oxygen supply should accelerate the natural precipitation processes. By injection of 29,000 kg $KMnO_4$ into 17 wells and piezometers the soluble As (III) species were oxidized to As (V) species, which were precipitated as $FeAsO_4$ or $Mn_3(AsO_4)_2$ or co-precipitated with Mn- and Fe-hydroxides.

McBride, K.K., "Decontamination of Ground Water for Organic Chemicals: Selected Studies in New Jersey," *Proceedings of the Second National Symposium on Aquifer Restoration and Ground Water Monitoring*, 1982, National Water Well Association, Worthington, Ohio, pp. 105-114.

Ground water contamination in New Jersey is a direct result of industrialization and population density. Over the past decade, government has responded to the problem by developing regulations governing all discharges to ground water. Contamination by synthetic organic chemicals is the most severe and widespread ground water problem, and has necessitated installation of monitoring and decontamination systems throughout the state. Hydrogeology, geography and contaminant-type affect design and monitoring wells and the final decontamination program. Sophisticated municipal wastewater treatment facilities have been used to treat contaminated ground water in industrialized areas, enabling comparatively inexpensive cleanup. In more rural settings, on-site systems using air-stripping or granular activated carbon have been successful. Spilled petroleum products are now pumped routinely from the ground using recovery wells or trenches. Recently, in-ground treatment of petroleum hydrocarbons has been accomplished using microorganisms. Overall, decontamination systems have repeatedly achieved 50 to 90 percent contaminant reduction. Contaminant levels in degraded aquifers are known to fluctuate in response to recharge events and the nature of the source. Nonetheless, marked decreases have been noted when the sources are removed. Aquifer restoration is a prolonged process requiring years to obtain a significant improvement.

McGarry, F.J. and LaMarre, B.L., "Proposed Cleanup of the Gilson Road Hazardous Waste Disposal Site, Nashua, New Hampshire", *Proceedings of the National Conference on Management of Uncontrolled Hazardous Waste Sites*, 1982, Hazardous Materials Control Research Institute, Silver Springs, Maryland, pp. 291-294.

The Gilson Road Hazardous Waste Dump site is located in Nashua, New Hampshire, near the New Hampshire/Massachusetts State line. The site was developed as a sand borrow pit by the owners, with substantial quantities of sand having been removed over several years of operation. The sand excavation at many locations extended into the ground water table underlying the site. At some point in the late 1960s, the owner decided to discontinue the same mining operation and began the operation of an unapproved and illegal refuse dump with the apparent intent of filling the pit left by the sand mining. Household refuse and demolition materials were the initial components of the wastes dumped at the site. Eventually, chemical sludges and aqueous chemical wastes were also dumped there. The domestic trash and demolition material were usually buried while the sludges and liquid chemicals were disposed of in several ways: mixed with the trash and buried, placed in steel drums and either buried or stored on the surface, or dumped into a makeshift leaching field and allowed to percolate into the ground. It was estimated that the site was used for the disposal of hazardous wastes for approximately five years.

Cleanup activities began in 1980, as soon as legal access to the site could be obtained. During May and June of that year, 1,314 drums, which were accessible, were removed by a contractor and disposed of at approved secure landfills in the States of New York and Ohio. Immediately following removal of the drums, a ground water exploration and monitoring program was initiated to determine the magnitude and extent of the contamination. In July of 1981, an engineering report prepared by another consultant stated that there were high concentrations of heavy metals, volatile organics, and extractable organics in the ground water under the site. The metals and much of the extractable organics could be traced to the domestic refuse while the volatile organics likely originated from the makeshift leaching field used for the disposal of most of the aqueous wastes.

The concentration of organics, measured as total organic carbon, was found to be over 4,000 mg/l. Total metals concentrations, in the most contaminated portion of the plume, were in excess of 1,000 mg/l. Some of the major organic contaminants included methylene chloride, methyl ethyl ketone, toluene, benzene, chloroform, tetrahydrofuran, and acetone. The metals found at the site included iron, manganese, nickel, zinc, barium and arsenic. The primary remedial measure recommended by the initial consultant was the encapsulation of a 12.5 acre site with a slurry trench cut-off wall and an impermeable surface cap. For that portion of the ground water plume outside the containment wall, a treatment system would be installed which would intercept and treat the ground water. The treatment process would consist of air stripping, biological oxidation, and activated carbon adsorption. Chemical precipitation of the metal contaminants was considered to be an optional treatment step. Following cleanup of the aquifer outside the slurry wall, all further activities would be terminated and the site would be left with the slurry wall to permanently contain the 12.5 acres of contaminants.

McLin, S.G. and Tien, P.L., "Hydrogeologic Characterization of Seepage from Uranium Mill Tailings Impoundments in New Mexico", *Proceedings of the Second National Symposium on Aquifer Restoration and Ground Water Monitoring*, 1982, National Water Well Association, Worthington, Ohio, pp. 343-358.

Seepage of waste liquids into shallow outcroppings of an interbedded sandstone-shale sequence of the Upper Gallup Sandstone aquifer was examined in order to characterize the potential for extensive off-site subsurface contaminant migration. These waste liquids are from a uranium mining and acid leach milling operation located near Gallup, New Mexico. Preliminary water balance studies from evaporation holding ponds indicated that as much as 80,000 gallons of rafinnate liquid containing up to 60,000 mg/l TDS had entered the aquifer each day during an 18 month period. Ground level and photogeologic mapping efforts depicted the presence of a major regional subsurface fracture trace adjacent to the tailings disposal site. Surface electrical resistivity, borehole geophysical and core fractographic analyses confirmed the presence of subsurface fractures, in addition to extensive contaminant movement. Formation hydraulic testing indicated at least some anisotrophy in hydraulic conductivity that may be unrelated to rock fracturing. Approximately 200 monitoring wells were eventually utilized in mapping a subsurface contaminant plume nearly 1.5 miles long and 0.75 miles wide. A system of pumping wells was designed and installed in order to partially control further extensive plume movement. Monitoring efforts are continuing in order to assess the feasibility of total containment on a long-term basis.

McMillion, L.G., "Ground Water Reclamation by Selective Pumping", *Society of Mining Engineers, AIME - Transactions*, Vol. 250, Mar. 1971, pp. 11-15.

A field project to develop and demonstrate a method for alleviating problems of highly mineralized ground water where they occur as isolated zones or pockets in fresh-water aquifers was conducted in the Ogallala aquifer of eastern New Mexico. Removal of contaminated water is by a large-capacity well and the pumped poor-quality water is used locally for secondary recovery of oil. Benefits of the operation are threefold in that the aquifer is being reclaimed for future fresh-water production, contaminated water is being used beneficially, and fresh water which otherwise would have been used for waterflooding is available for other purposes. Ground water quality in the site has improved markedly after two years of operation.

Michalovic, J.G., et al., "System for Applying Powdered Gelling Agents to Spilled Hazardous Materials", EPA-600/2-78-145, July 1978, U.S. Environmental Protection Agency, Cincinnati, Ohio.

Research had been conducted to develop a blended material that would optimally immobilize a wide range of liquid chemicals detrimental to the environment. The product of this research was Multipurpose Gelling Agent (MGA), a blend of four polymers and an inorganic powder. When applied to a chemical spill, MGA turns the hazardous liquid into a gelatinous mass which can be easily removed by shovel or other mechanical means. The MGA research program has also included design, fabrication, testing, and demonstration of a mobile, self-powered, mechanical system for dispensing powdered MGA to spill target areas in a safe and effective manner. A prototype of the Mobile Dispensing System (MDS) was constructed and tested. The MDS unit incorporates an auger-fed pneumatic conveyor system and a trailer that can be towed to remote spill sites. After initial testing to determine the MGA range and dispersal pattern, the MDS unit was tested against both small- and large-scale simulated spills.

Mido, K.W., "An Economical Approach to Determining the Extent of Ground Water Contamination and Formulation of a Contaminant Removal Plan", *Ground Water*, Vol. 19, No. 1, Jan.-Feb. 1981, pp. 41-47.

If past ground water movement in the vicinity of contamination could be shown pictorially, determination of the extent of contamination would be facilitated significantly. Furthermore, if pictorial representation of future ground water movement under presumed conditions reflecting alternative plans of contaminant removal or containment could be provided rapidly, an effective plan could be developed speedily and thus economically. A flow path and arrival time plot, which is developed through the use of a flow system kinematics mathematical model, provides such visualization of ground water movement. Contrary to the popular notion that the kinematics model requires much more geohydrologic data than the finite difference or element models, it was found that meaningful kinematics models could be developed based on relatively limited data. Presented in this paper are: a brief description of the development of flow path and arrival time plots, information obtainable from such plots, an example of the use of the plots in determining the extent of contamination of ground water, and an example of the use of the plots in formulating a contaminant removal plan.

Miller, D.G., Jr., "In Situ Treatment Alternatives and Site Assessment", *Proceedings of the National Conference on Management of Uncontrolled Hazardous Waste Sites*, 1983, Hazardous Materials Control Research Institute, Silver Spring, Maryland, pp. 221-225.

Remedial action alternatives at sites contaminated by waste management practices should be evaluated on the basis of technical feasibility, cost and long term effectiveness. One set of alternatives, in situ treatment, is attractive because it can result in minimal exposure and eliminate off-site transportation of the waste. In situ treatment, however, is questionable with regard to physical and chemical uniformity of treatment and long term chemical stability. Experience with numerous projects has demonstrated the need for adequate physical and chemical characterization of the system to be treated. In situ treatment as considered in this paper is restricted to the introduction of chemicals into the subsurface, rendering contaminants less hazardous to the environment. The process has also been referred to as neutralization/detoxification or detoxification and stabilization. In this paper, the author addresses in situ treatment of inorganic contaminants, primarily heavy metals, although similar concepts and physical characteristic concerns may be applicable to in situ biodegradation. In spite of demonstrations of physical feasibility by full scale tests, current use is limited because of the concerns of overall reliability and costs associated with demonstration and implementation. An understanding of these concerns is necessary to properly evaluate in situ treatment as an alternative.

Millet, R.A. and Perez, J.Y., "Current USA Practice: Slurry Wall Specifications", *Journal of the Geotechnical Engineering Division - ASCE*, Vol. 107, No. GT8, Aug. 1981, pp. 1041-1056.

To establish diaphragm and cutoff slurry wall design criteria and, thus, specifications, the designer must clearly establish the objectives or end results that are to be obtained. This paper considers the critical design criteria and the resulting specifications for both slurry trench diaphragm walls and cutoff walls. Diaphragm and cutoff walls are initiated with a common process. This process is the excavation of a narrow trench without the use of significant lateral support other than that provided by a bentonite-water slurry which is pumped into the trench so that the slurry level is maintained at or near the top of the trench throughout the excavation process. In the diaphragm wall process, after completion of a segment of excavated trench, either cast-in-place concrete (tremie process) or precast panels are used to displace the stabilizing mud and construct a load-bearing wall (vertical and lateral). There are two types of slurry cut-off walls: (1) soil-bentonite; and (2) cement-bentonite. In the soil-bentonite cutoff process, the bentonite slurry is displaced by a soil-bentonite mixture similar in consistency to high slump concrete. The soil backfill forms a low permeability highly plastic cutoff wall. In the cement-bentonite cutoff process, cement is added to a fully hydrated bentonite-water slurry. The cement-bentonite-water slurry is then used to both stabilize the slurry wall during excavation, and upon setting of the cement, for the permanent cutoff wall itself.

The critical design criteria for the two types of slurry walls are: (1) diaphragm walls—structural strength and integrity, permanence, and permeability; and (2) cutoff walls—permeability, deformability, and permanence. End-result (performance-guaranteed) specifications are acceptable if the owner, engineer, and contractor truly understand and can agree on what end results are to be obtained.

Molsather, L.R. and Barr, K.D., "Retrofit Leachate Collection System for an Existing Landfill", *Proceedings of the Second National Symposium on Aquifer Restoration and Ground Water Monitoring*, 1982, National Water Well Association, Worthington, Ohio, pp. 316-322.

This paper describes methods and special considerations involved in design and construction of a leachate collection system installed adjacent to an existing landfill site. While the authors use a recent project to illustrate the design methods employed and to discuss certain site-related problems, they are of the opinion that the methods used and the solutions to certain problems are generally applicable and should be of value in relation to planning and installation of facilities on other sites. Basic modeling techniques were employed in evaluating watershed yield, in predicting variations in seepage collection rate during normal operating conditions and construction and in evaluating the effects of the collection system on the site. Practical items discussed in the paper include the methods considered for collecting toe seeps, construction procedures used for installation of facilities in peat and other unstable soils, selection of construction materials and certain safety considerations related to construction adjacent to an existing landfill site.

Muller, B.W., Brodd, A.R. and Leo, J., "Picillo Farm, Coventry, Rhode Island: A Superfund and State Fund Cleanup Case History", *Proceedings of the National Conference on Management of Uncontrolled Hazardous Waste Sites*, 1982, Hazardous Materials Control Research Institute, Silver Spring, Maryland, pp. 268-273.

The cleanup of unauthorized hazardous waste sites presents varied operational and environmental problems. In this paper, the authors provide insight into the chronology, operational techniques, air monitoring aspects, water monitoring and problem areas associated with the abatement of such a site situated in Rhode Island. Field and off-site disposal analyses proved to be a problem in terms of turnaround time and procedural techniques. Particular attention must be paid to the analytical requirements so that they: (1) keep pace with excavation efforts, and (2) provide the required disposal analyses in a timely fashion. One of the primary operational constraints involves reducing the risk of injury and adverse health effects to personnel working on site and the surrounding population. A key aspect of this risk reduction strategy involved monitoring the atmosphere to assess the degree of hazard in the following areas: explosion, oxygen deficiency and exposure to contaminants. Funding of a project of this scope must around for the significant unknown factors associated with hazardous waste cleanup, i.e., what is the amount and type of material to be excavated and disposed of? The generation of contaminated soil poses a significant disposal problem. Successful efforts were made to limit this problem in Phase II and III in terms of the smaller amount of generated contaminated soil. However, at some point, a disposal or treatment option must be chosen to deal with this problem. Finally, and most importantly, all activities should be closely coordinated with any citizens organization or affected group that deals with the issues of hazardous wastes. An informed citizenry can aid procurement of funding and maintain the necessary interest in the issue that influences governmental decisions. Working closely with such organizations promotes cleanup effects, maintains good press relations and may accelerate completion of the job.

Muller-Kirchenbauer, H., Freidrich, W. and Hass, H., "Development of Containment Techniques and Materials Resistant to Groundwater Contaminating Chemicals", *Proceedings of the National Conference on Management of Uncontrolled Hazardous Waste Sites*, 1983, Hazardous Materials Control Research Institute, Silver Spring, Maryland, pp. 169-174.

In 1970, approximately 50,000 disposal sites were registered in the Federal Republic of Germany. Two percent of these seem to contain hazardous wastes and need appropriate treatment so that the ground water and the surroundings shall not be affected. Additionally, there are further unregistered intrusions from earlier times and from World War II such as bombed factories and oil or chemical storage. Clearly the best way to correct pollution at these sites is to dig them up and to refill the pits with clean material. However, in the case of very deep contamination or of intensively settled surroundings, the dig and refill procedure is neither technically nor economically possible. This has resulted in an intense search for alternative solutions in recent years. Progress has been made with encapsulation techniques. Although this concept can be successfully used, there are many special conditions that have to be considered; these are discussed in this paper.

Mutch, R.D., Jr., Daigler, J. and Clarke, J.H., "Clean-up of Shope's Landfill, Girard, PA", *Proceedings of the National Conference on Management of Uncontrolled Hazardous Waste Sites*, 1983, Hazardous Materials Control Research Institute, Silver Spring, Maryland, pp. 296-300.

Refuse and industrial waste from the Lord Corporation of Erie, Pennsylvania was dumped on the property of a maintenance department employee from the late 1950's to 1979. The landfill was closed in 1979 after having blanketed an area of about 4.5 acres. Very evident leachate problems at the landfill prompted the Pennsylvania Department of Environmental Resources to initiate discussions with the employee and the Lord Corporation. Lord Corporation, recognizing its role as generator of virtually all of the wastes in the landfill, accepted responsibility for investigation and remediation of the landfill. The selected remedial program encompasses a number of individual remedial components. The principle objective of the plan is the virtual elimination of recharge to the landfill, which is accomplished by the combination of the impermeable cap and the upgradient cut-off wall. The plan's major components are: (1) removal and off-site disposal of drums from the landfill surface and those encountered in regrading of the fill's side-slopes; (2) removal of standing pools of leachate and site drainage facilities; (3) surface water diversion and site drainage facilities; (4) an impermeable final cover (cap); (5) an upgradient (southern) cut-off wall; and (6) a short- and long-term monitoring program.

Myers, V.B., Domenico, D.D. and Hartsfield, B., "Remedial Activities at Florida's Uncontrolled Hazardous Waste Sites", *Proceedings of the National Conference on Management of Uncontrolled Hazardous Waste Sites*, 1982, Hazardous Materials Control Research Institute, Silver Springs, Maryland, pp. 295-298.

Florida's water resources are not limited to its expansive 1200 mile coast line and over 7000 freshwater lakes. It is also blessed with an abundant supply of ground water. The Floridan Aquifer underlies the majority of peninsular Florida and supplies most of the potable water needs of north and central Florida. The Floridan Aquifer is at the surface in many areas of central Florida. In coastal areas where the Floridan Aquifer is saline, surface aquifers are used for potable drinking water. The majority of the state has very pervious soils which allow for rapid transport of contaminants to the shallow surface aquifers; hence the ground water is extremely vulnerable to contamination. Toxic materials are finding their way into ground water from various sources: hazardous waste dumps, waste lagoons and pits, sanitary landfills, accidental spills, mining operations, and storm water runoff. A recent assessment funded by the USEPA showed over 70% of Florida industries using unlined impoundments for waste disposal. This is particularly alarming when 98% of the 6000 impoundments were located over useable drinking water aquifers. Florida's prominent position on EPA's Interim Priority List of uncontrolled hazardous waste sites is due, in part, to the susceptibility of the state's ground water to contamination.

This paper presents overviews of the preliminary remedial action planning and monitoring activities at seven of Florida's 16 Superfund sites. Included are discussions concerning a manufacturer of electrical solderless terminals, an inactive wood treatment (creosote) facility, a landfill, hydrocarbon spills at an airport, a drum recycling facility, a battery salvage company, and a series of oil pits.

Nebgen, J.W., et al., "Treatment of Acid Mine Drainage by the Alumina-Lime-Soda Process", EPA/600/2-76-206, Sept. 1976, U.S. Environmental Protection Agency, Cincinnati, Ohio.

The alumina-lime-soda process is a chemical desalination process for waters in which the principal sources of salinity are sulfate salts and has been field tested at the Commonwealth of Pennsylvania's Acid Mine Drainage Research Facility, Hollywood, Pennsylvania, as a method to recover potable water from acid mine drainage. The alumina-lime-soda process involves two treatment stages. Raw water is reacted with sodium aluminate and lime in the first stage to precipitate dissolved sulfate as calcium sulfoaluminate. In the second stage, the alkaline water (pH = 12.0) recovered from the first stage is carbonated to precipitate excess hardness. Following carbonation, product water meets USPHS specifications for drinking water. Alumina-lime-soda process economics are influenced most by the cost of sodium aluminate.

Neely, N.S., et al., "Survey of On-Going and Completed Remedial Action Projects", *Proceedings of the National Conference on Management of Uncontrolled Hazardous Waste Sites*, 1980, Hazardous Materials Control Research Institute, Silver Spring, Maryland, pp. 125-130.

The Resource Conservation and Recovery Act (P.L. 94-580) was promulgated in 1976 to control the ultimate disposal of solid and hazardous waste. Beginning in 1980, proper disposal and removal of hazardous substances became a major concern of the U.S. Environmental Protection Agency (EPA), other government agencies and private citizens. In an effort to determine the type and effectiveness of past corrective technology applied at uncontrolled sites, a nationwide survey of on-going and completed remedial action projects was implemented. The purpose of the survey was to provide examples of applied remedial action technologies to government officials, private facility owners/operators, and state regulators. Examples provided in the form of case histories identified problems, effectiveness, and cost related to implementing remedial action at uncontrolled hazardous waste disposal sites.

Remedial measures usually consisted of containment and/or removal of the hazardous wastes. The primary goal has been the prevention of further contamination of the environment rather than complete cleanup. Complete environmental cleanup of ground water or surface water generally requires higher technology, additional money and additional time. Therefore, a responsible party with sufficient funds and expertise must be located for complete cleanup to occur. In most cases sufficient funds have not been available for effective remedial action. It is difficult for the states and local governments to provide sufficient money for total cleanup, since any one site may require millions of dollars to correct its problems.

Based on the case studies and survey, the present state-of-the-art in applied remedial action does not look favorable when one considers that 46% of the time the applied remedial action was ineffective and only a portion of all uncontrolled sites have received some form of remedial action. In addition, remedial action applied at a site experiencing problems was only totally effective 16% of the time. It is emphasized that the percentage numbers are based on many assumptions, one of which is the subjective opinions of personnel interviewed. However, the numbers are an accurate representation of the tendency of applied remedial action. For example, it is apparent that there is a need for additional applications of corrective measures and that the state-of-the-art of remedial action needs improvement.

Neely, N.S., et al., "Remedial Actions at Hazardous Waste Sites: Survey and Case Studies", EPA 430/9-81-05, Jan. 1981, U.S. Environmental Protection Agency, Washington, D.C.

During the summer of 1980, a nationwide survey was conducted to determine the status of remedial actions applied at uncontrolled hazardous waste disposal sites. Over 130 individuals were contacted to obtain information on such remedial action projects. A total of 169 sites were subsequently identified as having been subject to corrective measures. Remedial actions were found to have been implemented at many kinds of hazardous waste disposal facilities including drum storage areas, incinerators, and injection wells, but most frequently landfill/dumps and surface impoundments. At the sites receiving such remedial actions, ground water was found to be the most commonly affected media, followed closely by surface water. Although several types of remedial measures were identified, remedial activities usually consisted of containment and/or removal of the hazardous wastes. Sufficient money was often not available for complete environmental cleanup (e.g., extraction and treatment). The survey determined that a lack of sufficient funds and/or selection of improper technologies were responsible for remedial actions having been applied effectively at only a portion of the uncontrolled hazardous waste disposal sites. Where they had been applied, remedial actions were found to be completely effective only 16 percent of the time. Nine sites were studied in detail to document typical pollution problems and remedial actions at uncontrolled hazardous waste disposal sites. Of these nine sites, remedial actions were completely effective at two and only partially effective at the other seven. Technologies employed at these nine sites included (1) containment, (2) removal of waste for incineration or secure burial, (3) institution of surface water controls, and (4) institution of ground water controls.

Niemela, V.E., Childs, K.A. and Rivoche, G.B., "Leachate Treatment and Mathematical Modelling of Pollutant Migration from Landfills and Contaminated Sites", *Proceedings of the National Conference on Management of Uncontrolled Hazardous Waste Sites*, 1982, Hazardous Materials Control Research Institute, Silver Spring, Maryland, pp. 437-441.

Pollution of both ground and surface water by toxic contaminants migrating from dangerous hazardous waste storage and disposal sites, and particularly from areas of land contaminated by various industrial activities, is a widespread problem. In this paper, the authors review and attempt to assess the current remedial technologies used in NATO countries, with special emphasis on control and treatment of leachate and of contaminated ground water. This overview includes the engineering design of various cut-off barriers, grout curtains, clay fill trenches, etc. required to contain contamination, but does not include an analysis of the physical and chemical effectiveness of such barriers. The focus of this paper is also on mathematical modelling, a proven tool in site-specific situations but perhaps a more controversial one from the viewpoint of general applicability.

Nightingale, H.I. and Bianchi, W.C., "Ground Water Chemical Quality Management by Artificial Recharge", *Ground Water*, Vol. 15, No. 1, Jan.-Feb. 1977, pp. 15-21.

The effectiveness of the Leaky Acres Recharge Facility in Fresno, California, for improving the regional ground water quality was studied as 65,815,000 m³ of high-quality surface water was recharged from 1971 through 1975. Observation wells at the facility showed some variability in chemical parameters associated with each recharge period. The long-term decrease in salinity could be described by a power law decay curve fitted by regression analysis. Recharge of Leaky Acres had noticeably decreased the ground water salinity for a distance of up to 1.6 km. in the direction of the regional ground water movement.

Ohneck, R.J., "Restoration of a Contaminated Aquifer Following an Accidental Spill of Organic Chemicals", *Proceedings of the Second National Symposium on Aquifer Restoration and Ground Water Monitoring*, 1982, National Water Well Association, Worthington, Ohio, pp. 339-342.

Current regulations prescribe efforts to be implemented to contain, monitor and clean up contaminated ground water. Accidental spills are one source of ground water contamination that is apparent and can be abated when actions are quickly and efficiently administered. New methods for containment, recovery and treatment of spills affecting ground water are being developed to satisfy these criteria. This paper contains a case history of a situation where a spill of 100,000 gallons of organic chemicals of various makeup contaminated a ground water system. The spill was responded to immediately, assessed, contained, recovered and treated on site. As a result of these efforts, the aquifer was restored to a usable condition. Innovative techniques were used, including hydraulic containment of the ground water by subsurface injection and recovery systems, mobile on-site treatment systems and in situ biological treatment of soil and ground water.

Opitz, B.E., Martin, W.J. and Sherwood, D.R., "Gelatinous Soil Barrier for Reducing Contaminant Emissions at Waste Disposal Sites", *Proceedings of the National Conference on Management of Uncontrolled Hazardous Waste Sites*, 1982, Hazardous Materials Control Research Institute, Silver Spring, Maryland, pp. 198-202.

The disposal of hazardous wastes, such as sanitary landfill leachate, chemical waste, or radioactive waste, is an increasing problem. Current practices call for waste disposal in earthen pits or repositories. However, these waste contaminants may eventually migrate so that potentially hazardous elements could be transported to the biosphere. Thus, reducing the transport of chemicals from waste disposal sites poses an immediate challenge.

Under sponsorship of the Department of Energy's Uranium Mill Tailings Remedial Action Program (UMTRAP), Pacific Northwest Laboratories (PNL) has investigated the use of engineered barriers for use as liners and covers for waste containment. A comparison of clay-soil mixtures and a gelatinous soil additive was performed to evaluate their use as a cover in reducing the escape of radioactive radon gas from abandoned uranium mill tailings piles. In addition, permeability measurements were performed to examine and compare the ability of gelatinous materials to reduce leachate movement and the migration of contaminants such as trace or heavy metals. The results of these studies led to the development of a low permeable, multilayer earthen barrier for effectively reducing contaminant emissions from waste disposal sites.

Osiensky, J.L. and Williams, R.E., "Ground Water Pump-Back System for a Uranium Tailings Disposal Site", *Proceedings of the Second National Symposium on Aquifer Restoration and Ground Water Monitoring*, 1982, National Water Well Association, Worthington, Ohio, pp. 30-37.

Seepage from unlined disposal ponds that contain uranium mill wastes has contaminated the ground water at several sites in the United States. Ground water "pump-back" systems are one alternative for controlling the migration of seepage at these sites. The installation and operation of an effective and efficient ground water pump-back system are dependent upon the adequate delineation of the site hydrogeology and the geometry of the seepage plume(s). Adequate delineation of the hydrostratigraphy and identification of the hydrostratigraphic unit(s) through which seepage is migrating is necessary to determine the proper locations, completion schedules and pumping rates for the pump-back wells. Disposal of mill wastes at one waste disposal site began in 1957. In 1980, a contaminated seep was discovered approximately 518 meters (1,700 feet) north of Tailings Pond No. 1. Hydrogeologic exploration at the site revealed an alluvium-filled paleo-channel. Precise delineation of the channel and the hydraulic properties of the alluvium were prerequisite to the implementation of a withdrawal scheme. A ground water pump-back system recently has been installed at the site to dewater the alluvium in the paleo-channel and return the contaminated water to a new tailings disposal facility designed so as not to leak.

Padar, F.V., "Identification and Control of Organic Chemicals in Sole Source Aquifers on Long Island, New York", *Proceedings of the Conference on Oil and Hazardous Material Spills: Prevention, Control, Cleanup, Recovery, Disposal*, 1979, Hazardous Materials Control Research Institute, Silver Spring, Maryland, pp. 140-150.

Widespread contamination of ground water, the sole source of drinking water for 2.8 million people on Long Island, New York, by industrial solvent discharges and underground gasoline spills has created unique public health and environmental protection problems. Ground water contamination by leaking of buried gasoline storage tanks has presented a series of technical problems including determining the spatial extent of aquifer contamination, notably from benzene, marshaling the best available technology for clean-up of spills, disposition of contaminated ground water pumped from recovery wells, development of field equipment for treatment of contaminated water, and determination of an appropriate standard or end point for restoration of ground water quality.

Paige, S.F., et al., "Preliminary Design and Cost Estimates for Remedial Actions at Hazardous Waste Disposal Sites", *Proceedings of the National Conference on Management of Uncontrolled Hazardous Waste Sites*, 1980, Hazardous Materials Control Research Institute, Silver Spring, Maryland, pp. 202-207.

Uncontrolled hazardous waste disposal sites produce adverse effects on public health and the environment as a result of the migration of hazardous waste constituents to off-site locations. Remedial actions are implemented as a means of mitigating the adverse effects produced by these sites. Before a final decision is made concerning the type of remedial actions to be implemented at a site, there should be discussions and investigations regarding the nature of the problem at the site and the efficacy of candidate remedial action techniques. At some point during this decision-making process, it will be necessary to develop preliminary designs and cost estimates for the remedial action programs being considered. In preparing the preliminary design, the investigator has to identify each of the steps that will be involved in implementation of the remedial action. Therefore, the exercise will provide officials a realistic picture of the extent of activity required by the remedial action. The cost estimate may provide useful input to the decision-making process. From a regulatory perspective it will also help to assess amount of damages that should be included in enforcement actions against responsible parties.

Parmele, C.S. and Allen, R.D., "Activated Carbon Adsorption with Nondestructive Regeneration—An Economical Aquifer Restoration Technology", *Proceedings of the Second National Symposium on Aquifer Restoration and Ground Water Monitoring*, 1982, National Water Well Association, Worthington, Ohio, pp. 99-104.

The use of activated carbon adsorption as an aquifer restoration technology has drawn widespread consideration. Often, critics point to the high cost of carbon replacement and/or thermal regeneration to justify claims of unreasonable cost. This is not the first time that the incentive to develop cheaper ways to regenerate activated carbon has been identified. IT Enviroscience has demonstrated that for process wastewaters nondestructive regeneration can be accomplished for one-third of the operating cost of thermal regeneration. IT Enviroscience believes that the economic incentive exists to evaluate activated carbon adsorption with nondestructive regeneration by itself and in conjunction with other technologies as solutions to the ground water contamination problem. This paper describes nondestructive regeneration techniques and applications. It shows how cost savings over other ground water treatment systems can be achieved.

Parmele, C.S., Alperin, E.S. and Exner, J.H., "Nondestructive Regeneration of Activated Carbon—A Viable Chemical Engineering Technology", Paper presented at the American Institute of Chemical Engineers Annual Meeting, Nov. 1981, New Orleans, Louisiana.

Solvent regeneration and steam regeneration were used to nondestructively regenerate activated carbon in a pilot plant study using six-inch adsorption columns. For solvent regeneration the pilot plant was operated on actual wastewater and also with wastewater spiked with a strong phenol >1%. In this demonstration, the study demonstrated that a stable working capacity of 0.15 phenol/lb carbon was established and maintained during 17 adsorption/desorption cycles. An effluent concentration of <1 mg/l was routinely achieved. Distillation of the acetone regenerant produced a product that could be reused in the process. Operating cost (exclusing of capital and labor) were $0.02/lb carbon regenerated while the value of the recovered phenol gave a credit of $0.03/lb carbon. Using a steam regeneration process in order to remove volatile organics such as ethylene dichloride (EDC) from process wastewater showed that a stable working capacity of 0.40 lb EDC/lb carbon was established with steam regeneration over 9 adsorption-desorption cycles, and that the EDC effluent concentration was ≤5 mg/l. Since the EDC feed concentration was 2000-7000 mg/l, EDC removal was >99.97%. Nondestructive regeneration processes will allow greater utilization of carbon adsorption by providing in-situ regenerator, adsorbate recovery and lower cost regeneration technology.

Parry, G.D.R., Bell, R.M. and Jones, A.K., "Degraded and Contaminated Land Reuse-Covering Systems", *Proceedings of the National Conference on Management of Uncontrolled Hazardous Waste Sites*, 1982, Hazardous Materials Control Research Institute, Silver Spring, Maryland, pp. 448-450.

As a result of proposals made by the United Kingdom in 1980, a NATO/CCMS pilot study on contaminated land has been established. A range of study areas and subjects for information exchange have been selected. Study Area C in this project has a UK lead, and will examine systems designed to prevent the migration of contaminants vertically or laterally or to prevent the ingress of surface or ground water into contaminated sites. In this paper, the authors discuss the desirable properties of covering systems and by example describe their use in the UK.

Pastrovich, T.L., et al., "Protection of Groundwater from Oil Pollution", Report No. 3/79, Apr. 1979, CONCAWE, The Hague, Netherlands.

Problems arising from ground water pollution by oil have gradually emerged during the past few years as an important area of concern for those involved in water resource exploitation and management. The most critical use of ground water is for domestic use. But in this case consumption of contaminated water in most cases poses little hazard to health because people and animals find the water unpalatable at levels of contamination far below those that would be harmful. The lower limit of detectability depends on the type of petroleum product and on the sensitivity of the person concerned. However, the rather low limits proposed by various authorities suggest that it is advisable to give recommendations and guidance on how to protect the subsoil drinking water resources from hydrocarbons and their biodegradation products. The "Oil Spill Manual" published by CONCAWE in August 1974 deals with the recovery of oil spilled onto the soil and onto surface waters. The main factors to be taken into account before implementing any remedial measures are also discussed in detail.

Pearson, F.H. and Nesbitt, J.B., "Combined Treatment of Municipal Wastewater and Acid Mine Drainage", Research Publication No. 73, undated, Institute for Research on Land and Water Resources, Pennsylvania State University, State Park, Pennsylvania.

Acid mine drainage (AMD), a serious water pollutant in many states, may be an economical source of ferrous iron for the chemical coagulation of municipal wastewater. AMD would then be neutralized by the alkalinity in wastewater. Samples of AMD and raw wastewater were collected from sites in Pennsylvania where AMD was found close to a wastewater treatment plant. The samples were mixed in varying ratios, then processed at controlled pH in a laboratory scale treatment plant that provided flocculation and sedimentation. The optimal pH was 8 for a maximum reduction in phosphorus, ferrous iron, and turbidity. At pH 8 the median reduction in total phosphorous was 95 percent when the molar ratio of ferrous iron to total phosphorous was 2. Ferrous iron was almost completely removed at pH 8. A cost analysis is given for the maximum distance for which it is economical to pump AMD for combined treatment with wastewater.

Pease, R.W., Menke, J.L. and Welks, K.E., "Management of Abandoned Site Cleanups: Wade Property, Chester, Pennsylvania", *Proceedings of the National Conference on Management of Uncontrolled Hazardous Waste Sites*, 1980, Hazardous Materials Control Research Institute, Silver Spring, Maryland, pp. 147-151.

In this paper, the authors describe the conditions at an illegal hazardous waste site in Chester, Pennsylvania and the steps taken for their resolution. Several of the "how to" aspects of the planning and management of an abandoned site cleanup are illustrated in the context of the Chester project. The site, known as the Wade Property, is an approximately three-acre parcel bounded by the Delaware River, the Commodore Barry Bridge, the Philadelphia Electric Company's deactivated gas storage tanks, and a railroad right-of-way. Historically, drums of wastes at the site were emptied either directly onto the ground or into trenches; this discharge has severely contaminated the soil at several locations. A fire occurred on the site in February, 1978 causing many of the stockpiled wastes to be destroyed. Large piles of debris containing exploded drums, building materials, tires, shredded rubber, and chemically contaminated earth litter the property. Approximately 150,000 gallons of waste chemicals remained after the fire; most of the material was contained in 2,500 55-gallon drums located inside fire-damaged buildings, although a large portion was stored in five bulk tankers in the front lot. The Pennsylvania Department of Environmental Resources (DER), through the aid of a contractor, was involved in the development and implementation of environmentally acceptable and cost-effective solutions to the problems of hazardous waste disposal and pollution abatement associated with the Wade site. This paper describes these solutions.

Pendrell, D.J. and Zeltinger, J.M., "Contaminated Ground-Water Containment/Treatment System: Northwest Boundary, Rocky Mountain Arsenal, Colorado", *Proceedings of the Third National Symposium on Aquifer Restoration and Ground Water Monitoring*, 1983, National Water Well Association, Worthington, Ohio, pp. 453-461.

The Rocky Mountain Arsenal was constructed by the Army just northeast of Denver, Colorado, in 1942 to manufacture mustard gas, incendiary bombs and other miscellaneous munitions. Following World War II the plant was leased to private industry for the manufacture of insecticides and herbicides. In the early 1950s the Army constructed an additional plant to manufacture nerve gas. Liquid chemical wastes from all of these operations were discharged into unlined basins, resulting in contamination of a shallow aquifer. A bentonite slurry-filled cutoff trench and a ground water treatment plant were installed at the north boundary of the arsenal in 1981. The system is successfully preventing migration of a plume of contaminated ground water off the arsenal to the north. A second plume is migrating northwesterly off the arsenal at the northwest boundary. Recent investigations have confirmed that the most feasible method to stop migration of the plume is a combined hydraulic-bentonite slurry trench barrier with a treatment plant. This barrier will consist of 11 withdrawal wells spaced 100 feet apart and a bentonite slurry-filled trench 1,300 feet long. The cutoff trench is required in a thinly saturated area of the shallow aquifer. The water pumped from the withdrawal wells will be treated in carbon columns and then injected back into the aquifer through a line of recharge wells located 600 feet downgradient from the hydraulic-bentonite slurry trench. The recharge line is necessary to establish reversal of the ground water hydraulic gradient and to assure no disruption in the quantity of ground water flow off the arsenal. The paper briefly discusses the north boundary system, but concentrates primarily on the recent study at the northwest boundary, including contaminants, aquifer analysis and design of the northwest boundary containment/treatment system.

Pennington, D., "Retardation Factors in Aquifer Decontamination of Organics", *Proceedings of the Second National Symposium on Aquifer Restoration and Ground Water Monitoring*, 1982, National Water Well Association, Worthington, Ohio, pp. 1-5.

> Movement of low-molecular weight halogenated organics in aquifer systems presents several problems for the investigation of contaminant control and aquifer restoration. Some of these organics are not appreciably absorbed in soils and move readily through aquifers. In contrast, other organics are relatively immobile. A measure of mobility of the organics is the retardation factor (ratio of the velocity of water flow to velocity of contaminant in aquifers). Retardation factors are listed for various organics including chloroform, 1,1,1-trichloroethane and chlorobenzenes. Because of retardation factors, dispersion and dilution processes will have different effects on selected organics and can result in variations of concentration of organics in the leachate plume. Therefore, the retardation factors can have a significant influence on abatement plans. Containment of some contaminants may be insignificant when an abatement plan assumes equal rates of movement of contaminants. Also, retardation factors have a significant effect on the calculation of the rate of flow of a leachate front and areal extent of the leachate plume. Case histories are utilized as examples of the significance of retardation factors.

Penrose, R.G. and Holubec, I., "Laboratory Study of Self-Sealing Limestone Plugs for Mine Openings", Sept. 1973, NUS Corporation, Pittsburgh, Pennsyvlania.

> Laboratory studies of self-sealing limestone plugs for mine openings were conducted to determine the optimum limestone material for such a treatment and sealant technique. Experimental results indicated that permeability, compressibility and strength of a limestone plug are primarily a function of the particle size distribution and density. Plug performance was most effective with high limestone placement density and smaller gradation of stone. Ferric waters were controlled most effectively. Additive effects were less significant throughout the tests.

Pye, V.I., Patrick, R. and Quarles, J., "Remedial Action and the Rehabilitation of Aquifers", *Groundwater Contamination in the United States*, 1st ed., University of Pennsylvania Press, Philadelphia, 1983, pp. 182-188.

This chapter of the book deals with a state-of-the-art review of the various aquifer restoration techniques. The most recent literature is reviewed for a number of topics including: in situ remedial alternatives; conventional alternatives of withdrawal, treatment and final disposal; and treatment options. Also included are a series of figures schematically depicting the various treatment technologies.

Quan, W., "The Effective Use of Resource Recovery in the Cleanup of Uncontrolled Hazardous Waste Sites - Based on the California Experience", *Proceedings of the National Conference on Management of Uncontrolled Hazardous Waste Sites*, 1981, Hazardous Materials Control Research Institute, Silver Spring, Maryland, pp. 380-386.

One of the most important environmental concerns today is the location and management of uncontrolled hazardous waste sites. Very often the most expeditious way of cleaning up a site is removal of the waste for disposal at a permitted hazardous waste disposal site; this is not necessarily the most prudent way of solving the problem. Furthermore, California and the rest of the country are now experiencing increasing difficulty in siting new hazardous waste disposal sites. According to a 1980 EPA contracted study of nine uncontrolled hazardous waste sites in the United States, the remedial action technologies usually employed were onsite containment and surface/ground water monitoring, and waste removal of landfill burial or incineration at permitted facilities. Only once was resource recovery employed as part of a cleanup program. In the long run, even though it usually takes a longer period of time for a site to be cleaned up through resource recovery, it is probably, in many instances, the soundest solution. Sometimes it may take a joint cooperative effort of everybody concerned for resource recovery to work. Furthermore, landfill disposal and/or on-site containment measures may be only temporary solutions. Four California case studies of uncontrolled hazardous waste sites are presented. For the purposes of this paper, an uncontrolled or abandoned hazardous waste site is defined as having one or more of the following characteristics: (1) is no longer used or is inactive and is no longer maintained; (2) is a location of unpermitted or improper waste disposal; (3) has no known owner; and (4) containment and monitoring techniques are inadequate.

Quince, J.R., "Monitoring, Recovery and Treatment of Contaminated Ground Water", *Proceedings of the Second National Symposium on Aquifer Restoration and Ground Water Monitoring*, 1982, National Water Well Association, Worthington, Ohio, pp. 58-68.

With ground water monitoring being implemented, it is becoming increasingly apparent that there exist many areas of ground water contamination. The degree of contamination varies from site to site, usually dependent on the character of the chemicals involved and the time elapsed since contaminant occurrence. In many cases, the contaminant plume has not migrated far from the source and remedial measures are possible. Contaminated ground water can be removed using a variety of techniques ranging from simple French drain systems to more complex well field configurations. Recovered ground water is treated on-site with current wastewater treatment technologies applied in numerous configurations. Treatment system effluent is commonly used for aquifer recharge creating a closed system of injection and recovery. The area of influence is isolated from the aquifer system using hydraulic control. Monitoring the physical and chemical changes in the ground water before, during and after implementation, is required to determine the system efficiency. Aquifer rehabilitation technology is evolving through direct application of current concepts.

Quince, J.R., "Subsurface Hydrocarbon Spill Recovery", *Proceedings of the Third National Symposium on Aquifer Restoration and Ground Water Monitoring*, 1983, National Water Well Association, Worthington, Ohio, pp. 47-50.

Hydrocarbon spills may well be the greatest threat to ground water resources. Hydrocarbons cover a wide range of commonly used oils (typically fuel oils and gasolines) and chemicals (such as benzene and toluene). These substances are used by almost everyone in all parts of the nation. Spills of hydrocarbons primarily occur in response to transportation accidents and storage system failures. In particular, the storage of hydrocarbons in subsurface tanks at both public and private facilities create a great potential for environmental contamination. This paper presents information regarding methods for monitoring underground storage systems, investigation of hydrocarbon spills affecting soil and ground water and remedial alternatives. A number of case histories are reviewed to outline the various elements to be considered during site investigations and to illustrate some options for cleanup.

Quince, J.R. and Gardner, G.L., "Recovery and Treatment of Contaminated Ground Water", *Proceedings of the Second National Symposium on Aquifer Restoration and Ground Water Monitoring*, 1982, National Water Well Association, Worthington, Ohio, pp. 58-68.

This paper presents information regarding underground recovery and treatment systems that have been applied to reclaim aquifers contaminated from a variety of sources. Aquifer rehabilitation is currently being considered, as an option for remedial action, at many locations across the United States. Published literature on this subject is limited, especially with regard to site-specific restorations. These systems have, however, been used in numerous cases for recovery and treatment of contaminated ground water following accidental spills of hazardous materials and contamination resulting from poor operational practice at various industrial facilities. Concepts for recovery and treatment of contaminated ground water are well established. It is the application of specialized techniques and equipment to real situations that has been limited. Recovery is accomplished using a variety of techniques, either singly or in combination, depending on site characteristics and contaminant(s) associated with the site. Effluent from the treatment process can be effectively used for recharge to the aquifer actively flushing contaminants to the recovery system. This method creates a closed loop of recovery, treatment and injection that can continually treat the ground water until removal of contaminants has been accomplished. Ground water and system monitoring is required to follow the progress of the project. Four case histories are discussed briefly regarding recovery system, treatment methods used, and system efficiency at selected monitoring points.

Raymond, R.L., "Reclamation of Hydrocarbon Contaminated Ground Waters", undated, U.S. Patent Office 3, 846, 290, Washington, D.C.

This process involves elimination of ground water contamination by providing nutrients and oxygen for natural in-situ microorganisms. Further, removing water from the contaminated area is carried out until hydrocarbons are reduced to an acceptable level. Near a contaminated well a second well is drilled for use as an injector well. The injection well is equipped with a nutrient mixing tank, an injection tube, an air pump and a sparger. The contaminated well is pumped to cause water flow from the injection site to the contaminated site. The injection well must be outside the perimeter of the contamination. Nutrients provided include nitrogen, phosphorus, and other inorganic salts. Liquid fertilizers are suggested in the 12-6-6 or 8-8-8 range. A concentration of 0.005-0.02 percent by weight of the formation water is suggested. The pH of the nutrient solution must be 6.5-7.5. To remove 1.25 tons of hydrocarbons per day 5 tons of air per day is required. If the above conditions are attained, and a growth rate of 0.02 grams per liter per day is attained, 90% of the hydrocarbons should degrade in six months. After the degradation is achieved the nutrient flow is stopped. This system is effective only in sites where the gasoline concentration is ≤ 40 ppm.

Rex Chainbelt, Inc., "Treatment of Acid Mine Drainage by Reverse Osmosis", Mar. 1970, Milwaukee, Wisconsin.

The report documents a study on the treatment of acid mine drainage by reverse osmosis. The objective of the study was to determine the feasibility of utilizing reverse osmosis to abate pollution due to acid mine drainage, and produce a water which could be used by industry or as a municipal water supply. A test site in Shickshinny, Pennsylvania was selected as a source of acid mine water for the study. A sample of this water was tested in a laboratory reverse osmosis unit to determine the design parameters for a 10,000 gallon per day demonstration unit. The results obtained during the demonstration period indicated that the reverse osmosis process has potential application in acid mine drainage treatment. A high quality water was produced which was suitable for use by industries or municipalities with a minimum of additional treatment. There are, however, operational problems which must be solved prior to utilizing reverse osmosis on a large scale. These include maintenance of high permeation rates through the membrane by reducing membrane fouling and determination of the optimum flow sheet for an acid mine treatment system utilizing reverse osmosis.

Rex Chainbelt, Inc., "Reverse Osmosis Demineralization of Acid Mine Drainage", Mar. 1972, Milwaukee, Wisconsin.

A two-phase study, involving both laboratory and field investigations, has demonstrated the feasibility of using reverse osmosis to provide potable water from acid mine drainage. The laboratory investigations involved the determination of methods for controlling iron fouling and the selection of a process flow sheet. During the field test, the process developed in Phase I was used to treat acid mine drainage from an underground aandoned anthracite coal mine. Treatment prior to reverse osmosis consisted of filtration (10 microns) followed by ultraviolet light disinfection. Brine from the RO unit was treated by neutralization, oxidation, and settling. Results obtained indicated that membrane fouling due to iron was satisfactorily controlled, but calcium sulfate fouling limited the recovery of product water to about 75%. Product water was of potable quality in all respects except for iron, manganese, and pH. Calcium sulfate precipitate on the RO membrane was successfully removed using a solution of ammoniated citric acid at pH 8.

Rich, C.A., "The Necessary Involvement and Responsibility of Separate Entities in Planning and Performance of Aquifer Restoration", *Proceedings of the Third National Symposium on Aquifer Restoration and Ground Water Monitoring*, 1983, National Water Well Association, Worthington, Ohio, pp. 3-9.

A practical problem that may misdirect or inhibit appropriate abatement of ground water contamination is the potential misunderstanding or lack of communication that can develop between various parties involved in segments of applied remediation. Advance decision making is oftentimes controlled by the problem forum, preconceived objectives, the need to posture, experience gained elsewhere, institutional guidelines and one's personal perception of the problem. As a result, those charged with a responsibility to coordinate, investigate, design and operate or litigate contamination/cleanup are confronted with interpreting technical information that may or may not be complete or properly understood. Parties involved in effecting remediation are likely to demonstrate varying levels of participation, responsibility, influence and interest toward in situ ground water conditions, off-site impacts, actual vs. projected costs and solution attainability in the real world. Certain parties will be allowed greater or lesser flexibility at different stages in the program. Others, possibly central to key concerns, may only be introduced in a peripheral capacity. For example, the hydrogeologist, usually relied upon to define the "nature and extent" of a problem, provides critically important interpretation that dictates technical approach, degree of expenditure and relative significance of future decisions by others. Because it is his duty, with the support of associated disciplines, to provide hard facts on local ground water conditions, his interpretation becomes seriously consequential in modifying determinations on the applicability, feasibility and site suitability of categorical remedial alternatives. In overview, early recognition that remedial action's success or degree of success is contingent upon the quality, experience and communicative skills of its participants and their ability to objectively receive and evaluate field conditions in a cooperative atmosphere is paramount to securing an effective program. Experiences with remedial action development at an abandoned site, in the backyards of manufacturing plants and during the course of emergency spill response are used as illustrative examples.

Rishel, H.L., Boston, T.M. and Schmidt, C.J., *Costs of Remedial Response Actions at Uncontrolled Hazardous Waste Sites*, First Edition, 1984, Noyes Data Publications, Inc., Park Ridge, New Jersey.

This book presents conceptual design cost estimates for remedial response actions at uncontrolled hazardous waste sites. Thirty-five unit operations, covering uncontrolled landfill or surface impoundment disposal sites, were costed in mid-1980 dollars for the Newark, New Jersey area; and upper and lower cost averages for the contiguous 48 states were also prepared. The data in the book are based on a review of pertinent literature with subsequent conversion to a consistent computational framework such that costs of remedial response options can be readily compared. Capital costs for each unit operation were totaled. Overhead and contingency allowances were then applied. Total capital costs were not amortized, but rather assumed to be fully incurred in the first year. Operating cost components were also identified over a period extending 10 years from the time of initial application. Inflation was included in future operating costs. These were then discounted back to the first year, and added to capital costs to yield total life cycle costs in present value dollars. The result of this costing exercise is a set of unit costs for each remedial action unit operation. These costs are in terms of the unit of application for the remedial action itself (e.g., square meters of wall face for slurry cut-offs). It should be noted that resultant costs are "base construction costs", and costs attributable to the safety and health of on-site workers and off-site receptors must be added to these.

Rishel, H.L., et al., "Costs of Remedial Actions at Uncontrolled Sites", *Proceedings of the National Conference on Management of Uncontrolled Hazardous Waste Sites*, 1981, Hazardous Materials Control Research Institute, Silver Spring, Maryland, pp. 248-254.

During 1980, the U.S. Congress enacted the Comprehensive Environmental Response Compensation and Liability Act (CERCLA, also known as Superfund, P.L. 96-510) which was proposed to provide funds for the U.S. Environmental Protection Agency (EPA) to assist in the mitigation of pollution problems at uncontrolled waste disposal sites through remedial actions. This study was conducted to provide technical information to support this process; it was done to review, compile, update and integrate existing data on the costs of such remedial actions, in terms of discrete unit operations which could then be combined to construct conceptual remedial action scenarios. This type of review-and-update approach was considered more appropriate than additional conceptual design efforts because much conceptual design work had already been done. The design work which exists, however, is scattered, incomplete and inconsistent in methodology; much of it is out of date and vague either about the methods used to arrive at a cost figure or about what components the cost figure included.

Through the review-and-update approach, a consistent methodology on the existing data in terms of scope, location, time frame and cost computations was imposed. In addition, the missing details were supplied and results presented in a uniform format, with a minimum of overlap between the individual unit operations. The resulting document presents these data in a framework of a broad and consistent methodology, with enhanced detail.

Rocky Mountain Arsenal Contamination Control Program Management Team, "Selection of a Contamination Control Strategy for Rocky Mountain Arsenal", DRXTH-SE-83206, July 1983, U.S. Army Toxic and Hazardous Materials Agency, Aberdeen Proving Ground, Maryland.

Rocky Mountain Arsenal (RMA) is an Army installation, 27 square miles in area, located northeast of the city of Denver, and adjacent to Stapleton Airport. It is surrounded by residential, business, and agricultural real estate. Since 1942, RMA has been used by both government and industry as a site for the manufacture, testing and packaging of various chemical agents and commercial chemicals. Several areas of the property have become contaminated with organic and inorganic chemical wastes as a result of these activities. In 1974, chemicals directly associated with RMA activities were found in ground water north of the Arsenal. The Army thereupon established the Contamination Control (CC) Program at RMA to identify and quantify the various types of contamination migrating across the boundaries, determine contamination sources, and develop and implement appropriate measures to assure compliance with Federal and State Environmental laws. This report documents the results of a two and one-half year study of potential contamination control strategies for RMA to ensure compliance with State and Federal statutes pertaining to the release of pollutants to the environment. The report deals with an extensive technical review and analysis of migratory pathways of hazardous contaminants and their sources, an assessment of applicable environmental laws, development of corrective strategies within available technology, screening and evaluation of alternative strategies and selection of a preferred strategy.

Ross, L.W., "Removal of Heavy Metals from Mine Drainage by Precipitation", Sept. 1973, Denver University, Denver, Colorado.

Heavy metals in mine drainage waters of the Rocky Mountains can be removed by a two-stage process consisting of (1) neutralization followed by (2) sulfide treatment. The first stage removes ferric and aluminum hydroxides, and the second (sulfide) stage precipitates the heavy metals that are most objectionable as pollutants, and that are of possible interest for economic recovery. The two-stage process has been demonstrated in the laboratory and in a field experiment.

Rott, U., "Protection and Improvement of Groundwater Quality by Oxidation Processes in the Aquifer", *Studies in Environmental Science: Quality of Groundwater*, Vol. 17, 1981, Elsevier Scientific Publishing Company, Amsterdam, The Netherlands, pp. 1073-1076.

Oxidation processes in an aquifer caused by injection of oxygen containing water can improve the ground water quality and protect the ground water against pollution. The so-called subterranean ground water treatment has been applied in several European countries for some years. Stream and transport mechanisms and chemical and biological reactions as well are described. The recharge system in a most practical manner uses the pumping well for injection of oxygenated water into the soil.

Rulkens, W.H., Assink, J.W. and Van Gemert, W.J., "Development of an Installation for On-Site Treatment of Soil Contaminated with Organic Bromine Compounds", *Proceedings of the National Conference on Management of Uncontrolled Hazardous Waste Sites*, 1982, Hazardous Materials Control Research Institute, Silver Spring, Maryland, pp. 442-447.

Soil contamination is one of the most pressing of the environmental problems existing in the Netherlands. Over the last few years particularly, much attention has been given to this subject. At the present time, a large number of hazardous waste sites that need to be cleaned up have been discovered. Current methods for treating highly contaminated soil mainly involve excavation of the soil followed by thermal treatment elsewhere. This method is rather expensive and furthermore cannot be applied in all cases. Thus there is a need for less costly methods of cleaning up contaminated soil. A large number of process alternatives for cleaning up contaminated soils can be given. Some examples of these alternatives are extraction, chemical conversion and biological degradation of the contaminations in the soil. With the exception of thermal treatment, however, no other method of treatment has been developed for practical use so far. Much research and development work still have yet to be carried out with the attendant practical problem that each type of soil contamination is different. This also means that the method of treatment has to be adapted in each particular case. The investigation described in this paper deals with the development of an on-site treatment method for a soil strongly contaminated with organic bromine compounds. The contaminated site is located in the neighborhood of the Dutch municipality of Wierden. The main problem is the potential danger of the contamination of ground water used for the production of drinking water.

Ryan, C.R., "Slurry Cut-Off Walls, Methods and Applications", Mar. 1980, Geo'Con, Inc., Pittsburgh, Pennsylvania.

Slurry cut-off walls are non-structural walls constructed underground to act as barriers to the lateral flow of water and other fluids. Principal applications are site dewatering, pollution control, and seepage barriers in the foundations of water retaining structures. In this paper, the two basic types of trench — soil bentonite (SB), and cement-bentonite (CB), and the principal kinds of slurry trenching equipment are discussed. There are examples of several recent projects with emphasis on the reasons behind the selection of the particular method. Slurry cut-off walls normally have a permeability in the range of 10^{-6} to 10^{-7} cm/sec. Recent advances in methods of analysis of slurry cut-off walls for the key factors of permeability and durability have provided much-needed assistance in the design process. The primary purposes of quality control are to check the continuity and depth of the wall, and to ensure a slurry and backfill which fall within workable limits while satisfying design criteria.

Sanning, D.E., "Surface Sealing to Minimize Leachate Generation at Uncontrolled Hazardous Waste Sites", *Proceedings of the National Conference on Management of Uncontrolled Hazardous Waste Sites*, 1981, Hazardous Materials Control Research Institute, Silver Spring, Maryland, pp. 201-205.

Many existing technologies, such as those currently being used for construction, hydrologic investigation, waste-water treatment, spill cleanup and chemical sampling and analysis, can be applied to uncontrolled hazardous waste sites. The minimization of surface infiltration will, in almost all cases, be an integral part of the remedial steps at those sites where the waste has been buried and the cost of removal is prohibitive. Minimizing surface infiltration typically consists of regrading, diverting surface water runoff

and preventing or eliminating infiltration. The effectiveness of surface sealing depends upon the contribution of surface infiltration to the total problem at the site. From a cost effectiveness standpoint and ease of applicability, minimizing surface infiltration poses marked advantages over other
types of remedial action unit operations. The Solid and Hazardous Waste
Research Division of EPA has been involved either directly, through actual
EPA funding or indirectly through technically supported efforts at several
sites where minimizing surface infiltration has been implemented. Two of
those sites are discussed in this paper.

Schmidt, K.D., "Limitations in Implementing Aquifer Reclamation
Schemes", *Proceedings of the Third National Symposium on Aquifer
Restoration and Ground Water Monitoring*, 1983, National Water Well
Association, Worthington, Ohio, pp. 105-109.

Often one of the first reactions to the discovery of polluted ground water
is the urge to remove it by pumping. In the Southwest, there are a number of
problems that may be encountered when attempting to pump polluted water.
These include land ownership, water rights issues, adverse impacts due to
pumping and disposal of treated water. In some cases, the land overlying a
considerable part of the plume and downgradient area is under separate
ownership from land where the pollution originated. Adjacent land owners
may not even allow access for monitoring purposes, let alone for the drilling
of new wells to pump polluted ground water. Thus it may not be possible to
site recovery wells in the hydrogeologically preferred locations, which
greatly limits the effectiveness of the reclamation program. In several states,
polluted ground water cannot be pumped without first obtaining a special
permit. In addition, in at least one case, the pumping of polluted water and
subsequent disposal to an evaporation pond was ruled to be a waste of
water, and hence this option was not possible.

Pumpage of polluted ground water from highly productive anisotropic
alluvial aquifers is inherently inefficient. That is, the pumped water is generally a mixture of polluted and unpolluted water. This is especially true at
sites where the vertical and horizontal extent of the plume have not been
precisely determined, prior to installation of recovery wells. At some operating reclamation projects, less than 20 percent of water pumped from recovery wells is polluted water. There can be large water-level declines in nearby
wells, and in some cases pumps and wells have gone dry. Due to changes in
the direction of ground water flow induced by pumping, substantial changes
in ground water quality can result that were unforeseen prior to their occurrence. Disposal of the pumped water can be a major problem in some cases.
Many plumes in the Southwest comprise tens or hundreds of thousands of
acre-feet of polluted water. If it is not possible to treat and immediately use
the pumped water, then there may be a substantial disposal problem.
Recharge of treated water is difficult in some situations due to unfavorable
soils and geologic conditions. Often, there are a substantial number of

potential pollutants in the pumped water. If attention is focused on only one or several of these pollutants and comprehensive chemical analyses are not performed, then disposal of the water by land treatment can create additional ground water pollution. An example would be land spraying of a pumped polluted water in order to volatilize a specific organic constituent. Although the levels of the organic constituents in percolating water may be adequately reduced by this process, ammonia and organic nitrogen in the pumped water could be oxidized to nitrate. This could pollute ground water beneath the land disposal site.

Another problem in some aquifer reclamation schemes in the Southwest is that substantial amounts of pollutants are present in the vadose zone. Even after polluted water is removed from the aquifer, substantial pollution may occur due to slow drainage of polluted water from the vadose zone over many years and decades. In addition, specific pollutants may be adsorbed on solid materials in the aquifer. Later, some of these could be desorbed, and pollute relatively unpolluted water then present in the former plume. The majority of the problems in implementing aquifer reclamation projects in the Southwest have been due to one or more of the following factors: (1) the hydrogeologic system was not adequately understood prior to implementing the reclamation project; (2) the extent of the plume had not been determined; and (3) all potential pollutants in the plume had not been identified. In turn, these factors often appear to be related to two other phenomena. First is the lack of ground water hydrologists at the forefront of many reclamation projects. Second are the unreasonable time constraints imposed by some regulatory agencies, often because of intense public pressure.

Schuller, R.M., Beck, W.W., Jr. and Price, D.R., "Case Study of Contaminant Reversal and Groundwater Restoration in a Fractured Bedrock", *Proceedings of the National Conference on Management of Uncontrolled Hazardous Waste Sites*, 1982, Hazardous Materials Control Research Institute, Silver Spring, Maryland, pp. 94-96.

Ground water contaminated with trichloroethylene (TCE) was discovered in the vicinity of a manufacturing plant in southeastern Pennsylvania where contamination had resulted from a series of spills. Subsequent to this discovery, the plant contracted with SMC Martin to study the extent and magnitude of the TCE contamination in the vicinity of the plant and to develop an appropriate method for cleanup. A rapid response to cleanup of the ground water was desirable due to the potential for contamination of residential wells in the vicinity of the plant. However, the study was constrained by the availability of locations for well placement and limited client resources. As is typical of most industrial projects, the client and the regulatory agency were interested in a cost effective and efficient solution to the problem, which is rarely possible in the complex hydrogeologic conditions of southeastern Pennsylvania. However, the uniquely simplistic fracture system controlling ground water flow in the area resulted in a well-defined contaminant migration path which permitted a direct approach to contaminant reversal and ground water restoration.

Schuurmans, R.A., Steinmetz, J.J. and van den Akker, C., "Multi-Purpose Solution for the Protection Against Pollution of a Groundwater-abstraction for Watersupply", *Studies in Environmental Science: Quality of Groundwater*, Vol. 17, 1981, Elsevier Scientific Publishing Company, Amsterdam, The Netherlands, pp. 1083-1088.

The Municipal Waterworks of Amsterdam (G.W.) founded a ground water pumping station in the year 1888 north of the town of Hilversum, 25 km south-east of Amsterdam. In 1901 another water supply company, the "Midden Nederland" Water Works (W.M.N.), started in the same area. Already in 1875 the municipality of Hilversum made infiltration works for sewage water east of town, in an area with small lakes with a distance of about 1500 m south-east of the pumping stations. Together with the natural flow to the north-west, the recharged sewage water flows to the pumping stations and can cause trouble. During the last century new residential and industrial areas were built to the north of the old town of Hilversum. Today the border of the town is close to the pumping stations. The ground water below the town is polluted by sewage water from the leaky sewerage and probably by direct infiltration of sewage. The water level in the sandy sub-soil is about 5 to 10 m below the land surface. From 1934 investigations were made to find a solution for the problem of pollution. In 1977 in the attracted ground water of G.W. a little amount of Trichloroethene (TCE) was discovered. This gave a start to a new approach to the problem.

Scott, R.E., Hill, R.B. and Wilmoth, R.C., "Cost of Reclamation and Mine Drainage Abatement: Elkins Demonstration Project", 1970, Robert A. Taft Research Center, Cincinnati, Ohio.

Acid mine drainage, discharging from coal beds, pollutes streams and rivers. An acid mine drainage reclamation project was established in the Roaring Creek-Grassy Run watershed near Elkins, West Virginia. The control costs established from the project are presented. The costs include those for clearing, grubbing, reclamation, and revegetation.

Sheahan, N.T., "Injection/Extraction Well System - A Unique Seawater Intrusion Barrier", *Ground Water*, Vol. 15, No. 1, Jan.-Feb. 1977, pp. 32-49.

A multiple-aquifer system in the bayfront area of Palo Alto, California, is being intruded with seawater from San Francisco Bay. In order to combat this potential degradation of the ground water supplies in the area, a sea water intrusion barrier is being constructed consisting of a series of injection wells used to inject 2.0 million gallons per day (7.6 x 10⁶ l/d) of reclaimed wastewater into a shallow aquifer. The injected water is subsequently removed by a similar system of extraction wells to avoid any possible degradation of the water-supply aquifers from this source, and to allow reuse of the reclaimed wastewater. The investigation phase included test drilling, aquifer testing and injection testing to determine the feasibility of the injection/extraction concept. The number, spacing and location of I/E doublets were optimized using a digital computer model. The double-cased, double screened wells were constructed using corrision-resistant materials, and were designed for ease of routine maintenance. In operation, injection and extraction will be computer controlled by sensing piezometric levels in a series of monitor wells. Water pumped from extraction wells will be sold for industrial and agricultural purposes.

Shepherd, W.B., "Practical Geohydrological Aspects of Ground-Water Contamination", *Proceedings of the Third National Symposium on Aquifer Restoration and Ground Water Monitoring*, 1983, National Water Well Association, Worthington, Ohio, pp. 365-372.

An understanding of some basic aquifer properties and fluid distribution principles is necessary in order to assess and resolve underground contamination problems. The varied effects of petroleum product and chemical spills on land are discussed, including the physical and chemical properties of both the spilled product and the soil. The effect of capillarity in the funicular or transition zone during percolation, and lateral movement of mobile product is discussed. Particular emphasis is placed on determination of the volumetrics of recoverable hydrocarbons as analyzed from monitoring well data and aquifer soil properties. The technical aspects of hydrologic determination of the lateral extent of underground contamination are discussed. Also considered is the endurance of taste and odor from contaminated ground water after years of percolating rainwater representing many pore volumes of displaced fluid. Aspects of contaminant flow are emphasized, including contaminant plume velocity versus ground water flow. Spill mitigative measures for phase-separated product plumes are discussed, with the emphasis on practical aspects and field determinations. In addition, some aspects of hydrologic containment of contaminants are compared to physical barrier emplacement.

Shipps, M.H., "A Cost Effective Approach to the Clean-Up of Groundwater Spills", *Proceedings of the Conference on Oil and Hazardous Material Spills: Prevention, Control, Cleanup, Recovery, Disposal*, 1978, Hazardous Materials Control Research Institute, Silver Spring, Maryland, pp. 160-163.

New developments related to ground water spill clean-up are providing a more cost-effective approach to clean-up. Currently, difficulties are being encountered with the use of costly excavation techniques and inefficient recovery devices without the capabilities of hydrocarbon/water separation. Recent improvements in recovery well installation and in the development of in-ground hydrocarbon separations, coupled with an increased awareness of hydrogeological principles have significantly improved the state of the art. This new technique results in pure product recovery and provides an efficient and cost effective clean-up of the ground water.

Shuckrow, A.J., et al., "Bench Scale Assessment of Technologies for Contaminated Ground Water Treatment", *Proceedings of the National Conference on Management of Uncontrolled Hazardous Waste Sites*, 1980, Hazardous Materials Control Research Institute, Silver Spring, Maryland, pp. 184-191.

Contamination from unsecured industrial waste storage and disposal sites is a widespread problem. Often, this contamination manifests itself in the form of hazardous leachates and contaminated ground and surface waters. These contaminant streams are diverse in terms of composition and concentration—varying from site to site, from location to location within a site and often over time at any given location. Some contaminant streams contain a broad spectrum of organic and inorganic constituents, while others have only a few compounds of concern. Regardless of whether contaminant streams are associated with active or abandoned sites, the need to detoxify/decontaminate these hazardous aqueous wastes sometimes arises. However, hazardous aqueous waste treatment for this application is not a routine operation. Little information on and/or experience with treatment technology applied to hazardous leachate or contaminated ground water exists.

In order to lessen this information gap, a project sponsored by the U.S. Environmental Protection Agency was undertaken to evaluate and verify selected concentration techniques for hazardous constituents of aqueous waste streams. This project entails literature search/data acquisition, desktop technology evaluations and experimental investigations to evaluate and adapt appropriate technologies for the applications of interest. The background of the study, and earlier efforts to compile data on hazardous waste problem sites and to screen pertinent treatment technologies have been reported elsewhere. Thus, the focus of this paper is upon the results of the desktop technology evaluations and upon ongoing experimental treatability studies. These treatability studies are being conducted at the Ott/Story site in Muskegon, Michigan.

Shukle, R., "Rocky Mountain Arsenal Ground Water Reclamation Program", *Proceedings of the Second National Symposium on Aquifer Restoration and Ground Water Monitoring*, 1982, National Water Well Association, Worthington, Ohio, pp. 366-368.

The most sizeable ground water contamination problem encountered in Colorado is the Rocky Mountain Arsenal. The Arsenal is a federal facility which has both manufactured and demilitarized chemicals and chemical-filled munitions plus leased facilities for pesticide manufacture. Generated wastes in the 1940s and 1950s were disposed of in several unlined lagoons. In 1975, the Department of Health issued orders to the Army and Shell Chemical Co. to respond to observed ground water contamination. Subsequent monitoring has identified unique organic contamination for more than 30 square miles. To date, the Army and Shell have been involved in monitoring, identifying contaminants, evaluating treatment alternatives, identifying hydrogeologic conditions, testing pilot systems and designing and constructing full-scale treatment facilities. The extended north boundary treatment system is 6,700 feet long, averages 25 feet deep, contains both dewatering and reinjection wells, functions with a bentonite barrier and handles about 400 gpm of water. The northwest boundary system is 1,700 feet long, wells range from 65 to 85 feet deep, initial flows were 1,600 gpm with projected flows at 700 gpm and functions with a hydrologic barrier, as opposed to a bentonite barrier. Both systems include activated carbon treatment for removal of organics.

Shultz, D.W. and Miklas, M.P., "Assessment of Liner Installation Processes", *Proceedings of the Sixth Annual Research Symposium on Disposal of Hazardous Waste*, 1980, U.S. Environmental Protection Agency, Cincinnati, Ohio, pp. 135-159.

Southwest Research Institute is conducting a study to identify current methods and equipment used to (1) prepare supporting subgrade and (2) place liners at various impoundments in the United States. Subgrade preparation and liner placement activities have been observed at fifteen sites to date. The sites selected have included landfills, wastewater impoundments, and potable water reservoirs. Information obtained during each site visit included: (1) methods and equipment used to prepare the subgrade upon which the liner is to be placed; (2) methods and equipment used to place the liner material; (3) special problems encountered and their solutions; and (4) important characteristics which must be considered during design and construction of an impoundment facility. Various aspects of subgrade preparation and liner placement are discussed herein. Photographs depicting construction and placement activities are also presented.

Shultz, D.W. and Miklas, M.P., "Procedures for Installing Liner Systems", *Land Disposal of Hazardous Waste: Proceedings of the Eighth Annual Research Symposium*, 1982, U.S. Environmental Protection Agency, Cincinnati, Ohio, pp. 224-238.

The field placement procedures used to construct a variety of generic types of liner systems for landfills and surface fluid impoundments are discussed. These include procedures to install flexible polymeric membranes, sprayed-on membranes, native materials, soil sealants and admixes. In addition to placement procedures, methods and equipment used to prepare and subgrades are discussed. Placement procedures used to install polymeric liner systems have certain commonalities, such as anchoring methods and approaches to material positioning. There are differences, however, with respect to seaming technology. For example, solvents and heat are two methods used to field seam the same material. Liner material placement procedures and subgrade conditions as recommended by manufacturers and installers are compared to field observations. Generally, field practice appears to follow these recommendations. The critical aspects of subgrade preparation and liner placement generally accepted by the industry as important to a good job include proper subgrade compaction, installation by knowledgeable, experienced crews familiar with the liner material, and a strong quality assurance program during installation.

Silka, L.R. and Mercer, J.W., "Evaluation of Remedial Actions for Groundwater Contamination at Love Canal, New York", *Proceedings of the National Conference on Management of Uncontrolled Hazardous Waste Sites*, 1982, Hazardous Materials Control Research Institute, Silver Spring, Maryland, pp. 159-164.

During the coming years, considerable effort and resources will be committed to the investigation and clean up of ground water contamination caused by hazardous waste disposal, spills and other activities. To optimize the limited resources available for the solution of these problems, and maximize the protection of public health and the environment, remedial actions must be adequately selected, designed and monitored. Ground water modeling is an excellent planning tool that can assist in the evaluation and selection of remedial actions. Modeling provides quantitative analyses of site hydrology and allows prediction of the effects of proposed remedial actions on the fate of subsurface contaminants. Simulation of remedial actions during the planning stages can provide decision makers with considerable insight for making decisions on appropriate actions to be taken at a site. Such a tool, applied from the outset of a remedial action, can reduce trial and error, costs, and time on a project. Modeling, in addition, provides the means to predict future consequences of planned actions.

Sills, M.A., Struzziery, J.J. and Silbermann, P.T., "Evaluation of Remedial Treatment, Detoxification and Stabilization Alternatives", *Proceedings of the National Conference on Management of Uncontrolled Hazardous Waste Sites*, 1980, Hazardous Materials Control Research Institute, Silver Sprin, Maryland, pp. 192-201.

Uncontrolled hazardous waste sites, such as Love Canal, New York, Woburn, Massachusetts and the Valley of the Drums, Kentucky, are symbolic of both a national problem and a national commitment. The environmental and public health hazards associated with these sites have emphasized the need to escalate regulatory control efforts, technical evaluations and the financial investments required for the identification, cleanup and remedial management of all existing uncontrolled hazardous waste sites. For those cases where excavation, removal and off-site disposal of hazardous materials from an uncontrolled site are employed, the technical, economic and environmental evaluations are normally straightforward and limited to the actual excavation, hauling and final disposal activity. In these cases, the private contractor with the lowest bid is usually the most cost-effective alternative. A more typical and difficult problem, however, is the case of in situ or in-place contamination of the subsurface soil matrix or ground water under an uncontrolled site, due to the leaching of hazardous substances. Subsequent contamination of adjacent surface water bodies by direct runoff of hazardous substances or recharge by contaminated ground water is an equally serious companion consideration. The evaluation of a wide range of remedial treatment, detoxification and stabilization alternatives for these situations require a systematic method of analysis in order to accurately compare the technical, economic and environmental impact aspects of each remedial management option under consideration. This paper describes the basic on-site treatment, detoxification and stabilization alternatives available for the remedial management of uncontrolled hazardous waste sites and outlines a systematic evaluation procedure that can be used to identify the most cost-effective, environmentally sound alternative for a given site. This procedure was initially developed for the evaluation of existing hazardous waste sites in New England.

Sims, R.C. and Wagner, K., "In Situ Treatment Techniques Applicable to Large Quantities of Hazardous Waste Contaminated Soils", *Proceedings of the National Conference on Management of Uncontrolled Hazardous Waste Sites*, 1983, Hazardous Materials Control Research Institute, Silver Spring, Maryland, pp. 226-230.

Uncontrolled disposal of hazardous wastes frequently results in the production of large quantities of contaminated soils. It is often cost prohibitive or impractical to excavate, haul and dispose of these soils in an approved landfill. In situ treatment of these soils was investigated in this project as an alternative. The approach to treating soils is based on: (1) fundamental soil chemical, physical and biological processes, (2) methods developed for land treatment of industrial wastes and (3) hazardous waste land treatment (HWLT) regulations recently promulgated in 40CFR, part 264, 1983. The goals of in situ management include treating contaminated soils until an acceptable level is achieved and protecting ground water and surface water resources without physically removing or isolating the contaminated soil from the contiguous environment.

Spencer, R.W., Reifsnyder, R.H. and Falcone, J.C., Jr., "Applications of Soluble Silicates and Derivative Materials in the Management of Hazardous Wastes", *Proceedings of the National Conference on Management of Uncontrolled Hazardous Waste Sites*, 1982, Hazardous Materials Control Research Institute, Silver Spring, Maryland, pp. 237-243.

Soluble silicates are currently being used in chemical based methods of waste disposal, but the authors feel they could be utilized much more. This report is a review by the authors of the literature pertaining to waste treatment for disposal, with emphasis on solidification of liquids and sludges. The chemical methods currently used, using both silicate-based, and those that use silicates indirectly, are described. Some theoretical papers with implications for this use are reviewed, and possible new ways of using silicates are described; solidifying solvents for transportation; grouting landfills to reduce permeability and divert or block subsurface flows; and modifications of other processes to improve leaching and physical properties.

Spooner, P.A., Wetzel, R.S. and Grube, W.E., Jr., "Pollution Migration Cut-Off Using Slurry Trench Construction", *Proceedings of the National Conference on Management of Uncontrolled Hazardous Waste Sites*, 1982, Hazardous Materials Control Research Institute, Silver Spring, Maryland, pp. 191-197.

Over the last two decades, cut-off walls emplaced by slurry trenching have been installed at many solid and hazardous waste facilities. This paper presents the interim results of a project for USEPA's Office of Research and Development to develop a technical handbook on the use of slurry trenching techniques to control pollution migration. Slurry trenching is a method by which a continuous trench is excavated (by backhoe or clam-shell grab) under a slurry of bentonite and water. This slurry, or more correctly, colloidal suspension, supports the trench walls and allows excavation to greater depths with no other means of support. Once a trench has been excavated under slurry to the required depth and length, it is backfilled to form a continuous, nearly impermeable barrier, or cut-off wall, to ground water flow. In some cases, the slurry is a mixture of water, cement, and bentonite which hardens in place to form the final barrier. In other instances, this trenching technique is used to place soil-bentonite cut-off walls, where the backfill is composed of the excavated soil with small amounts (1 to 4%) of bentonite added. In some instances, borrowed soil materials, such as additional fines, or gravel may have to be added to meet the requirements of lower permeability or higher strength. Another variation of this technique, seldom used for pollution migration control, is the use of pre-cast or cast-in-place concrete diaphragm walls, which are installed when both pollution migration cut-off and structural support are required.

Spooner, P.A., et al., "Compatibility of Grouts with Hazardous Wastes", EPA-600/S2-84-015, Mar. 1984, U.S. Environmental Protection Agency, Cincinnati, Ohio.

A study was conducted to determine existing information on the compatibility of grouts with different classes of chemicals. The data gathered can be used as a guide for testing and selecting grouts to be used at specific waste disposal sites with various leachates.

The 12 types of grouts used in this study were chosen because of their availability and use in waterproofing and soil consolidation projects. These grouts are bitumen, Portland cement Type I, Portland cement Types II and V, clay, clay-cement, silicate, acrylamide, phenolic, urethane, urea-formaldehyde, epoxy, and polyester. Sixteen general classes of organic and inorganic compounds are also identified as being the types most likely to be found in leachate from a hazardous waste disposal site. The known effects of each chemical class on the setting time and durability of each grout are identified and presented in a matrix. These data were based on a review of the available literature and contact with knowledgeable persons in industries, universities, and government agencies. The physical and chemical properties, reaction theory, and known chemical compatibility of each grout type are discussed. Since compatibility data are not complete for each grout type, predictions are made where possible for the silicate and organic polymer grouts based on their reaction theory. These results are also presented in a matrix.

To establish the compatibility of chemicals with grouts, a series of laboratory tests should be performed. The two grout properties that must be addressed are permeability of the grouted soil and set time of the grout. No established testing procedures are identified in the literature for determining the effects of chemicals on these grout properties. Fixed-wall and triaxial permeameters, which are used for soil testing, can be used for measuring the effects of chemicals on permeability. No single procedure applies to all grout types for determining set time. Visual observation is the easiest method, though somewhat subjective. The selection of a grout for a specific waste site depends on its injectability, durability, and strength. These factors relate site hydrology, geochemistry, and geology to grout physical and chemical properties.

Srivastava, V.K. and Haji-Djafari, S., "In Situ Detoxification of Hazardous Waste", *Proceedings of the National Conference on Management of Uncontrolled Hazardous Waste Sites*, 1983, Hazardous Materials Control Research Institute, Silver Spring, Maryland, pp. 231-236.

Uncontrolled disposal of hazardous wastes usually results in contamination of soil and rock followed by ground water contamination. The removal of contaminated ground and emplaced waste is costly and often impractical. Open excavation requires careful transportation of contaminated mixtures and disposal in a safe landfill. An alternative to contaminated ground removal is in situ treatment or detoxification. Although many of the fundamental concepts of in situ detoxification are known, and others are being studied at several laboratories, the feasibility assessment has not been collectively considered for different applications, mainly because these technologies are either in the conceptual or early infancy stage. Recently, the USEPA began a research program at the Municipal Environmental Research Laboratory, Cincinnati, Ohio, to investigate in situ treatment of hazardous wastes. A literature search has been conducted to identify the technologies that either exist or have the potential for further investigation. The most common technologies considered for in situ treatment include: immobilization; biodegradation; neutralization; solvent mining; and oxidation-reduction.

One significant limitation of in situ treatment is that it is waste specific. Heavy metals, pesticides, residues, organic solvents, inorganic salts, explosives, etc., have their own chemical properties and require different treatment. Proper identification of wastes and their physical, chemical and toxicological properties is vitally important to determine the best strategy planning for in situ treatment to avoid undesirable reactions or production of additional unwanted toxic species. The applications of in situ detoxification technologies for organic and inorganic contaminants are identified and their applications are assessed with particular regard to their: availability—commercially developed, pilot plant or laboratory scale; feasibility—design, construct and operation feasibility; cost-effectiveness—design, construction and operational cost; advantages and disadvantages—unwanted chemical reactions or end products; environmental impact—assess environmental hazards, spread of contamination; and safety method—personnel safety.

St. Clair, A.E., McCloskey, M.H. and Sherman, J.S., "Development of a Framework for Evaluating Cost-Effectiveness of Remedial Actions at Uncontrolled Hazardous Waste Sites", *Proceedings of the National Conference on Management of Uncontrolled Hazardous Waste Sites*, 1982, Hazardous Materials Control Research Institute, Silver Spring, Maryland, pp. 372-376.

Many uncontrolled hazardous waste sites across the nation are currently being "cleaned up" or are slated for remedial action implementation in the near future. An initial step in any remedial action plan at hazardous waste sites is identification of potential remedial alternatives, followed by the selection of the most appropriate one. The formality required in this identification/selection phase may be dependent on the source of funding for the remedial action. For example, CERCLA requires that remedial action conducted using Superfund monies be demonstrated to be the cost-effective alternative that adequately protects human health and the environment. Whether state agencies or private contractors use a cost-effectiveness assessment per se, some procedure is necessary to select the most appropriate plan from a list of potential alternatives.

In the first part of this paper, the authors discuss the concepts related to cost-effectiveness and their applicability to assessment of remedial action alternatives. After reviewing these concepts, the development of a specific methodology for systematic, accurate assessments of potential remedial action plans it outlined. In developing a methodology for assessing cost-effectiveness, the objectives and criteria which must be met should be well defined. One major purpose of the cost-effectiveness framework is to promote consistency in decision-making while maintaining applicability to widely varying situations. The methodology must be simple to apply, requiring only minimal instructions on its use. In this regard, the analysis should be based on the smallest possible number of independent variables that address all relevant concerns. Because of the nature of the problem at uncontrolled hazardous waste sites, the methodology must not be dependent on large amounts of information, either on site conditions or on the remedial alternatives being considered. However, the method must readily allow for consideration of newly obtained information if it becomes available. In addition, it should not be overly quantitative or impose a precision on the analysis that is inconsistent with the degree of knowledge about the problem or expected results. Finally, the methodology should incorporate a means for determining the sensitivity of the analysis to judgments made in applying it.

Stief, K., "Long Term Effectiveness of Remedial Measures", *Proceedings of the National Conference on Management of Uncontrolled Hazardous Waste Sites*, 1982, Hazardous Materials Control Research Institute, Silver Spring, Maryland, pp. 434-436.

The effectiveness of widespread remedial actions can only be proved in the field by a decrease in contamination, e.g., decreasing ground water pollution. The aim of assessment of long-term effectiveness is to find the point in time, at which a remedial action must be repeated. Constructional sealing measures can be expected to be effective for a maximum period of 50-100 years, which is accepted as a reasonable depreciation period in civil engineering. Whether particular chemical or microbiological strains at a contaminated site causes a shorter useful life time, must be checked by monitoring. If the contaminants, which contained by capping or barrier-system are not changed within this "calculated" lifetime, so that the hazardous contamination, present at the beginning of remedial action, has decreased to an acceptable threshold, a new remedial action scheme has to be designed. However, the degree of contamination present at this time may require less stringent remedial methods. The objective of monitoring these facilities is to check the accuracy of the assumption of long-term effectiveness, so that the remedial measures can be renewed before if needed at the appropriate time.

Stover, E.L., "Removal of Volatile Organics from Contaminated Ground Water", *Proceedings of the Second National Symposium on Aquifer Restoration and Ground Water Monitoring*, 1982, National Water Well Association, Worthington, Ohio, pp. 77-84.

Indiscriminate disposal of industrial wastes has become one of the most important environmental issues facing the United States today. A serious threat from these activities is contamination of drinking water supplies by a variety of organic compounds. Contamination of ground water supplies by organic compounds, including volatile organics commonly found in many industrial cleaning solvents, has recently received widespread media coverage. The feasibility of removing volatile organic compounds from contaminated ground water was evaluated in parallel activated carbon adsorption and air-stripping studies. The removal of 1,1,1-trichloroethane detected at levels of concern, as well as other volatile organics detected in lower concentrations, were investigated in pilot scale studies. The 1,1,1-trichloroethane could be effectively air-stripped to less than 10.0 ppb with other volatile organics stripped to less than 1.0 ppb.

Sudat, M.M., Mateo, M. and Farro, A., "Management Plan for Hazardous Waste Site Cleanups in New Jersey", *Proceedings of the National Conference on Management of Uncontrolled Hazardous Waste Sites*, 1983, Hazardous Materials Control Research Institute, Silver Spring, Maryland, pp. 413-419.

A Management Plan for hazardous waste site mitigation has been prepared by the Department of Environmental Protection (DEP), Division of Waste Management, of the State of New Jersey, for the period of 1983-1986. The purpose of the plan is to develop a systematic approach to remedial action at hazardous waste sites, to coordinate cleanup and enforcement actions and to identify future funding needs and sources. There are a total of 106 hazardous waste sites identified in the plan. These include: 65 sites on the National Priorities List (NPL) issued by the USEPA on Dec. 20, 1982 and eligible for Federal Superfund monies and 41 sites which were not included on the NPL. Thirty-four drum dump sites listed in the draft plan approved by USEPA on Apr. 11, 1983, were cleaned up in 1983. Those sites not on the NPL will be addressed through New Jersey's Spill Compensation Fund and Hazardous Discharge Fund as were the recently cleaned up sites. While cleanup projects are expected to be initiated within the period of the plan, completion of cleanups at all scheduled sites will be a seven- to eight-year effort with some sites requiring maintenance for many years beyond that time. The plan also lists the 68 hazardous waste sites with over 10,000 drums cleaned up by DEP since 1980 at a cost of approximately $3 million. Including the $26 million spent on cleanup of Chemical Control in Elizabeth and $6 million on Goose Farm in Plumsted Township, the state's cost for remedial work at these 70 hazardous waste sites is over $35 million to date.

Swinnerton, C.J., "Protection of Groundwater in Relation to Waste Disposal in Wessex Water Authority", *Studies in Environmental Science: Quality of Groundwater*, Vol. 17, 1981, Elsevier Scientific Publishing Company, Amsterdam, The Netherlands, pp. 1089-1095.

Wessex Water Authority as one of the ten water authorities in England and Wales has the responsibility of liaising the waste disposal authorities which are the County Councils to ensure that both ground water and surface water resources are not polluted by waste disposal activities. All new waste disposal sites require both planning permission and a site license, and both these authorisations are referred to the water authority for comment. Conditions can be requested for inclusion on both so that adequate protection is afforded. Failure for agreement on conditions necessary can be referred to the Secretary of State for decision. Several water authorities have developed aquifer protection policies in which the aquifers are specified as either requiring various degrees of protection or in some cases as being of minimal resource value and therefore not justifying protection.

Within Wessex no aquifer protection policy as such has been developed although the general policy adopted in waste disposal liaison is to discourage waste disposal sites on aquifers. However, taking a realistic view and bearing in mind the need to balance the various interests involved, as required by Government, each site is looked at on its individual merits and disadvantages, and sites on aquifers are permitted, subject to adequate safeguards, if the need is sufficiently great. In addition, preference is given to the "contain and treat" approach rather than the "dilute and disperse" unless adequate evidence is available to support the latter.

Of the 9918 km² total Wessex area, 36% is outcrop of a wide range of aquifers found in Wessex, and these provide water for approximately half of the total public water supply demand which is around 900 megalitres per day at present. Clearly with ground water contributing such a high proportion of potable supplies it is essential that due regard is given to its protection in siting waste disposal sites. The practical approach of looking at sites individually and accepting sites on aquifers where sufficient reliable evidence is provided to show that no harmful effects to ground water will result has been well accepted by the counties, this being demonstrated by the fact that only one site has been referred to the Secretary of State for decision due to failure to agree between Wessex and a county.

Tewhey, J.D., Sevee, J.E. and Fortin, R.L., "Silresim: A Hazardous Waste Case Study", *Proceedings of the National Conference on Management of Uncontrolled Hazardous Waste Sites*, 1982, Hazardous Materials Control Research Institute, Silver Spring, Maryland, pp. 280-284.

> The Silresim Chemical Corporation's chemical waste reclamation facility in Lowell, Massachusetts was abandoned in Jan. 1978; approximately one million gallons of hazardous material were left behind in drum and bulk storage. The five-acre reclamation facility, established in 1971, had been accepting approximately three million gallons of oil wastes, solvents, chemical process wastes, plating wastes, heavy metal containing sludges and other materials yearly. The facility was designed and licensed for the ultimate disposal or recycle of these chemical wastes. Site investigations conducted in 1977 revealed license violations; the license was revoked when Silresim declared bankruptcy later that year. The Commonwealth of Massachusetts, Department of Environmental Quality Engineering (DEQE) initiated efforts to clean up the site and by Sept. 1981 all stored materials had been removed. In Oct. 1981, a two-part study was initiated to characterize the nature and extent of soils and ground water contamination caused by the hazardous materials and to recommend actions to remedy the contamination problem.

Texas Research Institute, Inc., "Final Report, Underground Movements of Gasoline on Groundwater and Enhanced Recovery by Surfactants", D743 (1-3) - F, Sept. 1979, American Petroleum Institute, Washington, D.C.

> A combination of anionic and nonanionic surfactants (Richonate YLA [77030-90-1] and Hyonic PE 90 [65035-40-7]) was most effective in displacing gasoline from 1-dimensional sand-pack models of an aquifer. A 2-dimensional model simulating a soil section from a gasoline spill zone to a recovery well showed that the surfactants acted by 2 mechanisms: simple displacement of the gasoline and draining of the capillary zone which allowed immobile gasoline to drain to a narrower zone and flow to a recovery point.

Texas Research Institute, Inc., "Enhancing the Microbial Degradation of Underground Gasoline by Increasing Available Oxygen", 8081 (FR) WLM, Feb. 1982, American Petroleum Institute, Washington, D.C.

> The objective of this work was to develop means of enhancing the microbial degradation of gasoline in the subsurface soil strata by increasing the available oxygen. The project consisted of two phases—a literature survey,

followed by the experimental work. Several papers were found concerning the biodegradation of gasoline in ground water, but none with regards to the subsurface soil strata. The first aim of the literature review was to develop a thorough characterization of the hydrocarbon oxidizers and, from this, propose "ideal" degradation conditions. Focus was then brought on methods, both physical and chemical, for oxygenating the biodegradation environment. Only two physical means were proposed. Roto-tilling was deemed feasible for aeration of shallow contaminated areas and soil venting with liquid oxygen or air was a possibility for deeper contamination. Chemical techniques cited included decomposition of peroxides and superoxides, as well as peroxyacids and salts. Considering all possibilities, three approaches were deemed feasible. These were the injection of liquid oxygen or air, the injection of oxygen releasing compounds along with nutrients, and simple venting, if the soil was porous. The chemical costs were listed with regards to the cost per pound of oxygen produced. Liquid oxygen, compressed oxygen, and 35% hydrogen peroxide were the three cheapest chemicals. All things considered, the injection of H_2O_2, along with the necessary nutrients was deemed the most beneficial method.

The experimental phase examined the toxicity of H_2O_2 to the hydrocarbon utilizers in a liquid medium under various H_2O_2 and microbial concentrations. It was found that H_2O_2 was toxic at concentrations of ≥ 100 ppm; new cultures (>2 days old) could withstand concentrations as high as 10^3 to 10^4 ppm. In fact, it was seen that bacterial growth, at times, exhibited an increase, probably due to H_2O_2 decomposition. As the concentration of microorganisms increased, a higher concentration of H_2O_2 was required to be toxic. The final phase of the experimental work was an attempt to look at the toxic effects under constant H_2O_2 concentrations in a sand packed column. Problems resulted when bubble formation in the column caused a blockage of the H_2O_2 flow, even at concentrations as low as 100 ppm.

Thompson, S.N., Burgess, A.S. and O'Dea, D., "Coal Tar Containment and Cleanup, Plattsburgh, New York", *Proceedings of the National Conference on Management of Uncontrolled Hazardous Waste Sites*, 1983, Hazardous Materials Control Research Institute, Silver Spring, Maryland, pp. 331-337.

A gas generating plant was operated along the banks of the Saranac River in the City of Plattsburgh, New York, from 1896 to 1960. Coal tar, a by-product of the gasification process, was intermittently disposed of in unlined ponds adjacent to the river. Throughout the years, the coal tar has migrated through the substrata resulting in periodic coal tar release into the river. Acres American Incorporated (Acres), Buffalo, New York, was retained in 1979 to: identify the extent of coal tar contamination; define the physical and chemical properties of the contaminant; define the mode of contaminant transport; define and assess the corrective remedial alternatives; perform final design; and provide construction supervision and management. The work was undertaken in a multi-phase approach from 1979 to 1982. In this paper, the authors discuss the site investigation, remedial alternatives and final design and construction undertaken at the site.

Threlfall, D. and Powiak, M.J., "Remedial Options for Ground Water Protection at Abandoned Solid Waste Disposal Facilities", *Proceedings of the National Conference on Management of Uncontrolled Hazardous Waste Sites*, 1980, Hazardous Materials Control Research Institute, Silver Spring, Maryland, pp. 131-134.

The selection of remedial options at abandoned solid waste disposal sites is a function of economic and technical feasibility. A site investigation to determine subsurface conditions and waste characteristics is necessary to choose the best option. Under certain circumstances, limited remedial measures may be sufficient. In others, though, extensive measures may be required despite higher costs. The case study presented herein demonstrates that a thorough understanding of subsurface conditions and the waste characteristics determined the selection of viable remedial options. The use of natural barriers, such as the local soil cover and underclay, to control leachate and ground water flow direction reduced engineering costs without neglecting environmental concerns. The case study also shows that further laboratory and field investigations may be required to determine what remedial options are feasible. In this case, tests of compacted cover material had to be performed to determine if required low permeability values could be attained. By utilizing the results of the detailed site investigation and conducting permeability studies, the most economical yet technically and environmentally effective remedial measures to protect ground water were selected.

Thurrott, J.C., et al., "Trihalomethane Removal by Coagulation Techniques in a Softening Process", EPA-600/52-83-003, Mar. 1983, U.S. Environmental Protection Agency, Cincinnati, Ohio.

This research program investigated various potable water treatment processes in combination with lime softening to effect maximum removal of trihalomethane precursor compounds. A study of the literature was used to guide the initial test work. Bench-scale jar tests investigated various combinations of coagulants that earlier studies indicated would be promising. The test work evaluated the relative effectiveness of lime softening, alum coagulation, ferric coagulation, and clay coagulation with respect to their ability to remove THM precursors by themselves and in various combinations with lime softening. The bench-scale test work was followed by a series of eight pilot-plant test runs using the U.S. Environmental Protection Agency's trailer-mounted pilot facility at the 45.4-MLD (12-MGD) Water Treatment Plant in Daytona Beach, Florida. The raw water studied is a moderately colored, high-hardness ground water emanating from the Floridan aquifer. Various treatment processes studied included single-stage coagulation/lime softening, two-stage coagulation/lime softening, lime softening/ coagulation, bentonite clay with lime softening, and polymeric coagulant/ clay coagulation/lime softening. Extensive analytical data were collected and summarized on raw water for test samples, pilot-plant process waters, treatment plant samples, and Daytona Beach distribution system samples. These data were used to evaluate the effectiveness of each process studied, as well as to compare pilot-plant performance with full-scale plant results.

Tolman, A.H., et al., "Guidance Manual for Minimizing Pollution from Waste Disposal Sites", EPA-600/2-78-142, Aug. 1978, U.S. Environmental Protection Agency, Cincinnati, Ohio.

The purpose of this manual is to provide guidance for municipal officials and engineers in the selection of available engineering technology to reduce or eliminate leachate generation at existing dumps and landfills. The manual emphasizes remedial measures for use during or after closure of landfills and dumps which do not meet current environmental standards. All of these measures must be designed and engineered by competent professionals for each specific site. Some of these measures are passive, that is, they require little or no maintenance once emplaced. Others are active and require a continuing input of manpower or electricity. Since the emphasis of this manual is on techniques of reducing leachate generation or controlling its movement, leachate treatment processes per se are not detailed.

Most of the techniques discussed deal with the reduction or elimination of infiltration into landfills in one of five categories: surface water control, passive ground water management, active ground water or plume management, chemical immobilization of wastes, and excavation and reburial. The technology presented is widely used in construction but has not necessarily as yet been applied to landfill closure. Surface water control measures include contour grading, surface sealing, and revegetating the landfill. These methods reduce infiltration of precipitation through the landfill surface, involve standard engineering procedures, and provide means of finishing the site. Passive ground water control techniques are used to minimize infiltration of ground water into the fill. They involve more technically advanced engineering procedures designed to provide an underground barrier between the ground water and the landfill. They are generally more costly and are useful for isolating a landfill when applied in conjunction with surface sealing methods. Plume management procedures involve actively altering the course of leachate movement by either adding or removing water from around the landfill. These methods can effect greater changes in water table elevations than can passive barriers but require continued maintenance and energy supplies and are therefore more costly.

Truett, J.B., Holberger, R.L. and Sanning, D.E., "In Situ Treatment of Uncontrolled Hazardous Waste Sites", *Proceedings of the National Conference on Management of Uncontrolled Hazardous Waste Sites*, 1982, Hazardous Materials Control Research Institute, Silver Spring, Maryland, pp. 451-457.

> Problems associated with contaminated lands — lands used directly as waste disposal sites or contaminated as a result of industrial or other activities — are common throughout the NATO alliance and many other industrial nations. These problems range from imminent health and environmental hazards posed by contaminants, to the increasing pressures to reclaim the land for safe, beneficial use.
>
> The North Atlantic Treaty Organization's (NATO) Committee on the Challenges of Modern Society (CCMS) is studying in situ treatment of contaminated lands. Portions of the study are being conducted in several nations. The Netherlands are evaluating in situ treatment applications in gas works and other chemically contaminated sites. The United Kingdom has completed a grouting feasibility study and is evaluating an experimental study of shallow-depth grouting problems. The Federal Republic of Germany has treated arsenic-contaminated soils and ground water by injection of permanganate solution, and has investigated the effects in ground water of silicate-gel injections used to consolidate foundation soils. The United States has recently completed a study on in situ techniques of solidification/ stabilization using one actual site as a scenario for evaluating the feasibility of such treatment. These studies represent a substantial effort in the evaluation and application of in situ treatment techniques. Certain of the in situ treatment techniques were not only to reduce the pollution impacts from contaminated soils, but also to improve the properties of the soils for certain end uses of the site. Selected grouts, for example, not only immobilize specific pollutants, but can increase the load-bearing properties of the soil in a contaminated area intended for eventual use as a building site.

Trussell, R.R., Trussell, A. and Kreft, P., "Selenium Removal from Ground Water Using Activated Alumina", EPA-600/2-80-153, Aug. 1980, U.S. Environmental Protection Agency, Cincinnati, Ohio.

> Selenium is a contaminant found in trace quantities in some ground- and surface-waters in the United States. Two species of inorganic selenium, with valence states of $+4$ and $+6$, are typically found in selenium-contaminated waters. Se(IV) and Se(VI) act very different, chemically. Se(IV) occurs as $HSeO_3^-$ in the pH range of 2.7 to 8.5. Se(VI) occurs as SeO_4^- above pH 1.7. The valence of either of these species is thought to be determined by the oxidation-reduction potential of a water at a certain pH. Knowing the oxidizing state of a water, one can predict whether Se(IV) or Se(VI) should be present.

Initial batch studies indicated that Se(IV) was preferentially adsorbed over Se(VI) in side-by-side tests. The isotherm capacity of activated alumina for Se(IV) was roughly three times the capacity of Se(VI). Other studies indicated that while bicarbonate mildly interfered with Se(IV) removal, both bicarbonate and sulfate heavily interfered with Se(VI) adsorption. Initial column studies with a three-inch deep bed helped delimit the amounts of NaOH and H_2SO_4 to be used during regeneration of Se(IV)-saturated alumina. Other items addressed in the three-inch column studies included how varied concentrations of NaOH and H_2SO_4 affected regeneration capabilities and how the varied concentrations affected alumina degradation. Deeper (9-inch) column studies showed that capacities for Se(IV) decreased with increasing influent water pH. pH 5 showed the highest capacity for Se(IV) adsorption. The kinetics of regeneration were the most important factors in determining the capacity of activated alumina for Se(IV), with pore diffusion seeming to be the rate-limiting step. Slow 0.5% NaOH flow rates (0.5 gpm/ft^2 or less) are necessary to effectively recover a high percentage of Se(IV) removed during a previous treatment run.

Tsand, C.F., Mangold, D.G. and Doughty, C., "A Study of Contaminant Plume Control in Fractured Porous Media", *Proceedings of the Third National Symposium on Aquifer Restoration and Ground Water Monitoring*, 1983, National Water Well Association, Worthington, Ohio, pp. 446-452.

A generic study has been carried out in the control and manipulation of a contaminant plume in an aquifer by production and injection procedures. The calculation is based on a numerical simulator CPT developed at the Lawrence Berkely Laboratory. The simulator employs an integrated finite-difference (IFD) numerical scheme and calculates coupled thermohydrologic flows with simple chemical advective transport in porous media with discrete fractures. It is a three-dimensional code that includes the effects of density and viscosity variations of the fluid, gravitation or buoyancy effects, aquifer heterogeneity and complex boundary conditions. The current version is derived from an early code PT which calculates heat and mass flows and which has been widely applied and validated in geothermal and non-isothermal hydrological problems. The paper applies this simulator to the study of a contaminant plume in four different types of media: (1) homogeneous porous medium; (2) porous medium with two regions of different permeabilities; (3) porous medium with a major fracture (or fault); and (4) multiple-fractured medium where the fractures can be represented statistically. In this generic exploratory study, two production-injection schemes are calculated for each case to investigate the transport and shape change of the contaminant plume. General discussions on the effectiveness and limits of such cleanup schemes are also made.

Tyco Laboratories, "Electrochemical Treatment of Acid Mine Waters", Feb. 1972, Waltham, Massachusetts.

Synthetic acid mine drainage (AMD) water was prepared by draining tap water through waste coal, and the resulting AMD was treated on a laboratory scale by an electrolytic oxidation process. Tests of fluidized bed, packed-bed, and annular flow prototype reactors demonstrated the packed-bed reactor to be most efficient. Oxidation of Fe(2+) to Fe(3+) takes place on a carbon electrode at a mass transport limited rate, while hydrogen evolution occurs on a polished 316 stainless steel cathode, limited by a slower electrochemical kinetic step. Preliminary economic analyses, using a packed-bed reactor are given. The electrolytic process is free from both the safety hazards associated with radio isotope-induced oxidation and the temperature dependence of biological oxidation methods.

Underwood, E.R. and Thornton, J.C., "Contaminated Ground-Water Recovery System Analysis, Design and Construction at a Waste Management Facility Located in a Gulf Coastal Plain", *Proceedings of the Third National Symposium on Aquifer Restoration and Ground Water Monitoring*, 1983, National Water Well Association, Worthington, Ohio, pp. 142-147.

A shallow, saline, nonpotable saturated zone consisting of silty sand was slightly contaminated under a waste management facility located in a coastal zone. A 29-well ground water recovery system, using submersible pumps, showed a somewhat limited success in recovery and in effecting a cone of depression to halt the forward progress of the plume of contamination since low yield capacity of the saturated zone was not sufficient to continuously supply pumps at their minimum pumping rates. A detailed technical study was conducted to develop a high-yield recovery system to form a significant cone of depression and effect complete recovery of contaminated water. Hydrological data was gathered at all wells on-site by physical measurements and bailing tests. Field data and data from previous reports and literature was entered into a finite-difference ground water model. Once calibrated, the model was used to evaluate various recovery systems including jet-eductor pump systems, a French drain at the base of the saturated zone, recharge pits ahead of the plume and combination systems. The most effective system found was a combination of a 520-foot-long French drain and jet-eductor pumps capable of pumping at low formation yield located strategically in the plume of contamination. A French drain and four jet-eductor pumps were designed and installed. Pumping rates indicate a potential yield greater than expected. Recent monitoring indicates a positive change in ground water contours and formation of a cone of depression although the system is not operating continuously due to water-handling limitations. A water treatment system is being developed to treat the estimated volume of water to be recovered on a continuous basis.

Uniroyal, Inc., "Use of Latex as a Soil Sealant to Control Acid Mine Drainage", June 1972, Wayne, New Jersey.

Acid formation in a mine cavity can be prevented by keeping water (one of the reactants) out of the mine. This might be accomplished by forming a waterproof seal over the mine cavity to prevent the seepage of surface water into the mine. In laboratory tests using reconstructed soil columns, rubber latex showed good sealing efficiency. The ideal situation in which latex would coagulate in a narrow zone two to three feet below the surface by reacting with acidic or metallic constituents of the soil was not attained. Rather, in field tests the latex was deposited progressively. Latex stability appears to be a more critical property than latex particle size in controlling penetration. Addition of excess anionic surfactants to latex improved its penetration into the soil. Costs are discussed.

Unterbery, W., Stone, W.L. and Tafuri, A.N., "Rationale for Determining Prioroties and Extent of Cleanup of Uncontrolled Hazardous Waste Sites", *Proceedings of the National Conference on Management of Uncontrolled Hazardous Waste Sites*, 1981, Hazardous Materials Control Research Institute, Silver Spring, Maryland, pp. 188-197.

The Comprehensive Environmental Response, Compensation and Liability Act of 1980, PL 96-510, requires revision of the National Contingency Plan with a section known as the National Hazardous Substance Response-Plan and requires that the plan establish procedures and standards for responding to releases of hazardous substances, pollutants, and contaminants. One aspect of this new plan is defined as: "105 (3) methods and criteria for determining the appropriate extent of removal, remedy, and other measures as required by this Act." In this paper the authors attempt to develop a rationale for determining "appropriate extent of removal" for uncontrolled hazardous waste sites. This involves three types of decisions: (1) determination of cleanup priorities—to what area or areas to direct the cleanup effort when time or resources are limited and/or the affected area is too extensive to clean up the entire release; (2) evaluation of alternative cleanup methods and selection of optimum methods—once the priority cleanup areas have been selected, (a) to evaluate cleanup alternatives considering availability and cost of manpower and equipment, effectiveness and speed of deployment, effect on the environment and other applicable parameters; and (b) to select the optimum method or combination of methods for the priority cleanup operations; and (3) determination of extent of cleanup (how clean is clean)—once the optimum cleanup methods have been selected, how far and how long should cleanup proceed? The desirable "extent of removal" depends on (a) reaching acceptably low levels of residuals, (b) the (usually rising) cost of removal per unit mass of contaminant as lower and lower contamination levels are achieved, (c) environmental impact of cleaning methods themselves; and other suitable parameters. The authors deal, in order, with these three questions and provide decision-making methodologies in these areas to assist those charged with directing cleanup operations.

U.S. Environmental Protection Agency, "Surface Impoundments and Their Effects on Ground Water Quality in the United States - A Preliminary Survey", EPA 570/9-78-004, June 1978, Washington, D.C.

Sections VIII and IX of this report address ground water pollution prevention techniques and the associated costs, respectively. The techniques are grouped into direct and indirect methods. Direct methods include: (1) installation of impermeable membranes; (2) installation of layer of impermeable material; (3) collection of contaminated water seeping from impoundment (infiltration galleries, wellpoint systems, and wells); (4) return of collected water back to impoundment; (5) physiochemical immobilization of waste material; (6) ground water cutoff wall (slurry trench, and grout cutoff); (7) capping of impoundment surface; and (8) treatment of contaminated water (equalization, biological treatment, activated carbon adsorption, heavy metals removal, dissolved solids removal) and treated water discharge. Indirect methods include development of a new source of water supply, and treatment of contaminated ground water prior to use. Size-independent cost figures for each of the methods are presented. The figures are based on available data and needed assumptions. Some of the resulting costs are actual numbers while some are cost relationships depicted graphically.

Villaume, J.F., Lowe, P.C. and Unites, D.F., "Recovery of Coal Gasification Wastes: An Innovative Approach", *Proceedings of the Third National Symposium on Aquifer Restoration and Ground Water Monitoring*, 1983, National Water Well Association, Worthington, Ohio, pp. 434-445.

Abandoned city coal gasification plants, which operated from the mid-1800s until just before World War II, are increasingly being identified as potential environmental troublespots because of the poor waste management practices which were used at the time. By 1920 there were more than 900 of these plants in existence in the U.S. One such abandoned plant site, which was the first site in the nation to receive emergency Superfund money for containment action, is located in Stroudsburg, Pennsylvania, along what was at one time a world famous trout fishery. A 1981 investigation of the site identified an eight-acre area contaminated with up to 1.8 million gallons of coal tar wastes from the old plant. Low-temperature coal tar is a complex misture of microscopic free carbon particles and some 10,000 individual organic compounds, only about 600 of which have ever been identified. Some of these compounds, such as anthracene and benzo(a)pyrene, are known or suspected human carcinogens. A large accumulation of stratigraphically trapped coal tar is currently being recovered by the present property owner through an in-ground pumping operation which relies on the particular physical properties of the material. In this operation ground water is withdrawn at a constant rate to form a shallow cone of depression in the ground water table. Because of the density difference between the two fluids, this causes an upwelling of the coal tar, which in turn stresses the system and forces the coal tar to flow into the well more rapidly than its extremely low transmissivity would otherwise allow. Proper screening and gravel packing of the well casing assure a virtually water- and sediment-free product which can be sold as either a chemical feedstock or supplemental fuel.

Wallace, J.R., Badalamenti, S. and Ogg, R.M., "Price Landfill: Interim and Long-Term Remedial Actions", *Proceedings of the National Conference on Management of Uncontrolled Hazardous Waste Sites*, 1983, Hazardous Materials Control Research Institute, Silver Spring, Maryland, pp. 358-361.

The Price Landfill site near Atlantic City, New Jersey, is an inactive landfill that operated from the mid-1960s to the mid-1970s. Initially, it served as a sand and gravel pit, but when the contents had been excavated to within approximately 2 ft of the ground water level, gravel operations ceased. At that time the site began receiving private and municipal wastes. For approximately 18 months, the landfill reportedly received liquid industrial wastes, both in 55-gal drums and bulk liquids that were poured directly onto the ground. Landfill operations ceased in the mid-1970s, and at that time the landfill was covered with sand and gravel. Ground water generally flows northeast and east towards the ocean. Between the site and the ocean lie a number of privately owned drinking wells and the Atlantic City Municipal Utilities Authority (ACMUA) wellfield, which supplies drinking water to Atlantic City.

Before the summer of 1982, several actions were taken to ensure a safe supply of water until long-term remedial actions were evaluated. A water conservation program was initiated by Atlantic City. This program was very successful, and the anticipated peak demand during the summer of 1982 of 20 mgd was never reached. The actual peak demands were approximately 15 mgd. Piping and structural modifications were installed at two ACMUA production wells to accommodate portable, truck-mounted, granular activated carbon adsorption units. These units could be used for well-head treatment in the event that contamination reached the upper Cohansey wells located near the water treatment plant. After well-head treatment, the water would be treated in the ACMUA plant. An interconnecting pipe was installed to allow Atlantic City to receive water from the NJWCo distribution system. This water would be added directly to the ACMUA clearwell and could be used to augment ACMUA treatment capabilities. Plant piping modifications were installed to allow the ACMUA to bypass the water treatment plant with the Kirkwood Aquifer supplies if the plant's hydraulic capacity became a limiting factor in supplying sufficient quantities of water to Atlantic City. The raw water quality of the Kirkwood Aquifer is sufficient to allow discharge directly into the clearwell for chlorination only.

Two long term remedial action alternatives have been proposed for the site. Alternative 13 involves relocating 13.5 mgd of ACMUA water supply capacity north of the reservoir with a 2 mgd plume abatement well. Alternative 14b involves relocating 13.5 mgd north of the reservoir, a 2 mgd plume abatement well and a slurry wall around the landfill site. Both alternatives provide: an adequate drinking water supply for Atlantic City that should not be affected by the plume from Price Landfill; positive measures to control and remove the contaminated plume; and positive measures to control the source at Price Landfill.

Walsh, J.J., Lippitt, J.M. and Scott, M., "Costs of Remedial Actions at Uncontrolled Hazardous Waste Sites — Impacts of Worker Health and Safety Considerations", *Proceedings of the National Conference on Management of Uncontrolled Hazardous Waste Sites*, 1983, Hazardous Materials Control Research Institute, Silver Spring, Maryland, pp. 376-380.

In December 1980, the U.S. Congress passed legislation entitled "The Comprehensive Environmental Response, Compensation and Liability Act". This Act provides the USEPA with the legislative mandate and the money to assist in the elimination of public health hazards posed by uncontrolled hazardous waste sites. Several studies have been conducted to evaluate the types of remedial actions and associated costs applicable to Superfund sites (i.e., sites for which Superfund monies have been allocated) and other hazardous waste sites. In these studies, costs associated with health and safety of workers were either not included or not uniformly identifiable as separate cost items. As a result, the project presented in this paper was designed to specifically address the additional costs of protecting worker health and safety on a hazardous waste site. These costs do not include costs associated with addressing concerns of the public health and safety in the vicinity around an uncontrolled hazardous waste site. However, the controls and costs associated with protection of workers on the site should reflect much, if not all, of the additional costs of protecting the public in areas removed from the source of contamination (i.e., the hazardous waste site itself). The objectives of this project were: (1) to identify categories of health and safety costs; (2) collect and compile health and safety cost estimates and determine a range of costs which can be encountered on hazardous waste sites, (3) calculate percentage incremental health and safety cost adjustment factors; and (4) identify factors which impact health and safety costs and should be considered for future study and evaluation.

Wardell, J., Nielson, M. and Wong, J., "Contamination Control at Rocky Mountain Arsenal, Denver, Colorado", *Proceedings of the National Conference on Management of Uncontrolled Hazardous Waste Sites*, 1981, Hazardous Materials Control Research Institute, Silver Spring, Maryland, pp. 374-379.

In response to requirements to contain or clean up contamination at Rocky Mountain Arsenal (RMA), Denver, Colorado, the Department of the Army (DA), began a program to mitigate the contamination problem at that installation in 1975. The approach initially taken by DA was to contain contamination at the installation's boundaries. In 1978, a pilot wastewater treatment system began operations at the northern arsenal boundary to remove organic contamination from ground water. The pilot system proved successful. It has been expanded to intercept all contaminated ground water crossing that boundary. A similar system is also being designed to intercept all contaminated water crossing the northwest boundary. This paper describes the history and operation of these containment systems.

Weinstein, N.J., "A Successful Underground Water Decontamination System: Its Meaning for Future Projects", *Proceedings of the Second National Symposium on Aquifer Restoration and Ground Water Monitoring*, 1982, National Water Well Association, Worthington, Ohio, pp. 87-93.

This paper describes the problems encountered in an industrial plant where underground water was contaminated with 1,1,1-trichloroethane and where this contamination threatened a public water supply. The steps taken to define and resolve these problems will be described, including the development, scale up, design, installation and start up of an air-stripping system which has resulted in 99.9+ percent cleanup. Although proprietary process information is not disclosed, assessment of results are discussed in detail. The application of air-stripping to similar problems are discussed, including alternative air-stripping systems and combinations of air-stripping and adsorption as a general tool for purifying organic chemical contaminated aquifers.

Werner, J.D., Yang, E.J. and Nagle, E., "Remedial Action Management and Cost Analysis", *Proceedings of the National Conference on Management of Uncontrolled Hazardous Waste Sites*, 1983, Hazardous Materials Control Research Institute, Silver Spring, Maryland, pp. 370-375.

Superfund presents a classic economic and management dilemma — how to clean up the seemingly unlimited number of problem hazardous waste sites using limited resources. Congress' recognition of this dilemma is reflected in the Comprehensive Environmental Response, Compensation and Liability Act of 1980, by the specific statutory requirement for selection of the most "cost-effective" remedial alternative at Superfund sites. Cost effectivenes, as out-lined in section 300.68, subpart F of the National Contingency Plan, does not mean that public health concerns should be subordinated but rather that the least cost alternative should be selected from among adequately effective options. However, carrying out this mandate for cost effective remedies requires first, that accurate cost information be available for estimating the relative costs of remedial alternatives and second, that this information be used to implement the remedial alternative as efficiently as possible through effective planning and management. To help provide this information to the EPA's Superfund remedial action program, the Environmental Law Institute (ELI) performed detailed case studies on remedial actions at 23 hazardous waste sites across the United States. This study is the most comprehensive compilation of actual expenditure data for remedial actions that has been performed to date. For 16 of the 23 case studies, ELI worked in conjunction with JRB Associates, which compiled the technical information at the sites. This paper summarizes the findings on the costs, the planning and the management of the cleanups at these sites.

Whittaker, K.F. and Goltz, R., "Cost Effective Management of an Abandoned Hazardous Waste Site by a Staged Cleanup Approach", *Proceedings of the National Conference on Management of Uncontrolled Hazardous Waste Sites*, 1982, Hazardous Materials Control Research Institute, Silver Spring, Maryland, pp. 262-267.

The Bridgeport Rental and Oil Services (BROS) site is a former oil processing and reclamation facility in southern New Jersey. The site has an overall area of approximately 26 acres, contains an 11.5 acre unlined lagoon and over 88 tanks and storage vessels, and is bounded on three sides by two fresh water ponds, a peach orchard and marshland (Cedar Swamp). A small creek (Timber Creek) running through the swamp passes in the near vicinity of the lagoon and eventually discharges to the Delaware River about 3 miles from the site. The lagoon has an average depth of 10 to 15 ft but depths as great as 60 ft in places have been reported. A thick layer of heavy oil floats on the surface. Large quantities of construction debris, trash, and several large tank trucks are partially submerged in the lagoon and all are coated with oil. Large numbers of floating drums are also present. The storage vessels on site range in capacity from a few thousand gallons to tanks with volumes of over 300,000 gal. The volume of material in each tank is not consistent. Initial site surveys have shown the majority of the tanks are either empty or contain only small quantities of bottom sludges. Two of the seven major tanks (i.e., greater than 300,000 gal) contain large quantities of liquid material.

The site poses a threat to both surface and ground water in the area. Private drinking wells in the area have been contaminated with varying amounts of volatile organic compounds. Previous breaching of the dike surrounding the lagoon has led to discharge of material to Timber Creek and the deforestation of 8-10 acres of land adjacent to the lagoon. There is visual evidence of seepage of oily materials into the surrounding fresh water ponds.

A treatment system for waters drawn from the waste oil storage lagoon. Treatment processes were recommended on the basis of laboratory and full-scale treatability tests. The most cost effective treatment solution is the application of a combined flocculation/sedimentation oil removal, activated carbon adsorption treatment system.

Williams, D.E. and Wilder, D.G., "Gasoline Pollution of a Ground-Water Reservoir - A Case History", *Ground Water*, Vol. 9, No. 6, Nov.-Dec. 1971.

An estimated 250,000 gallons of gasoline seeped into an underground reservoir between 1968 and 1971. Methods for cleanup include "skimming" wells to produce a high gasoline/water ratio and on-site treatment facilities. Tests on the aquifer showed a semiconfined aquifer with the confining layers being somewhat leaky. The result is multiple aquifer systems. Seventy wells were drilled to define the problem area and expedite removal. The principle removal technique consisted of packing off the lower portion of the aquifer and pumping clean water from this zone. Meanwhile, gasoline was removed from the upper zone by a smaller pump. The skimming wells pumped into separator tank assemblies. These separators were composed of density stratification tanks coupled to a flotation tank. The gasoline was stored until disposal. Only 50,000 gallons of gasoline were recovered. The authors suggest that this was because of an error in original estimates of aquifer porosity. Alternately they suggest that the lower recovery could be due to a reduction in the relative permeability of the formation to one of the fluids relative to the saturation of the other fluid. (This is current oil field technology). This was never proved by lab experiments with filter sand. Since gasoline fills smaller pores than water, clean water passed through the uncontaminated pores and produced cleaner water (when the aquifer is flushed with clean water). The final idea addressed by the authors is bacterial degradation. They state that slime is a problem in degradation. They did not analyze degradation at this site although they note that bacteria are evident at the site.

Williams, E.B., "Contamination Containment by In Situ Polymerization", *Proceedings of the Second National Symposium on Aquifer Restoration and Ground Water Monitoring*, 1982, National Water Well Association, Worthington, Ohio, pp. 38-44.

A unique method of containment, by polymerizing a contaminant in the ground, was successfully applied on a recent pollution control project. Approximately 4,200 gallons of acrylate monomer leaked from a corroded underground pipeline at a small industrial plant adjacent to a suburban area.

A preliminary investigation of soil borings and sample analyses detected migration of the contaminant on two separate levels. The plume spread initially through a surficial zone of cinders and then passed downward through a storm sewer trench into a lower glacial sand and gravel, where the direction of migration was reversed by local ground water drainage. Below this lower level was another clay layer, which was in turn underlain by the major regional aquifer of glacial outwash. The hydrogeologic conditions did not favor recovery by conventional techniques. On the lower level, conductivity of formation materials was moderate to low and saturated thickness was only two or three feet. The upper level, although porous, was largely unsaturated. Winter weather and continuous traffic were practical considerations also. As an alternative, field testing confirmed that injection of catalysts could be used to immobilize the contaminant by polymerization and an injection program was subsequently conducted on both levels. A comparison of soil borings from before and after injection visually showed substantial polymerization and soil extractions indicated as much as a 99 percent reduction of monomer may have been obtained.

Wilmoth, B.M., "Salty Ground Water and Meteoric Flushing of Contaminated Aquifers in West Virginia", *Ground Water*, Vol. 10, No. 1, Jan.-Feb. 1972, pp. 99-105.

Salty ground water is commonly encountered at relatively shallow depths of 100 to 300 feet beneath the major stream channels in the western half of West Virginia. Because of the wide distribution of salty ground water and connate brine at various depths, it is difficult to distinguish natural contamination from that caused by subsurface industrial activities. Natural changes in quality apparently are minor. The available historical data indicate no large-scale natural variations in salt content during the period of record. Histories of some water well developments show unnatural large-scale increases in salt content from various industrial activities that affect the fresh water zones. Some records also reveal decreases in salt content after the source of the salt was eliminated or after the subsurface activity responsible for artificial migration of the salt water was stopped.

Artesian brine contaminated a fresh water aquifer in Fayette County. Chloride content changed from 53 mg/l to more than 1,900 mg/l in a period of 5^1/$_2$ years. When pumping was stopped, chloride content decreased to 55 mg/l in 10 years. Heavy pumping of well fields in Charleston during 1930 to 1956 accelerated migration upward of salt water. Chloride content increased from less than 100 mg/l to more than 300 mg/l in some wells and to more than 1,000 mg/l at individual wells. Pumpage has declined greatly since 1956 and chloride content has decreased below 200 mg/l at some of the contaminated wells. In an oil field of Kanawha County, a water well was contaminated by salt water accelerated by subsurface activities. Chloride content increased from less than 100 mg/l to more than 2,900 mg/l within 2 months. After the oil-field activity was curtailed, chloride content decreased to 190 mg/l in about 2¹/₂ years. Road salt piles contaminated a carbonate aquifer in Monroe County. Chloride concentrations in wells located 1,500 feet from the piles increased from 185 mg/l to 1,000 mg/l in 5 years. The greatest change was 1,000 mg/l in 1969 to 7,200 mg/l in 1970 when the salt storage area was enlarged. All salt piles were removed in late 1970 and within 2 months chloride content decreased to 188 mg/l.

Willmoth, B.M., "Procedures to Reduce Contamination of Groundwater by Hazardous Materials", *Proceedings of the 1978 National Conference on Control of Hazardous Material Spills*, 1978, Hazardous Materials Control Research Institute, Silver Spring, Maryland, pp. 293-295.

Protection of ground water aquifers and their zones of recharge by hazardous materials is as important as the protection of surface water bodies. The zone of recharge to an aquifer can be quite vulnerable to contamination by a spilled material that presents no threat to a stream or lake. Although the source of most contaminants is surface spills, underground facilities also may leak and cause enormous damage. Thousands of small underground tanks and lines are presently exempt from regulations for prevention of spills — including such common facilities as service stations and small heating oil tanks and distribution lines. The effects of even a very small undetected leak over an extended period can be devastating. Most often water wells nearest the spill are affected first. It is obvious then that the public should be alerted to the need for effective spill prevention regulations and contingency response procedures to reduce the serious and long-lasting degradation of ground water. New conservation and education programs should be initiated to alert the industrial and business community to the need for immediate and thoroughly effective response actions to spills of hazardous materials that may damage local aquifers. The program must show and define the critical importance of protecting the local zones of recharge to aquifers even when no threat to a stream is evident.

Wilmoth, R.C., "Limestone and Limestone-Lime Neutralization of Acid Mine Drainage", June 1974, U.S. Environmental Protection Agency, Rivesville, West Virginia.

The critical parameters affecting neutralization of ferric-iron acid mine waters were characterized by the U.S. Environmental Protection Agency in comparative studies using hydrated lime, rock-dust limestone, and a combination of the two as neutralizing agents. The advantages and disadvantages of each of these neutralizing agents were noted. On the ferric-iron test water, combination limestone-lime treatment provided a better than 25 percent reduction in materials cost as compared to straight lime or limestone treatment. Significant reduction in sludge production was noted by the use of rock-dust limestone and by the use of combination treatment as compared to hydrated-lime treatment. Emphasis on optimizing limestone utilization efficiencies resulted in an increase from approximately 35 percent to 50 percent utilization. Studies using limestone that had been ground to pass a 400-mesh screen resulted in utilization efficiencies near 90 percent.

Wilmoth, R.C., "Limestone and Lime Neutralization of Ferrous Iron Acid Mine Drainage", EPA-600/2-77-101, May 1977, U.S. Environmental Protection Agency, Rivesville, West Virginia.

> The U.S. Environmental Protection Agency conducted a 2-year study on hydrated lime and rock-dust limestone neutralization of acid mine drainage containing ferrous iron at the EPA Crown Mine Drainage Control Field Site near Rivesville, West Virginia. The study investigated optimization of the limestone process and its feasibility in comparison with hydrated lime treatment. Operating parameters, design factors, and reagent costs for both processes were determined. Effluent quality was considered of prime importance in these investigations. Coagulants were considered essential to successful thickener operation for both lime and limestone treatment. Effluent iron, suspended solids, and turbidity values could be maintained below 3 mg/l, 10 mg/l, and 10 JTU, respectively, using coagulant addition. Although the limestone process was demonstrated to be technically effective in ferrous iron treatment situations, the process was judged to be less efficient overall in comparison with lime neutralization.

Wilmoth, R.C., Mason, D.G. and Gupta, M., "Treatment of Ferrous Iron Acid Mine Drainage by Reverse Osmosis", *Proceedings of the 4th Symposium on Coal Mine Drainage*, 1972.

> In previous research, reverse osmosis treatment of a ferrous iron mine discharge at Mocanaqua, Pennsylvania was studied. Using a tubular system, a loss of 90 percent of the original product flow in 400 hours of operation due to iron fouling was observed. Though sodium hydrosulfide flushes were successful in flux restoration, fouling immediately recurred upon resumption of operation on the ferrous water. To investigate the Mocanaqua fouling phenomena, field studies were undertaken. Reported are the results of the tests and comparisons with a spiral wound and hollow fiber reverse osmosis unit studied at the same time.

Wilmoth, R.C., Scott, R.E. and Hill, R.D., "Combination Limestone-Lime Treatment of Acid Mine Drainage", *Proceedings of the 4th Symposium on Coal Mine Drainage*, 1972.

> The research investigated treatment methods for ferric iron acid mine drainage. One of the most promising techniques studied was combination limestone-lime neutralization. To date, neutralization has been shown to be the most economical treatment method to remove iron, aluminum, and

acidity. Lime treatment has received the majority of research attention. However, limestone treatment offers several distinct advantages over lime; namely higher density, lower volume sludges, cheaper raw material costs, easier handling. Conversely, the biggest disadvantages to limestone are its relatively inefficient reaction rate.

Wilson, J.L., Lenton, R.L. and Porras, J., "Ground Water Pollution: Technology, Economics and Management", TR 208, Jan. 1976, Massachusetts Institute of Technology, Cambridge, Massachusetts.

A review of the technical, economic and management aspects of ground water pollution is presented. The ground water pollution problem is described and several of its important characteristics are pointed out. A description of the physical, chemical and biological aspects of ground water pollution is given. The technology of ground water pollution detection and observation is reviewed, including elements in the design and operation of a ground water quality monitoring system. The causes, types and extent of ground water pollution are described. These include agricultural, industrial, domestic and urban, radiological, and natural or induced sources of pollution. Methods of control for each are given. Methods of analysis as applied to ground water pollution management are described within the context of total resource management and long-range planning. Criteria for choosing among alternative plans, systems analysis, and optimization and simulation techniques are discussed. A case study of integrated ground water-surface water management with specific ground water quality considerations is described and discussed.

Wilson, L.W., Matthews, N.J. and Stump, J.L., "Underground Coal Mining Methods to Abate Water Pollution: A State of the Art Literature Review", Dec. 1970, West Virginia University, Morgantown, West Virginia.

The report reviews published information concerning the abatement of harmful drainage from underground coal mines. Although much has been written on mine water management, very little literature is available on the specific area of preventing the formation of acid water. The references used in this report include mining engineering and hydrology studies and spans the period of time when water quantity rather than quality was the major consideration. Physical approaches to the problem of interdicting water entry into coal mines, beyond removal and treatment, are (1) land management for surface and subsurface water diversion, (b) the exploitation of water carrying strata, and (c) new mining methods. Chemical approaches to abatement include (a) the use of silica gel solutions underground to prevent acid formation, (b) the use of inert gas in active mines, and (c) the use of new and refinement of known grouting agents.

Winston, A., Kirchner, D.G. and Rosthauser, J.W., "Functional Polymers for Removal of Heavy-Metal Pollutants from Water", WRI-WVU-80-01, 1980, West Virginia University, Morgantown, West Virginia.

This study involved designing polymeric ion-exchange resins specific for certain metal ions, with particular attention being given to systems for selectively removing iron from water. The resins are vinyl polymers bearing hydroxamic acids, a functional group having a particularly strong affinity for iron. A series of polymers was prepared in which the spacing of the hydroxamic acid groups were varied. The stability constants for the iron complexes showed that the spacing of the hydroxamic acid groups was important both in complex stability and in the selectivity of the resin for iron. The polymer having the largest binding constant for iron had an 11-atom spacing between hydroxamic-acid groups. This polymer was prepared in a cross-linked insoluble form and tested as an ion-exchange medium. The polymer removed iron from water to less than 0.5 ppm. The column could be regenerated by washing with sodium hydrosulfite or EDTA. Copolymers bearing hydroxamic acid groups undergo cross-linking in the presence of iron, resulting in large increases of viscosity and in flocculation. Polymers for binding copper and mercury were synthesized and evaluated. The functional group for copper was the tripeptide glycylglycylhistidine. For mercury, polymers containing sulfide were studied. Preliminary evaluation showed that the polymers formed complexes with copper and mercury, respectively. Various applications of these resins in environment, health, and industry were suggested.

Wood, P.R. and DeMarco, J., "Treatment of Groundwater with Granular Activated Carbon", *Journal of the American Water Works Association*, Vol. 71, No. 11, Nov. 1979, pp. 674-682.

Removal of halogenated organic compounds, total organic compounds, trihalomethane formation potential and several high molecular weight compounds from raw, lime softened and finished water by using granular activated carbon (GAC) has been studied for a three year period at the Preston Water Treatment Plant in Hialeah, Florida. Ground water is the sole source of drinking water in this region. Total organic carbon (TOC) of the ground water is about 10 μg/l and trihalomethane formation potential (THMFP) ranges from 650 to 950 μg/l, THMFP and halogenated organic compounds of the finished water in the 300 μg/l and 330 μg/l range respectively.

The effectiveness of four types of GAC in removing these organics was evaluated. Carbon 1 was tested in beds of 0.75, 1.5, 2.25 and 3 m deep with empty bed contact times (EBCT) of 6.2, 12.4, 18.6 and 24.8 min. Carbons 2, 3 and 4 were tested in beds of 0.75 m deep with an EBCT of 6.2 min. each. Chloroform removal from finished water showed that equal volumes of the four types of GAC removed 78% (carbon 4), 53% (carbon 1), 48% (carbon 3) and 40% (carbon 2). On an equal weight basis, carbon 4, 3, 1 and 2 removed 78, 57, 56 and 36 percent respectively. The removal for eighteen other purgeable halogenated organic compounds rated the carbons in a similar order. Removal of THMP substances from finished water showed that, during a 59-day test period, an equal volume of four types of GAC removed 27 (carbon 2), 26 (carbon 1), 16 (carbon 3) and 8 (carbon 4) percent of total THMFP entering the water. On an equal weight basis carbons 1, 2, 3 and 4 removed 26, 23, 18 and 8 percent, respectively.

The level of total trihalomethanes reaching a consumer's tap was simulated by rechlorinating water from GAC column to 3 μg/l of free chlorine and bottle-aged two days. The four types of GAC were evaluated at equal depths of 0.75 m with a bed life of ten days. At an equal volume basis, carbons 1, 2, 3 and 4 removed 54, 35, 14 and 9 percent of the THMFP substances. TOC removal showed a similar order of removal. The water was also spiked with high molecular weight industrial and agricultural chemicals for evaluating the adsorption capacities of the GAC. Chlorinated pesticides exhibited early low level breakthrough. Overall, more than 80 percent of the spiked high molecular weight compounds were removed. Removal of 2-nitrobenzene, ethyl benzoate, naphthalene, 1-chloro-2-nitrobenzene, dimethylphthalate and diisobutylphthalate ranged from complete to poor.

Yaniga, P.M., "Alternatives in the Decontamination of Hydrocarbon Contaminated Aquifers", *Proceedings of the Second National Symposium on Aquifer Restoration and Ground Water Monitoring*, 1982, National Water Well Association, Worthington, Ohio, pp. 47-57.

Hydrocarbon contmaination of ground water is a widespread problem of increasing concern. Hydrocarbons in ground water affect both domestic and public water supplies, resulting in significant economic losses, both to the water owner and the hydrocarbon owner. The severity of the problem can range from the objectionable tastes and odors imparted by small amounts of dissolved hydrocarbons to the destruction of aquifer usability through the accumulation of a considerable thickness of hydrocarbons on the top of the water table. The application of sound hydrogeologic principles is the key to rehabilitating hydrocarbon-degraded aquifers. The extent and magnitude of contamination are defined on the basis of observation wells, isopach maps and other devices. One of the major problems in analzing the problem is to distinguish between the real thickness and apparent thickness of the hydrocarbon lens or free product. Groundwater Technology Inc. has successfully applied several techniques to the recovery of lost product and the abatement of residual, dissolved hydrocarbons. Among these are: (1) a special two-pump system that utilizes water/hydrocarbon-sensitive probes and oleophilic/hydrophobic filters that efficiently retrieve the free product that floats on water; and (2) processes that involve air-stripping, granular activated carbon filters and biochemical processes that efficiently lower the dissolved hydrocarbon concentration of ground water. The successful application of these techniques is described in the context of case histories of several projects completed by Groundwater Technology Inc.

Yaniga, P.M. and Demko, D.J., "Hydrocarbon Contamination of Carbonate Aquifers: Assessment and Abatement", *Proceedings of the Third National Symposium on Aquifer Restoration and Ground Water Monitoring*, 1983, National Water Well Association, Worthington, Ohio, pp. 60-65.

> The words assessment and abatement in the past few years have become terms of major significance in the yet-to-be-compiled encyclopedia of ground water contamination. That same encyclopedia, when compiled, will contain under the lead "ground-water contamination", a category designated "hydrocarbon losses to ground water: definition and cleanup". Under that section title will doubtless appear subheadings labeled: (1) geologic considerations in contamination assessment and abatement: consolidated vs. unconsolidated rock types; (2) structural influence and control on hydrocarbon movement in the subsurface: fracture and joint set control; and (3) lithologic considerations in fractured bedrock porosity and permeability: carbonate vs. noncarbonate rock types. This paper discusses one case history in which each of those noted headings came to play a significant role in the assessment and abatement of a hydrocarbon loss to ground water. The loss of a leak of No. 2 fuel that occurred over a prolonged period through a subsurface fuel transfer line. The loss area was underlain by a consolidated bedrock unit of regional extent that is locally tapped as an aquifer. The geologic unit is a steeply dipping, locally well-jointed carbonate rock. This structure and lithology restricted the plume configuration and its direction of movement to a narrow, linear band. Apparent product thicknesses in observation wells within the plume appear joint-set and solution-channel controlled, showing values approaching 60 feet. Comprehensive evaluation and assessment of the geologic controls affecting the plume culminated with the development and enactment of a program of water-table manipulation that is controlling product movement and retrieving the fugitive product for reuse.

Zaval, F.J. and Robins, J.D., "Water Infiltration Control to Achieve Mine Water Pollution Control—A Feasibility Study", Jan. 1973, West Virginia Department of Natural Resources, Charleston, West Virginia.

> The study determined the feasibility of conducting a full-scale demonstration to document the effectiveness of land reclamation at mined-out areas in establishing surface water infiltration control to prevent acid mine water pollution. The study site was the Dents Run Watershed, Monongalia County, West Virginia. Investigative measures included: investigation of each mine area and opening; a detailed description of each site; sampling and analysis of all receiving streams and discharge points to determine the

severity of acid mine water pollution; and evaluation and selection of weir structures, monitor enclosures and instruments to be placed in unattended areas to provide a continuous record of stream conditions. Recommendations and cost estimates are presented for reclamation at each site and for the installation of monitoring facilities.

INDEX

acid mine drainage 349-358,407,417,
423,425,437,441-443,469,476,
481-482,489,507-508,516-517,522
acrylonitrile 4,147-148,392,394,404,
439
activated carbon adsorption 4,89,
96-102,363-369,374,376-377,382,
387,392-394,399,415,419,426,440,
446-447,463,474-475,492,499,
509-510,514,519
activated sludge process 103-108,149,
394
adsorption 100,148,426,434,448,512
adsorption isotherms 100
Air Quality Risk Assessment System
290
air stripping 89-95,106-107,121,
148-149,171,363-366,391-393,
398-400,415,447-448,450,453,463,
499,512,520
anaerobic degradation 131-132,421
aquifer protection plan 458
arsenic 460-462,505

bacterial degradation 135,392,514
barrier 381,402,419,424,431,490,504
batch studies 506
bench-scale testing 398,439,453,503
See also pilot plant studies
bentonite barrier 68,84,183,189,
376-379,404,413-414,438,443,456,
477,492,495
biological treatment 89,102-106,110,
130,349,363-369,398,426,444,453,
463,486,509
bioreclamation process 132,138-139

biostimulation process 137,140,149,
387-390,393,444
block displacement 412
bottom seals 424

capping 5,51,54,63
carbon adsorption 121,354,398,400,
411,430
case studies 401,414-415,429,431,440,
455,469,478-481,503,513,521
cement-bentonite (C-B) trench 77-78,
80,83,424,465-466,486
chemical precipitation 89,110,
112-113,117,363,441,453,463
citizen committees 324-326,330,467
clay barriers 402-403,406
clay caps 416,426
clay liners 402,411
clay slurry wall 378,434
Clean Air Act 265
Clean Water Act amendments 265
coal mine waters 406-407,426,502,509
coke tray aerators 90-91
communications process models 303,
305
community contacts 319
comparisons of alternatives 205
conflict management and resolution
295,299-300,332-336
containment 173
cost-benefit analysis 197
cost-effectiveness analysis 197
costs 483,509,511,513
countercurrent packed columns 90-91
covering systems 438,445,472,475
cross-flow towers 90-91
cutoff walls 409,456,468

decision matrix 206,252-253
decision-tree analysis 163,269,424
Delphi technique 242-247,256
dewatering wells 378,434,492
dichloropentadiene (DCPD) 373-374
diisomethylphosphonate (DIMP) 373-382
diffused aeration 90
digital-transport modelling 377,379
dispersants 142
dissolved gasoline components 399-400

Environmental Evaluation System (EES) 233-242,251-253
environmental impacts 197,205,214, 227-229,232-233,238,247,252, 256-257,284,291
environmental mediation 331-335
environmental risk analysis 264, 270-272
environmental risk assessment 264, 272
evaluation of alternatives 206,218
event tree 267-269

field studies 397,437,448
final coefficient matrix 257
French drain systems 408,416,453, 480,507
Freundlich isotherm 96-97
functional curves 251-253,256

gasoline 1,4,149,414,448,473,501,514
gasoline-contaminated aquifers 135-138
genetically engineered microorganisms 143-144
Gilson Road hazardous waste site 349,358,361-370,452,462
goals identification matrix 178
granular activated carbon 149,461, 520
ground water modeling 409,493
grouting 13,66-69,74-75,170,183,186, 194,197,436-437,442,449,456,458, 471,495-496,505,509,518

Hazard Ranking System 278,282,284
heavy metals 4-5,15,112,115,127,134, 354,360-361,363,369,435,452,462, 465,472,485,497,501,509
hybrid cutoff wall 403,413
hydraulic containment 378,383,472
hydrocarbon-contaminated aquifers 131,139
hydrocarbon recovery systems 33,384
hydrocarbons 33,133,135,145, 182-183,385,388-391,399-400,453, 480-481,520
hydrogen peroxide 139,149
hydrogen sulfide 455

identification of publics 295,305
impermeable liners 183,185,449,468
importance weighting 205-209, 212-219,227-228,232,237-238,244, 252-253,255,257
infiltration galleries 509
injection trenches 444
injection wells 127-128,136,391,489
in situ aeration wells 444
in situ biological treatment 2,130,465, 472
in situ chemical fixation 2
in situ chemical treatment 127,130, 438
in situ treatment 184,186,450,465, 494,497,505
Installation Restoration (IR) Program 421-422,434-435
institutional constraints 173
interactive group methods 319
interceptor trenches 2-4,13,41-46, 49-50,182,185,408
interceptor wells 406,416
ion exchange 113,115,121,354,441, 443,447,519
iron 352-353,406-407,417,423, 426-428,435,442,447,462,476,478, 482,485,507,516-517,519

jar test studies 112
jet-eductor pumps 507

laboratory studies 105,368-369,389, 434,437,442-443,476,507-508,514
land use patterns 173

Milton Keynes UK
Ingram Content Group UK Ltd.
UKHW021927071024
449327UK00022B/1716

9 780367 451714